Neurotoxicity of Environmental Metal Toxicants

Neurotoxicity of Environmental Metal Toxicants

Editors

Richard Ortega
Asuncion Carmona

MDPI • Basel • Beijing • Wuhan • Barcelona • Belgrade • Manchester • Tokyo • Cluj • Tianjin

Editors
Richard Ortega
LP2i Bordeaux, CNRS
University of Bordeaux
Gradignan
France

Asuncion Carmona
LP2i Bordeaux, CNRS
University of Bordeaux
Gradignan
France

Editorial Office
MDPI
St. Alban-Anlage 66
4052 Basel, Switzerland

This is a reprint of articles from the Special Issue published online in the open access journal *Toxics* (ISSN 2305-6304) (available at: www.mdpi.com/journal/toxics/special_issues/neurotoxicity_metal_toxicants).

For citation purposes, cite each article independently as indicated on the article page online and as indicated below:

LastName, A.A.; LastName, B.B.; LastName, C.C. Article Title. *Journal Name* **Year**, *Volume Number*, Page Range.

ISBN 978-3-0365-5180-7 (Hbk)
ISBN 978-3-0365-5179-1 (PDF)

Contents

About the Editors

Richard Ortega

Dr. Richard Ortega is a CNRS research director at LP2iB (Laboratory of Physics of the 2 Infinites) from the University of Bordeaux in France. His research focus is on the imaging and speciation of metals in cells and proteins, applied to neurosciences and neurotoxicology. His main research projects include the role of biological metals in synapse formation, neurotoxicology of environmental metals, and metal dyshomeostasis in neurodegenerative diseases. He obtained a Chemical engineer and Master's Degree in Chemistry from the University of Marseille (1991), and a Ph.D. in Physics at the University of Bordeaux (1994). He was a Postdoctoral fellow with Pr. Bibudhendra Sarkar at the Research Institute of the Hospital for Sick Children, Canada (1995–1996). He obtained a permanent research position at CNRS in 1996 at CENBG, today LP2i, where he leads the group of Chemical Imaging and Speciation. He was an invited scientist at the European Synchrotron Radiation Facility (2001) and gained an award from the French Society of Chemistry in Analytical Chemistry (2004). He was vice-president of the program review committee 'Biology and Health'at SOLEIL Synchrotron, France (2015–2017). He is member of the Advisory Boards of Metallomics, NeuroToxicology and of the Journal of Analytical Atomic Spectrometry.

Asuncion Carmona

Dr. Carmona is a research associate at CNRS (Centre National de la Recherche Scientifique) and Bordeaux University in France since 2008. She obtained a Master's Degree in Physics from the University of Cordoba in Spain in 2001. And, she got her Ph.D. in Physics at the University of Seville in 2007 in Spain.

Currently, she is working at LP2IB in France where she conducts interdisciplinary research. She is interested in understanding the role of metals in neurodegenerative pathologies and environmental toxicology using analytical methods based on synchrotron radiation (X-ray fluorescence, X-ray absorption spectrometry) and ion beam techniques.

toxics

MDPI

Editorial

Neurotoxicity of Environmental Metal Toxicants: Special Issue

Richard Ortega * and Asuncion Carmona *

University of Bordeaux, CNRS, LP2I Bordeaux, UMR 5797, F-33170 Gradignan, France
* Correspondence: richard.ortega@u-bordeaux.fr (R.O.); asuncion.carmona@u-bordeaux.fr (A.C.)

Environmental exposure to metallic neurotoxicants is a matter of growing concern, since it may have very significant consequences for human health, from impairing neurodevelopment in children to the neurodegeneration processes involved in aging. The evaluation of the risks associated with the release of metals in the environment, the speciation analysis in environmental and biological samples, and the definition of relevant biological models to assess neurotoxicity are important research objectives. The aim of this Special Issue on the "Neurotoxicity of Environmental Metal Toxicants" is to provide a broad overview of the current work being performed in the field of the neurotoxicology of metallic contaminants, from the identification of emerging toxic compounds and the assessment of environmental exposures and associated risks to the description of the molecular mechanisms involved in neurotoxicity. Five original research articles and five review articles are reported in this Special Issue.

The impact of Alzheimer's disease (AD) on human health is a matter of great concern. The causes of AD are still unknown and exposure to certain environmental metals may contribute to the etiology of this disease. In particular, the association of exposure to environmental arsenic and AD is an emerging field of research and the literature on this very important topic is reviewed by Rahman et al. [1]. Long-term exposure to arsenic has been associated with the loss of memory and of cognitive functions in humans, including early indicators of AD. Arsenic can induce oxidative stress, neuroinflammation, mitochondrial dysfunction, endoplasmic reticulum stress, and impaired calcium signaling. The authors also discuss possible prevention strategies such as using zinc or selenium supplementation.

Citation: Ortega, R.; Carmona, A. Neurotoxicity of Environmental Metal Toxicants: Special Issue. *Toxics* **2022**, *10*, 382. https://doi.org/10.3390/toxics10070382

Received: 21 June 2022
Accepted: 7 July 2022
Published: 9 July 2022

Publisher's Note: MDPI stays neutral with regard to jurisdictional claims in published maps and institutional affiliations.

The mechanisms of copper neurotoxicity and their presentation of commonalities with the known mechanisms of neurodegeneration in AD are reviewed in the article from Patel and Aschner [2], particularly concerning the relationship between beta-amyloid plaques and copper. Beta-amyloid plaques have two copper-binding sites that may be involved in the generation of reactive oxygen species. Copper ions have a high affinity for the metal-binding site of the Aβ peptide, which increases the proportions of β-sheet and α-helix structures in Aβ aggregations, thus forming plaques. In addition, exposure to excessive concentrations of copper may induce brain inflammation, which in combination with Aβ aggregation would contribute to the progression of AD.

There are multiple sources of exposure to environmental metals. One of these sources of exposure is food. The review article from Spencer and Palmer [3] examines the neurotoxic potential of metal/metalloids in plants and fungi used for food, dietary supplements, and herbal medicine. Some fungi are particularly likely to accumulate neurotoxic metals and metalloids such as arsenic, lead, cadmium, or mercury. Air pollution and industrial effluent discharges contaminated with metals lead to the accumulation of these metals in plants, and ultimately, in food. As reviewed by the authors, the neurotoxicity of metals through plant consumption can be direct or indirect by modulating chemical species known to produce neurotoxic effects.

Neurotoxic metals have a wide variety of intracellular targets in neurons where they alter physiological functions. One of the main targets of metal-induced neurotoxicity is the mitochondria, as shown in the review article from Cheng et al. [4]. Mitochondrial dysfunction has been associated with most neurodegenerative diseases, such as AD, Parkinson's disease, Huntington's disease, and amyotrophic lateral sclerosis, but also with autism. For example, arsenic-induced mitochondrial dysfunction in neurons is a central mechanism of neurotoxicity. The effects of arsenic, aluminum, copper, cadmium, mercury, lead, zinc, iron, and manganese on mitochondrial functions and their consequences on neurodegeneration are reviewed by the authors.

Synapses are another potential target of metal-induced neurotoxicity, as reviewed by Carmona et al. [5]. Neurotoxic metals can interact with neurotransmitter receptors, as is the case for arsenic, cadmium, manganese, and lead, through interactions with the expression or activity of several neurotransmitter receptors, mainly the glutamatergic receptors, but also dopamine, GABA, and acetylcholine receptors. Another molecular mechanism of metal-induced neurotoxicity is the modification of the synaptic structure targeting important scaffolding proteins such as SHANK3 or cytoskeleton proteins (e.g., F-actin and tubulin). The molecular mechanisms involved may be oxidative stress or competition with zinc- or copper-binding sites in synaptic scaffold and/or cytoskeleton proteins.

Two research articles reported in this Special Issue used the 2011–2014 National Health and Nutrition Examination Survey (NHANES) data. Barahona et al. determined the relationship between blood and urinary manganese levels and cognitive function using the NHANES data [6]. This is the first study to examine associations between blood and urinary manganese levels and cognitive function in an elderly population. Blood and urinary manganese levels were inversely associated with the CERAD (Consortium to Establish a Registry for Alzheimer's Disease) after adjusting for all covariates; these associations were no longer significant after adjusting for the medical history. The findings suggest that increased blood and urinary manganese levels are associated with poorer cognitive function in an elderly population.

Metal neurotoxins not only affect cognition and memory but also neuromotor functions. This is well illustrated in the article from Gbemavo and Bouchard [7], who examined the relationship between blood concentrations of various metals and hand grip strength using the NHANES 2011–2014 dataset. They show that blood concentrations of mercury and manganese are not associated with grip strength, whereas lead is associated with a weaker grip strength in women, but not in men, even at low levels of exposure. In addition, low blood selenium levels were associated with weaker grip strength.

Methylmercury is an environmental toxicant of high concern, and chronic methylmercury exposure is associated with diabetes mellitus and metabolic syndrome. The study by Crawford et al. addressed methylmercury toxicity by examining its possible role in metabolic diseases using *C. elegans* [8]. The worms fed a cholesterol supplemented diet were more sensitive to methylmercury toxicity. The authors also examined whether the feeding habits of *C. elegans* were related to specific neuronal activity. The results showed that diet did not affect neurotoxicity pathways, such as dopaminergic dysfunction, in response to methylmercury exposure; however, diet did affect the feeding rate.

Another major concern in recent years has been cobalt-induced neurotoxicity after hip arthroplasty with cobalt chromium alloy prostheses. Gómez-Arnaiz et al. studied the mechanisms of cobalt neurotoxicity in rats treated with low concentrations of cobalt [9], which resulted in cobalt levels in the blood similar to those of patients with hip implants. A significant accumulation of cobalt in various organs of the rats, including the brain, was found. The authors measured the gene expression in neural tissue and revealed that the most up- or down-regulated genes were located in the choroid plexus, which is in direct contact with neurotoxicants at the blood–cerebrospinal fluid barrier.

Manganese is a well-known neurotoxic element, but the mechanisms of this toxicity are still poorly understood. Hernández et al. used SH-SY5Y neuroblastoma cells differentiated with retinoic acid to compare the toxicity of manganese chloride and manganese citrate [10]. Both chemical species of manganese induced similar toxicity governed by the disruption of protein metabolism, but with some differences. Manganese chloride impaired amino acid metabolism while manganese citrate inhibited the E3 ubiquitin ligase–target protein degradation pathway.

In conclusion, the scientific community will face many challenges in identifying and preventing the adverse effects of environmental metal exposure on brain health. We hope that this Special Issue will help to increase the visibility of this field of research, intensify collaborations, and increase the exchange of information between the different scientific communities concerned by this research topic.

Funding: This research received no external funding.

Conflicts of Interest: The authors declare no conflict of interest.

References

1. Rahman, M.A.; Hannan, M.A.; Uddin, M.J.; Rahman, M.S.; Rashid, M.M.; Kim, B. Exposure to Environmental Arsenic and Emerging Risk of Alzheimer's Disease: Perspective Mechanisms, Management Strategy, and Future Directions. *Toxics* **2021**, *9*, 188. [CrossRef] [PubMed]
2. Patel, R.; Aschner, M. Commonalities between Copper Neurotoxicity and Alzheimer's Disease. *Toxics* **2021**, *9*, 4. [CrossRef] [PubMed]
3. Spencer, P.S.; Palmer, V.S. Direct and Indirect Neurotoxic Potential of Metal/Metalloids in Plants and Fungi Used for Food, Dietary Supplements, and Herbal Medicine. *Toxics* **2021**, *9*, 57. [CrossRef] [PubMed]
4. Cheng, H.; Yang, B.; Ke, T.; Li, S.; Yang, X.; Aschner, M.; Chen, P. Mechanisms of Metal-Induced Mitochondrial Dysfunction in Neurological Disorders. *Toxics* **2021**, *9*, 142. [CrossRef] [PubMed]
5. Carmona, A.; Roudeau, S.; Ortega, R. Molecular Mechanisms of Environmental Metal Neurotoxicity: A Focus on the Interactions of Metals with Synapse Structure and Function. *Toxics* **2021**, *9*, 198. [CrossRef] [PubMed]
6. Barahona, A.J.; Bursac, Z.; Veledar, E.; Lucchini, R.; Tieu, K.; Richardson, J.R. Relationship of Blood and Urinary Manganese Levels with Cognitive Function in Elderly Individuals in the United States by Race/Ethnicity, NHANES 2011–2014. *Toxics* **2022**, *10*, 191. [CrossRef] [PubMed]
7. Gbemavo, M.C.J.; Bouchard, M.F. Concentrations of Lead, Mercury, Selenium, and Manganese in Blood and Hand Grip Strength among Adults Living in the United States (NHANES 2011–2014). *Toxics* **2021**, *9*, 189. [CrossRef] [PubMed]
8. Crawford, N.; Martell, M.; Nielsen, T.; Khalil, B.; Imtiaz, F.; Nguidjo, E.; Newell-Caito, J.L.; Bornhorst, J.; Schwerdtle, T.; Caito, S.W. Methylmercury-Induced Metabolic Alterations in Caenorhabditis elegans Are Diet-Dependent. *Toxics* **2021**, *9*, 287. [CrossRef] [PubMed]
9. Gómez-Arnaiz, S.; Tate, R.J.; Grant, M.H. Cobalt Neurotoxicity: Transcriptional Effect of Elevated Cobalt Blood Levels in the Rodent Brain. *Toxics* **2022**, *10*, 59. [CrossRef] [PubMed]
10. Hernández, R.B.; de Souza-Pinto, N.C.; Kleinjans, J.; van Herwijnen, M.; Piepers, J.; Moteshareie, H.; Burnside, D.; Golshani, A. Manganese-Induced Neurotoxicity through Impairment of Cross-Talk Pathways in Human Neuroblastoma Cell Line SH-SY5Y Differentiated with Retinoic Acid. *Toxics* **2021**, *9*, 348. [CrossRef] [PubMed]

toxics MDPI

Review

Exposure to Environmental Arsenic and Emerging Risk of Alzheimer's Disease: Perspective Mechanisms, Management Strategy, and Future Directions

Md. Ataur Rahman [1,2,3,4,*], Md. Abdul Hannan [5], Md Jamal Uddin [4,6], Md Saidur Rahman [7], Md Mamunur Rashid [8] and Bonglee Kim [1,2,*]

1 Department of Pathology, College of Korean Medicine, Kyung Hee University, 1-5 Hoegidong Dongdaemungu, Seoul 02447, Korea
2 Korean Medicine-Based Drug Repositioning Cancer Research Center, College of Korean Medicine, Kyung Hee University, Seoul 02447, Korea
3 Global Biotechnology & Biomedical Research Network (GBBRN), Department of Biotechnology and Genetic Engineering, Faculty of Biological Sciences, Islamic University, Kushtia 7003, Bangladesh
4 ABEx Bio-Research Center, East Azampur, Dhaka 1230, Bangladesh; hasan800920@gmail.com
5 Department of Biochemistry and Molecular Biology, Bangladesh Agricultural University, Mymensingh 2202, Bangladesh; hannanbmb@bau.edu.bd
6 Graduate School of Pharmaceutical Sciences, College of Pharmacy, Ewha Womans University, Seoul 03760, Korea
7 Department of Animal Science & Technology and BET Research Institute, Chung-Ang University, Anseong 17546, Korea; shohagvet@gmail.com
8 Department of Clinical Pharmacology and Therapeutics, Seoul National University College of Medicine and Hospital, Seoul 03080, Korea; mamun@snu.ac.kr
* Correspondence: ataur1981rahman@hotmail.com (M.A.R.); bongleekim@khu.ac.kr (B.K.)

Citation: Rahman, M.A.; Hannan, M.A.; Uddin, M.J.; Rahman, M.S.; Rashid, M.M.; Kim, B. Exposure to Environmental Arsenic and Emerging Risk of Alzheimer's Disease: Perspective Mechanisms, Management Strategy, and Future Directions. *Toxics* **2021**, *9*, 188. https://doi.org/10.3390/toxics9080188

Academic Editor: Michael Aschner

Received: 14 July 2021
Accepted: 11 August 2021
Published: 14 August 2021

Abstract: Alzheimer's disease (AD) is one of the most prevailing neurodegenerative diseases, characterized by memory dysfunction and the presence of hyperphosphorylated tau and amyloid β (Aβ) aggregates in multiple brain regions, including the hippocampus and cortex. The exact etiology of AD has not yet been confirmed. However, epidemiological reports suggest that populations who were exposed to environmental hazards are more likely to develop AD than those who were not. Arsenic (As) is a naturally occurring environmental risk factor abundant in the Earth's crust, and human exposure to As predominantly occurs through drinking water. Convincing evidence suggests that As causes neurotoxicity and impairs memory and cognition, although the hypothesis and molecular mechanism of As-associated pathobiology in AD are not yet clear. However, exposure to As and its metabolites leads to various pathogenic events such as oxidative stress, inflammation, mitochondrial dysfunctions, ER stress, apoptosis, impaired protein homeostasis, and abnormal calcium signaling. Evidence has indicated that As exposure induces alterations that coincide with most of the biochemical, pathological, and clinical developments of AD. Here, we overview existing literature to gain insights into the plausible mechanisms that underlie As-induced neurotoxicity and the subsequent neurological deficits in AD. Prospective strategies for the prevention and management of arsenic exposure and neurotoxicity have also been discussed.

Keywords: arsenic; Alzheimer's disease (AD); environmental risk factor; mitochondrial dysfunction; proteostasis; apoptosis; phytochemicals

1. Introduction

Alzheimer's disease (AD) is the most prevalent neurodegenerative disease and the prime cause of dementia among the elderly. The main pathological features are intracellular aggregates comprising phosphorylated tau protein, which forms neurofibrillary tangles (NFTs), as well as extracellular deposition of amyloid-β (Aβ) leading to the formation of senile plaques (Figure 1) [1–4]. About 70% of AD risk has been considered to be inherited, and several genes are commonly involved, while the actual causes along with molecular mechanisms have not yet been well understood [4,5]. It has been found that brains are affected by AD mostly via activated microglia, reactive astrocytes, oxidative damage, and altered proteostasis [6]. Recent observations suggest that exposure to several environmental factors may enhance the prospective risk of AD [3], although the authentic etiology of AD is not yet clear. Several investigations have emphasized that environmental risk factors may play a significant role in accelerating or decelerating the progression of AD. Such environmental risk factors include particulate air, pesticides, metal-containing nanoparticles, and numerous metalloids such as arsenic, aluminum, lead, cadmium, and mercury [3,7,8]. Arsenic has been found to be the most toxic metalloid responsible for the neurotoxicity associated with AD development in the brain, impairing cognitive function [7], although the role of As exposure in AD development has been poorly elucidated.

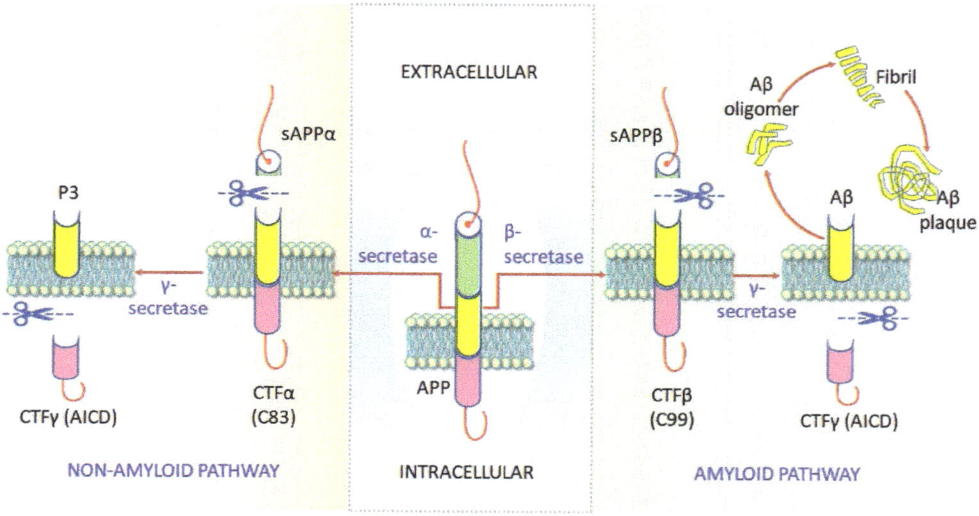

Figure 1. Mechanism of AD pathogenesis. The amyloid pathway is important to accumulate neurotoxic Aβ plaque via releases of β-secretase, which produces extracellular sAPPβ and C99. Cleavage of the C99 fragment by γ-secretase releases Aβ oligomer, which subsequently produces Aβ peptide. By the action of α-secretase, two fragments of sAPPα and C83 are produced by the non-amyloid pathway. The non-amyloid pathway cleavages APP via α-secretase to produce two fragments: C83, an 83 amino acid intracellular C-terminal fragment, and extracellular sAPPα, soluble amyloid precursor protein α. Cleavage of the C83 fragment by γ-secretase yields a P3 peptide and CTFγ.

According to epidemiological investigations, populations around the world are increasingly recognizing that childhood is the most vulnerable stage during which a low level of As exposure has a detrimental effect on AD development [8]. Exposure has been found to be connected to learning deficiency, variations of neurotransmitter release, and behavioral deficits [9]. Inorganic arsenic (iAs) exposure through drinking water leads to the bioenergetic impairment of 3xTgAD, in an AD model, compared to counterparts, suggesting that As exposure exacerbates AD pathophysiological progression [10]. Additionally, inorganic and organic As exposures increase APP and sAPPβ expression in cholinergic SN56.B5.G4 cells [9]. Moreover, prolonged arsenic exposure causes an increase in tau phosphorylation and insoluble tau aggregates [11]. These observations suggest that As might have the capacity to enhance Aβ accumulation and tau phosphorylation to initiate senile plaque formation, leading to increased susceptibility to AD. Therefore, an exhaustive literature review has been conducted to better understand the relationship between As toxicity and AD development along with the pathomechanism in redefining the health risks of As.

2. Environmental Sources of Arsenic Exposure

The most abundant inorganic arsenic found in the air is arsenic trioxide (As_2O_3); however, it may also be found in water, soil, or food, with the most prevalent being inorganic arsenic (AsO_3) or arsenites (AsO_2) [12]. Gallium arsenide (GaAs) is an inorganic arsenic compound that has serious human health effects because of its widespread use in microelectronics [13]. Arsenic and other metals generally are the major source, with seafood, grains, mushrooms, and poultry as the predominant dietary types. Although seafood contains more arsenic than other meals, it is mainly called arsenobetaine in an organic form, far less dangerous than other types. The most common causes of arsenic poisoning include occupational exposure, contaminated wine or moonshine, or malicious delivery [14]. Recently, traditional Chinese herbal arsenic-enhanced remedies have been found to represent a considerable risk to health [15]. In cosmetic color, pigments used to make eye shadows, toxic substances such as arsenic are often discovered. Eyelash skin is particularly sensitive, and eczema can be caused by applying eye-shadows [16]. Arsenic particles that are water-soluble can be absorbed by wet skin. When arsenic enters the circulatory system at high amounts through percutaneous absorption, there is a risk of carcinogenesis [17]. It was advised that the cosmetic products comprise fewer than 5 ppm of metal contaminants based on recognized toxicity research. Arsenic, which is common in water, soil, and food, can quickly enter the human body when swallowed. The majority of dust parts are placed on the lining when the air containing arsenic dust is inhaled [18]. The material that goes through the skin into the body is very small, which limits internal exposure to arsenic; therefore, there is highly unlikely that it will lead to arsenic poisoning [19]. A simple diffusion mechanism enables the majority of arsenic to enter the body in the trivalent inorganic form As(III) [20]. Only a limited amount of pentavalent inorganic arsenic can move through cell membranes via an energy-dependent transport process prior to being converted into trivalent arsenic. As arsenic is excreted through urine, both organic and inorganic, most inorganic arsenic could be removed in a few days, although some will linger for months. Organic arsenic is normally removed faster than inorganic arsenic within a few days [21]. Arsenic and other metals in soil have a seriously detrimental effect on health in various places of the world. In Bangladesh and West Bengal, India, some of the worst occurrences of arsenic poisoning were observed where close to 43 million people drank arsenic-laden water [22]. WHO suggested the arsenic limit for water is 0.001 μg/L, although a level up to 0.05–3.20 μg/L has been found [23].

Epidemiological data show that environmental arsenic ranges from 7 to 18 μg/L in topsoils are certainly associated with the mortality and prevalence of AD [24]. In an epidemiological study with 434 human participants, it was found that low-level arsenic exposure was related to poor neuropsychological action [25]. Nevertheless, an additional study showed a positive relationship between cognitive ability and serum arsenic, which suggests that consumption of seafood arsenic (docosahexaenoic acid) plays an important role in delaying AD pathogenesis [26]. Another study has indicated that in rat cerebellar granule neurons, As exposure led to apoptosis and neurotoxicity via activation of JNK3 and p38 MAP kinase signaling pathways [27]. Importantly, animal and human studies have found that air pollution is also a source of exposure and can worsen the pathology of neurodegenerative disease, which may involve the development of neurotoxicity [28]. Thus, experimental, clinical, observational, and epidemiological studies have described that As causes AD pathogenesis. The relationship between environmental arsenic and incidence of AD is presented in Table 1.

Table 1. Environmental arsenic factors highly related with AD pathogenesis.

Dose and Level of Arsenic	Study Model	Effects/Molecular Mechanism	References
13–15 mg/kg.	Mortality data by WHO, epidemiological and geological data.	Induce AD and other dementias as a composite morbi-mortality index.	[24]
Sodium arsenite (10 μM).	Cerebellar granule neurons of rats.	Activation of p38 and JNK3 MAP kinases cause cerebellar granule neurotoxicity and apoptosis.	[27]
Groundwater long exposure of 240.15 ± 182.96 μg/L.	Longitudinal epidemiological human study.	Low and long As exposure linked to global cognition function.	[25]
Drinking water (10 μg/L).	Rat and human brain.	Tau hyperphosphorylation and APP over transcription.	[28]

3. Prevalence of Arsenic Exposure and Potential Risk of AD Development

While inorganic arsenic (As) is a well-known neurotoxic metalloid with adverse neurological and cognitive impacts, the consequences on the elderly have received less attention. Only a few investigations have looked at As exposure as a risk factor for developing Alzheimer's disease [8]. According to a study conducted on rural and elderly people in Texas (Project Frontier), poor cognitive ability and memory (after being adapted to confounders such as ApoEε4) revealed early indicators of Alzheimer's disease, which are associated with long-term exposure to low arsenic levels (3–15 μg/L As in water) [29]. Occupational arsenic exposure was associated with memory loss and cognition function in humans [30]. In Thailand, India, Bangladesh, Mexico, Taiwan, and mainland China, chronic arsenic exposure in the air or drinking water has consistently been found to be related to memory reduction as well as intellectual capabilities in adolescents or children [7]. It has been documented that there was a potential connection between arsenic exposure in drinking water and cognitive dysfunction among adults living in a rural area of Cochran County, Texas [25]. Furthermore, arsenic exposure in childhood might be linked with lower education levels and AD. Therefore, arsenic exposure appears to have a more potent effect on cognitive function in AD throughout the lifetime, consistent with the lifelong-exposure hypothesis for AD.

4. Molecular Basis of Arsenic Toxicity and Its Implication in AD Pathobiology

Arsenic exposure has a direct impact on its toxic effects, but its exact molecular mechanism of action is not fully understood yet. There are several hypotheses: one is the generation of reactive free radicals, which oxidize cellular components such as lipids, proteins, and DNA, ultimately causing oxidative stress and subsequent damage to cells (Figure 2). Arsenic-induced oxidation of DNA leads to a reduction in the antioxidant capability of rodents' brains and protein thiols in the hippocampus, striatum, and cortex, resulting in downregulating ATP-synthase and encouraging peroxidation of lipids in rat brains [31]. In drinking water, 0.005 to 0.02 ppm as well as 0.01 to 0.05 ppm environmental arsenic exposure in humans and mice, respectively, enhances oxidation of DNA and proteins, and oxidative damage with inflammatory responses [32]. All this evidence suggests that arsenic is a possible etiologic factor for the oxidative stress hypothesis in AD pathogenesis, which could anticipate that levels of oxidized metabolites of RNA, DNA, fatty acids, and proteins increase in AD brains [9,33,34].

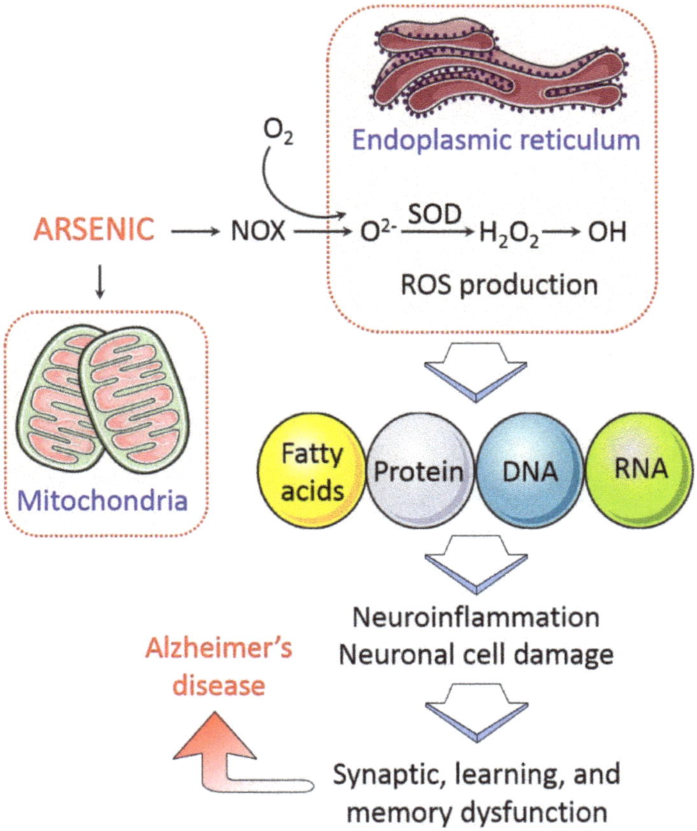

Figure 2. Arsenic-induced mitochondrial dysfunction and ROS generation in AD. As induces ROS production by mitochondrial dysfunction. NADPH oxidase (NOX) contributes to generate superoxide anions, which leads to the release of ROS in cells. ROS generation impairs fatty acids, proteins, and DNA, inducing neuroinflammation and subsequently causing cognitive dysfunction in AD.

Arsenic exposure has been found to reduce memory as well as learning ability in animal models where offspring of mice treated with 0.75 mg/kg of arsenic during pregnancy displayed neurobehavioral retardation in fetal origin [35]. Inorganic As (20 mg/L) induced severe spatial memory losses in mice during pregnancy and early postnatal life [7]. The modification of the amyloid pathway is a plausible explanation of cognitive and memory problems due to As exposure [3,36]. It has been reported that incubation of cholinergic SN56.B5.G4 cells with organic dimethylarsinic acid (DMA) (5–10 μM/12–24 h) increased Aβ levels [9]. However, Tg2576 mouse neurons have been shown to have similar effects (a murine model that overexpresses a mutant form of APP most used in AD) [37]. The effects of DMA are presumed to be attributable to higher Aβ anabolism (greater APP explication), although the Aβ degradation process has not changed. The method by which As produces excess Aβ is not known, but the inflammatory response in the brain and the oxidative stress are associated, which is compatible with the hypothesis of Alzheimer's pathobiology (Figure 3) [33,38].

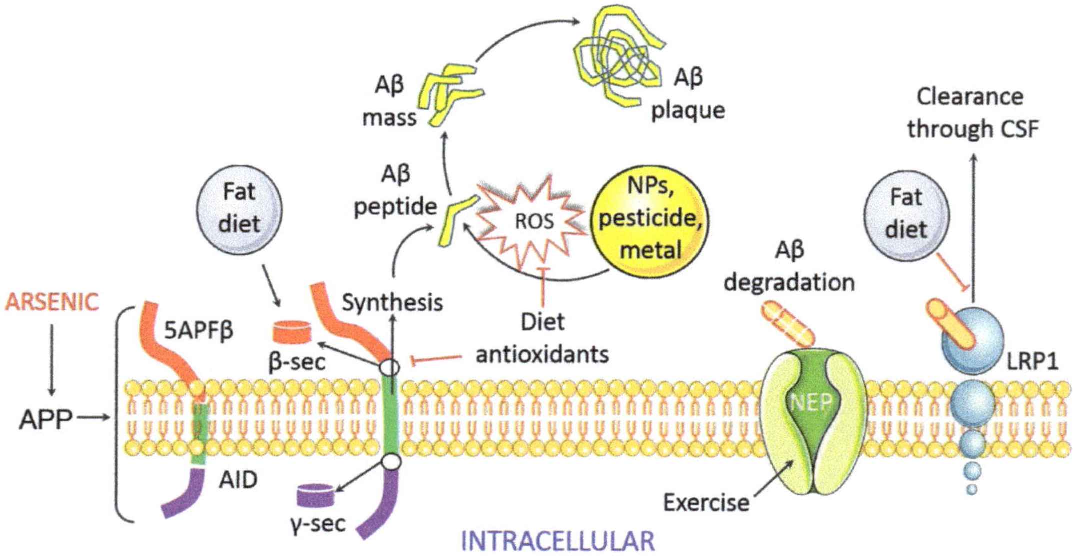

Figure 3. Arsenic induces Aβ accumulation during AD pathogenesis. APP cleaves through the enzymatic action of β- and γ-secretase, which subsequently produces Aβ plaque. Dietary antioxidants may hinder the formation of Aβ plaque and ROS, and thereby may prevent AD progression.

4.1. *Mechanism of Arsenic-Induced Neurotoxicity*

Human exposure to As is related to extensive neurological problems, including poor concentration, impaired memory, Guillain–Barre-like neuropathy, encephalopathy, verbal comprehension, Parkinson's disease, and peripheral neuropathy [34]. The exact mechanisms that underlie arsenic neurotoxicity are largely not yet known. However, experimental evidence suggests that pathological factors including oxidative stress, inflammation, mitochondrial dysfunctions, ER stress, apoptosis, and impaired protein homeostasis are supposed to be implicated in AD pathobiology [39].

4.1.1. Oxidative Stress

Oxidative stress plays a critical role in the pathobiology of various neurological disorders, including AD [40]. Oxidative stress is accompanied by an increase in reactive oxygen species (ROS) and lipid peroxidation, reduced levels of superoxide dismutase and glutathione (GSH) [41], and constitute the most important mechanism hypothesized for arsenic-mediated neurotoxicity. Intracellular ROS production appears to be the key mechanism of arsenic-induced neurotoxicity. Arsenic-methyltransferase, possibly a protein-binding substrate in the brain, is involved in the processing of inorganic arsenic (iAs) in the presence of GSH, monomethylarsonic acid (MMA), and DMA [42]. Depletion of GSH, therefore, hampers the metabolic processing of IAs. Furthermore, oxidative damage has been associated with the rise in BACE1 activity and consequently with Aβ levels in animal brains exposed to iAs over time [3]. Oxidative stress thus seems to play an essential role in arsenic-mediated neurotoxicity.

4.1.2. Neuroinflammation

Neuroinflammation is another crucial event that largely contributes to AD pathobiology [43]. Evidence suggests that arsenic plays a vital role in neuroinflammation via activation of microglia through secretion of pro-inflammatory cytokines, which could damage cognitive function, intellectual ability, and learning and memory functions [11,44]. Arsenic administration augmented ROS-mediated expression of pro-inflammatory cytokines in rats via activation of MAP kinase and protein kinase C through increased mRNA levels of pro-inflammatory markers such as TNFα, IL-1β, and IFNγ with protein expression of TNFα and IFNγ [45]. Intake of excess arsenic implicated MBP-activated autoimmunity and neuroinflammation in athymic nude mice with depleted T cell populations in peripheral lymphoid organs [46].

4.1.3. Mitochondrial Dysfunction

Mitochondria occupies a central position in cellular bioenergetics. Mitochondrial dysfunction and oxidative stress are closely linked and, together with inflammation, constitute a pathological triad [47]. Understanding molecular mechanisms that underlie mitochondrial dysfunction might be useful to progress therapeutic approaches to be effective against arsenic-mediated neurotoxicity. Several reports have described impairment of brain mitochondria during arsenic-mediated toxicity [48,49]. Mitochondrial ROS accumulation has been found after arsenic treatment in rat brains [49]. It has been indicated that Aβ lowered the activity of mitochondrial complex I in As-induced 3xTgAD mice. This study indicates that amyloid aggregates can hinder mitochondrial function. In this hypothesis, reductions in ATP generation are responsible for the endoplasmic reticulum (ER) stress leading to the accumulation of the malfunctioning proteins in its lumen [50]. Arsenic has shown its mutagenic response via disruption of mitochondrial function. As-induced perturbation to mitochondrial oxidation caused production of excess superoxide anions, which when reacted with nitric oxide generate highly reactive peroxynitrites [44]. Arsenic toxicity affects the mitochondrial membrane potential via generating ROS and DNA fragmentation, and ROS overproduction is associated with apoptosis induction by the release of cytochrome c, which activates the caspase pathway (Figure 4) [51]. Thus, mitochondria are the major target in arsenic-mediated neurotoxicity and the subsequent neurological deficits in AD.

4.1.4. Endoplasmic Reticulum (ER) Stress

Current evidence suggests a relationship between ER stress and arsenic-induced neurodegenerative diseases [3,52]. Arsenic has been found to induce endoplasmic reticulum (ER) stress via accumulation of misfolded proteins, which results in neurotoxicity and cell death [53], although the underlying molecular mechanisms are not well understood. Recently, it has been described that arsenic-mediated ER stress and neurotoxicity are associated with early neurodevelopment, which can be alleviated by microRNA-124 [52]. Additionally, in MIN6 cells, arsenic impaired mitochondrion function through diminishing mitochondrial membrane potential and decreased cytochrome c release, which cause mitochondrial ROS generation [54].

4.1.5. Apoptosis

Arsenic toxicity involves apoptosis as a common phenomenon of cell death. Arsenic-induced neurotoxicity has been involved in apoptosis induction in the cerebral neurons via activation of JNK3 and p38 mitogen-activated protein kinase (p38MAPK) pathways [27]. Recently, it has been found that arsenic prompts neuronal apoptosis through upregulating Bax levels as well as decreasing Bcl-2 protein [55]. Additionally, arsenic-induced keratinocyte apoptosis in the extrinsic apoptotic pathway involves Fas/FasL, which is correlated with modifications of AP-1 and NF-κB pathways [56]. Arsenic-induced neuronal cell death involves activation of autophagy-dependent apoptosis through inactivation of the Akt pathway and activation of the AMPK pathway [57].

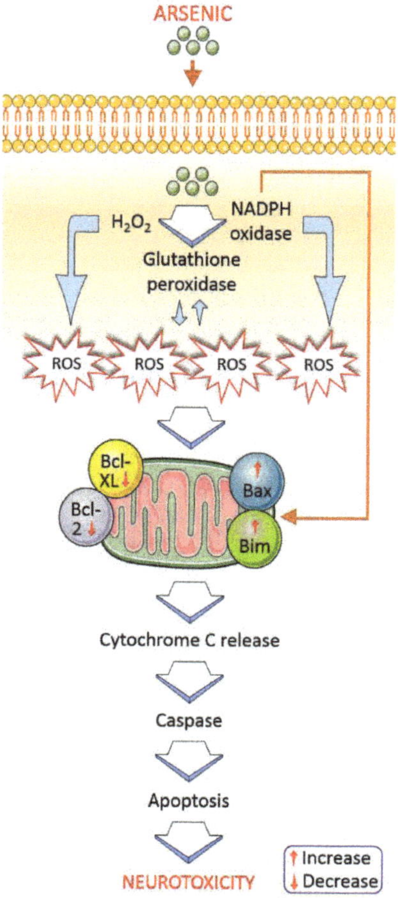

Figure 4. Arsenic exposure induces apoptotic factors from mitochondria to cause neurotoxicity. Mitochondria are the main target in arsenic-mediated neurotoxicity. Arsenic-mediated oxidative stress by hydrogen peroxide and NADPH oxidase generates ROS production, which can disrupt membrane potential resulting in cytochrome c release. Mitochondrial apoptotic markers, for example, Bax/Bim and Bcl-XL/Bcl-2, and numerous inflammatory markers (NF-kβ) alter and activate caspase activation, which causes cell damage and cell death via apoptosis.

4.1.6. Impaired Proteostasis

Protein quality control systems, also known as protein homeostasis or simply proteostasis, play a significant role in cellular physiological functions. Impairment in proteostasis leads to aberrant deposition of protein aggregates, which are characteristics of many neurodegenerative disorders such as Aβ aggregate in AD [58,59]. Compelling evidence from recent studies suggests that chronic exposure to inorganic arsenic can disrupt protein quality control and clearance systems, which contribute to the pathobiology of proteinopathic brain disorders such as AD. Arsenic-mediated post-translational modifications of proteins and disruptions of ubiquitination may culminate in impairment in proteostasis [60]. Genome-wide imaging screen analysis uncovers the molecular causes of arsenite-mediated protein aggregation and toxicity studied on *Saccharomyces cerevisiae* deletion mutants [61]. Recently, it has been found that three ER stress sensors viz. ATF6α, IRE1α, and PERK have an important role in ER stress-related autophagy, mitochondrial dysfunction, and unfolded protein response (UPR) in arsenic-induced malignancies to identify vital targets for therapeutics of neurodegenerative and cancer prevention [62]. Furthermore, arsenic prevents SNARE complex formation via increasing SNAP29 O-GlcNAcylation, which perturb proteostasis, and transfection of O-GlcNAcylation-defective CRISPR-mediated SNAP29 knockout cells eliminates arsenic-induced autophagy inhibition [63]. Additionally, arsenic has been found to induce autophagy inhibition and ER stress induction, which suggests low arsenic-mediated recovery of oxidative stress, and restoring proteostasis by the autophagy pathway [64]. Particularly, arsenic-mediated UPR activation is related to the accumulation of p62 and LC3 protein aggregates in the sequestration of autophagic protein clearance [65], and activation of UPR as well as formation of aggregated protein might be targeted to the lysosomal degradation of proteostasis.

4.1.7. Impaired Calcium Signaling

Arsenic-induced mitochondrial dysfunction causes a decline in ATP generation, which is responsible for ER stress, leading to calcium build-up in intracellular compartments and impairment in calcium signaling, most likely due to a lack of recovery systems [50]. Altered calcium signaling can cause cognitive impairment or the development of tau hyperphosphorylation by activating protein kinases such as GSK-3. Hyperphosphorylated tau and amyloid aggregates interact with mitochondria, generating complex I shortages and leading to an establishment of a vicious cycle of energy deprivation and proteostasis (Figure 5) [66].

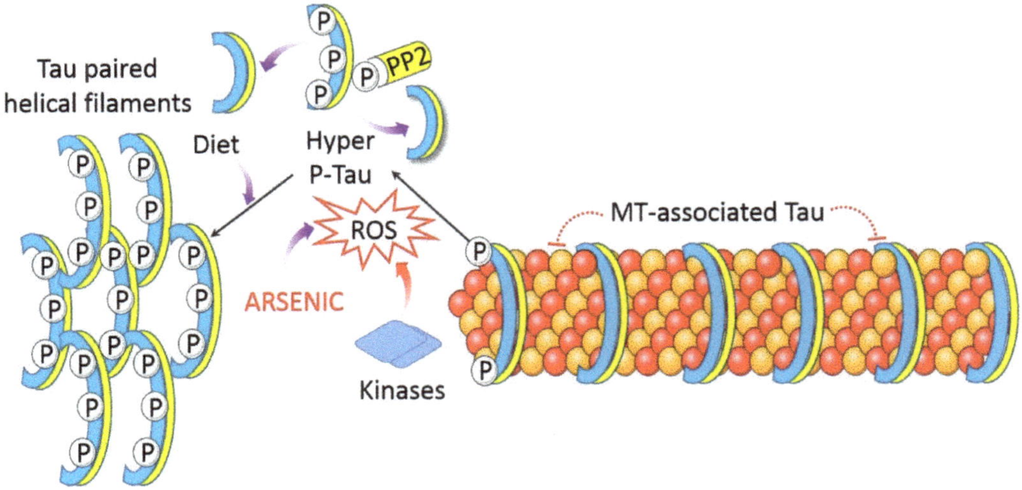

Figure 5. Arsenic influences tau hyperphosphorylation in AD. Hyperphosphorylated tau aggregates and acts on mitochondria, triggering energy deficit.

5. Management and Control of Arsenic-Induced Neurological Deficits

Although the epidemiological reports and experimental evidence clearly demonstrate that high exposure to arsenic results in abnormalities in the developing brain and cognitive deficits in adults, the specific management strategy that can effectively address arsenic-induced neurological deficits is yet to be developed. However, the approaches such as administration of biological trace elements (zinc and selenium), antioxidants and arsenic chelators, high-protein diets, and exposure to an enriched environment could be the possible strategies that can ameliorate arsenic-induced systemic deficits.

The forefront strategy that has been shown to be effective in alleviating arsenic-induced toxic effects is the dietary inclusion of trace elements such as zinc and selenium. Zinc can prevent arsenic-induced neurotoxicity in fish models by preserving the blood–brain barrier and attenuating apoptosis and autophagy dysfunction [67]. In several other studies with common carp, zinc supplementation also has been shown to be effective in protecting against arsenic-induced toxicity in the heart [68], kidney [69,70], liver [71], spleen [72], and pancreas [73]. In a rat model of chronic arsenic toxicity, zinc supplementation can protect against damages to the liver and kidney [74]. In a neuronal cell line, zinc was shown to alleviate arsenic-induced apoptosis (Figure 6) [75]. Early-stage zinc supplementation in pregnant women can prevent preterm birth induced by arsenic toxicity, as observed in a rural Bangladesh birth cohort [76]. Another vital trace element with antioxidant potential is selenium that has been reported to be protective against arsenic toxicity. Selenium co-administration can protect against arsenic-induced behavioral deficits in rats through a mechanism involving anti-inflammation, antioxidation, and anti-apoptosis (Figure 6) [77]. In another study by Samad and colleagues, selenium supplementation through drinking water ameliorated arsenic-induced anxiety/depression and memory deficits in rats [78]. In chickens exposed to arsenic at a subacute dose, selenium (up to 10 mg/kg) prevented oxidative damage, neurotransmitter disorders, and apoptosis in the brain [79]. Additionally, selenium was reported to be effective in ameliorating arsenic-induced damage to other tissues including the liver [80], reproductive organs [81], and other organs [82].

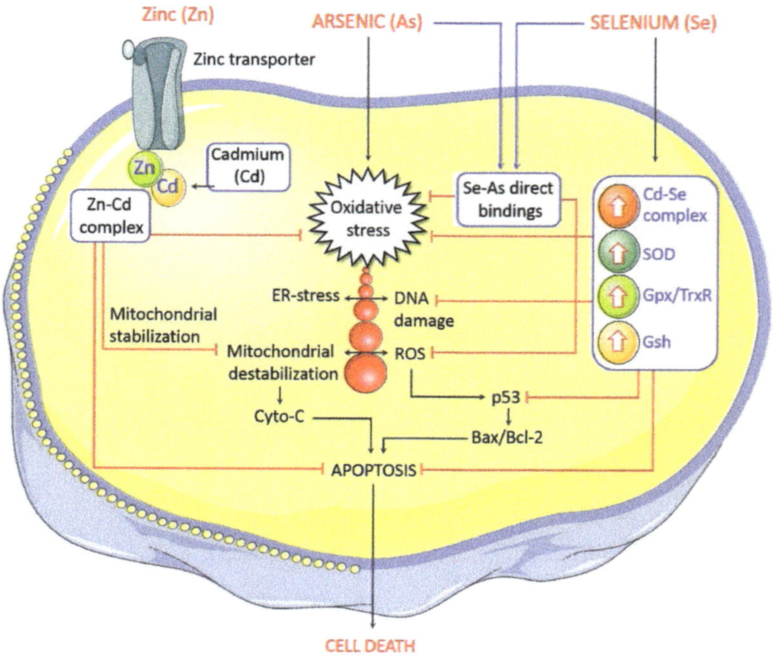

Figure 6. Possible protective mechanism of selenium and zinc against As toxicity. Arsenic induces oxidative stress, followed by DNA damage, ER stress, and mitochondrion dysfunction, which subsequently trigger apoptosis. Se–As complex inhibits oxidative stress, and Cd–Se prevents ROS production, DNA damage, and apoptosis. Additionally, Zn–Cd complex activates mitochondrial stabilization and inhibits oxidative stress and apoptosis.

In addition to biological trace elements, food-derived bioactive compounds that can attenuate oxidative stress and inflammation have protective potential against arsenic-induced tissue damage [83]. The most notable of the natural compounds are curcumin, quercetin, gallic acid, genistein, resveratrol, and thymoquinone, whose protective effects against arsenic toxicity have been supported by multiple studies in animals [84,85]. However, extrapolation of the beneficial effects of these compounds from preclinical evidence to clinical patients requires further validation. A limited number of studies of phytochemical effects on arsenic-induced toxicity in brain disorders have been reported. Green tea and vitamin C attenuated arsenic-induced lipid peroxidation in rat brains [86]. In addition, the leaf extracts of *Annona muricata* reduced arsenic-mediated neurotoxicity [87,88]. Besides, the treatment of arsenic intoxication is carried out mainly by chelation therapy using dimercaprol and aqueous garlic extract in A375 cells [89]. Garlic extract is a potential antidote to the toxic effects of sodium arsenite in mice [90]. Allicin, a bioactive compound of garlic, decreases arsenic-induced oxidative stress and toxicity in mice [91]. The aqueous extracts of garlic and its derivative, allicin, take part in chelation of arsenic [92]. Additionally, combined effects of various plant extracts have been recorded for arsenic-induced hematological, renal, and hepatic alteration in experimental animals [93–95]. Nanocapsulated quercetin prevented arsenic-induced damage in various organs, including brains of rats and mice [96,97].

Access to a high-protein diet also can help mitigate arsenic toxicity. Evidence suggests that ensuring an enriched environment can be a prospective strategy against arsenic-induced neurological problems, probably by alleviating depression and stress [7]. Moreover, conventional therapy that uses various chelating agents such as dimercaprol (BAL, British antilewisite), dimercaptosuccinic acid (DMSA), and penicillamine still may benefit arsenic-induced neurological deficits [98]. Beyond the management strategies outlined above, preventive measures such as access to arsenic-free drinking water, avoiding occupational exposure to arsenic, avoiding risk factors to neurological disorders, and provision of nutritious foods need to be considered for the ultimate prevention of As-induced health consequences, including neurological deficits [99].

6. Conclusions and Future Directions

Health consequences of As exposure represent one of the devastating setbacks of environmental pollution in human history. As reported in various epidemiological and experimental studies, arsenic extends its toxic effects to a number of vital organs, including the brain, where it causes neurodevelopmental abnormalities in childhood and cognitive deficits in adults. Although the underlying precise mechanisms of As-induced neurotoxicity have not yet been determined, the information from this review shows that the changes caused by arsenic exposure coincide with the pathological progression, clinical symptoms, and biochemical features of AD. While acute As toxicity can be managed by clinical use of specific antidotes, existing strategies to manage chronic As exposure are limited to supportive only. Extensive research involving animal models and appropriate human subjects is crucial to explore the detailed molecular mechanisms in order to design an effective therapeutic strategy against this menace.

Author Contributions: M.A.R.: conceptualization, data curation, writing—original draft preparation, writing—review and editing. M.A.H.: original draft preparation, writing—review and editing. M.J.U.: visualization, writing—original draft preparation, M.S.R.: data curation and prepared the figures, M.M.R.: writing—original draft preparation. B.K.: supervision, review and editing, funding. All authors have read and agreed to the published version of the manuscript.

Funding: This research was supported by Basic Science Research Program through the National Research Foundation of Korea (NRF) funded by the Ministry of Education (NRF-2020R1I1A2066868), the National Research Foundation of Korea (NRF) grant funded by the Korea government (MSIT) (No. 2020R1A5A2019413). M.J.U. is supported by National Research Foundation (No. 2020R1I1A1A01072879) and the Brain Pool program funded by the Ministry of Science and ICT through the National Research Foundation (No. 2020H1D3A2A02110924), Republic of Korea.

Institutional Review Board Statement: Not applicable.

Informed Consent Statement: Not applicable.

Data Availability Statement: Not applicable.

Conflicts of Interest: The authors declare no conflict of interest.

References

1. Rahman, M.A.; Rhim, H. Therapeutic implication of autophagy in neurodegenerative diseases. *BMB Rep.* **2017**, *50*, 345–354. [CrossRef]
2. Moya-Alvarado, G.; Gershoni-Emek, N.; Perlson, E.; Bronfman, F.C. Neurodegeneration and Alzheimer's disease (AD). What Can Proteomics Tell Us About the Alzheimer's Brain? *Mol. Cell. Proteom.* **2016**, *15*, 409–425. [CrossRef]
3. Rahman, M.A.; Rahman, M.S.; Uddin, M.J.; Mamum-Or-Rashid, A.N.M.; Pang, M.; Rhim, H. Emerging risk of environmental factors: Insight mechanisms of Alzheimer's diseases. *Environ. Sci. Pollut. Res.* **2020**, *27*, 44659–44672. [CrossRef]
4. Rahman, M.A.; Rahman, M.S.; Rahman, M.H.; Rasheduzzaman, M.; Mamun-Or-Rashid, A.N.M.; Uddin, M.J.; Rahman, M.R.; Hwang, H.; Pang, M.; Rhim, H. Modulatory Effects of Autophagy on APP Processing as a Potential Treatment Target for Alzheimer's Disease. *Biomedicines* **2021**, *9*, 5. [CrossRef]
5. Ghai, R.; Nagarajan, K.; Arora, M.; Grover, P.; Ali, N.; Kapoor, G. Current Strategies and Novel Drug Approaches for Alzheimer Disease. *CNS Neurol. Disord. Drug Targets* **2020**, *19*, 676–690. [CrossRef]
6. Clayton, K.A.; Van Enoo, A.A.; Ikezu, T. Alzheimer's Disease: The Role of Microglia in Brain Homeostasis and Proteopathy. *Front. Neurosci.* **2017**, *11*. [CrossRef]
7. Tyler, C.R.; Allan, A.M. The Effects of Arsenic Exposure on Neurological and Cognitive Dysfunction in Human and Rodent Studies: A Review. *Curr. Environ. Health Rep.* **2014**, *1*, 132–147. [CrossRef]
8. Ramos-Chavez, L.A.; Rendón-López, C.R.; Zepeda, A.; Silva-Adaya, D.; Del Razo, L.M.; Gonsebatt, M.E. Neurological effects of inorganic arsenic exposure: Altered cysteine/glutamate transport, NMDA expression and spatial memory impairment. *Front. Cell. Neurosci.* **2015**, *9*, 21. [CrossRef]
9. Zarazua, S.; Bürger, S.; Delgado, J.M.; Jiménez-Capdeville, M.E.; Schliebs, R. Arsenic affects expression and processing of amyloid precursor protein (APP) in primary neuronal cells overexpressing the Swedish mutation of human APP. *Int. J. Dev. Neurosci.* **2011**, *29*, 389–396. [CrossRef]
10. Nino, S.A.; Morales-Martínez, A.; Chi-Ahumada, E.; Carrizales, L.; Salgado-Delgado, R.; Pérez-Severiano, F.; Díaz-Cintra, S.; Jiménez-Capdeville, M.E.; Zarazúa, S. Arsenic Exposure Contributes to the Bioenergetic Damage in an Alzheimer's Disease Model. *ACS Chem. Neurosci.* **2019**, *10*, 323–336. [CrossRef]
11. Wisessaowapak, C.; Visitnonthachai, D.; Watcharasit, P.; Satayavivad, J. Prolonged arsenic exposure increases tau phosphorylation in differentiated SH-SY5Y cells: The contribution of GSK3 and ERK1/2. *Environ. Toxicol. Pharmacol.* **2021**, *84*, 103626. [CrossRef]
12. Chung, J.Y.; Yu, S.D.; Hong, Y.S. Environmental source of arsenic exposure. *J. Prev. Med. Public Health* **2014**, *47*, 253–257. [CrossRef]
13. IARC Working Group on the Evaluation of Carcinogenic Risks to Humans. Cobalt in hard metals and cobalt sulfate, gallium arsenide, indium phosphide and vanadium pentoxide. *IARC Monogr. Eval. Carcinog. Risks Hum.* **2006**, *86*, 1–294.
14. Emadi, A.; Gore, S.D. Arsenic trioxide—An old drug rediscovered. *Blood Rev.* **2010**, *24*, 191–199. [CrossRef] [PubMed]
15. Luo, L.; Wang, B.; Jiang, J.W.; Fitzgerald, M.; Huang, Q.; Yu, Z.; Li, H.; Zhang, J.Q.; Wei, J.H.; Yang, C.Y.; et al. Heavy Metal Contaminations in Herbal Medicines: Determination, Comprehensive Risk Assessments, and Solutions. *Front. Pharmacol.* **2021**, *11*, 595335. [CrossRef] [PubMed]
16. Borowska, S.; Brzoska, M.M. Metals in cosmetics: Implications for human health. *J. Appl. Toxicol.* **2015**, *35*, 551–572. [CrossRef]
17. Briffa, J.; Sinagra, E.; Blundell, R. Heavy metal pollution in the environment and their toxicological effects on humans. *Heliyon* **2020**, *6*, e04691. [CrossRef] [PubMed]
18. Wu, X.Y.; Cobbina, S.J.; Mao, G.H.; Xu, H.; Zhang, Z.; Yang, L.Q. A review of toxicity and mechanisms of individual and mixtures of heavy metals in the environment. *Environ. Sci. Pollut. Res.* **2016**, *23*, 8244–8259. [CrossRef] [PubMed]
19. Liang, J.; Schoenau, J.J. Development of Resin Membranes as a Sensitive Indicator of Heavy-Metal Toxicity in the Soil Environment. *Int. J. Environ. Anal. Chem.* **1995**, *59*, 265–275. [CrossRef]
20. Liu, Q.Q.; Leslie, E.M.; Le, X.C. Accumulation and Transport of Roxarsone, Arsenobetaine, and Inorganic Arsenic Using the Human Immortalized Caco-2 Cell Line. *J. Agric. Food Chem.* **2016**, *64*, 8902–8908. [CrossRef]
21. Middleton, D.R.S.; Watts, M.J.; Hamilton, E.M.; Ander, E.L.; Close, R.M.; Exley, K.S.; Crabbe, H.; Leonardi, G.S.; Fletcher, T.; Polya, D.A. Urinary arsenic profiles reveal exposures to inorganic arsenic from private drinking water supplies in Cornwall, UK. *Sci. Rep.* **2016**, *6*, 25656. [CrossRef]
22. Ahmad, S.A.; Khan, M.H.; Haque, M. Arsenic contamination in groundwater in Bangladesh: Implications and challenges for healthcare policy. *Risk Manag. Healthc. Policy* **2018**, *11*, 251–261. [CrossRef]
23. Mukherjee, A.; Sengupta, M.K.; Hossain, M.A.; Ahamed, S.; Das, B.; Nayak, B.; Lodh, D.; Rahman, M.M.; Chakraborti, D. Arsenic contamination in groundwater: A global perspective with emphasis on the Asian scenario. *J. Health Popul. Nutr.* **2006**, *24*, 142–163.
24. Dani, S.U. Arsenic for the fool: An exponential connection. *Sci. Total Environ.* **2010**, *408*, 1842–1846. [CrossRef] [PubMed]
25. O'Bryant, S.E.; Edwards, M.; Menon, C.V.; Gong, G.; Barber, R. Long-Term Low-Level Arsenic Exposure Is Associated with Poorer Neuropsychological Functioning: A Project FRONTIER Study. *Int. J. Environ. Res. Public Health* **2011**, *8*, 861–874. [CrossRef]
26. Baum, L.; Chan, I.H.S.; Cheung, S.K.K.; Goggins, W.B.; Mok, V.; Lam, L.; Leung, V.; Hui, E.; Ng, C.; Woo, J.; et al. Serum zinc is decreased in Alzheimer's disease and serum arsenic correlates positively with cognitive ability. *Biometals* **2010**, *23*, 173–179. [CrossRef] [PubMed]
27. Namgung, U.; Xia, Z. Arsenic induces apoptosis in rat cerebellar neurons via activation of JNK3 and p38 MAP kinases. *Toxicol. Appl. Pharmacol.* **2001**, *174*, 130–138. [CrossRef] [PubMed]

28. Gong, G.; O'Bryant, S.E. The Arsenic Exposure Hypothesis for Alzheimer Disease. *Alzheimer Dis. Assoc. Disord.* **2010**, *24*, 311–316. [CrossRef] [PubMed]
29. Bakulski, K.M.; Seo, Y.A.; Hickman, R.C.; Brandt, D.; Vadari, H.S.; Hu, H.; Park, S.K. Heavy Metals Exposure and Alzheimer's Disease and Related Dementias. *J. Alzheimers Dis.* **2020**, *76*, 1215–1242. [CrossRef]
30. Sharma, A.; Kumar, S. Arsenic exposure with reference to neurological impairment: An overview. *Rev. Environ. Health* **2019**, *34*, 403–414. [CrossRef]
31. Alboghobeish, S.; Pashmforosh, M.; Zeidooni, L.; Samimi, A.; Rezaei, M. High fat diet deteriorates the memory impairment induced by arsenic in mice: A sub chronic in vivo study. *Metab. Brain Dis.* **2019**, *34*, 1595–1606. [CrossRef]
32. Wai, K.M.; Umezaki, M.; Mar, O.; Umemura, M.; Watanabe, C. Arsenic exposure through drinking Water and oxidative stress Status: A cross-sectional study in the Ayeyarwady region, Myanmar. *J. Trace. Elem. Med. Biol.* **2019**, *54*, 103–109.
33. Butterfield, D.A.; Boyd-Kimball, D. Oxidative Stress, Amyloid-beta Peptide, and Altered Key Molecular Pathways in the Pathogenesis and Progression of Alzheimer's Disease. *J. Alzheimers Dis.* **2018**, *62*, 1345–1367. [CrossRef]
34. Singh, A.P.; Goel, R.K.; Kaur, T. Mechanisms pertaining to arsenic toxicity. *Toxicol. Int.* **2011**, *18*, 87–93. [PubMed]
35. Ma, L.; Zhang, C.; Liu, W.J. Effects of arsenic on the offspring development in mice. *Zhonghua Yu Fang Yi Xue Za Zhi* **1994**, *28*, 20–23. [PubMed]
36. Chin-Chan, M.; Navarro-Yepes, J.; Quintanilla-Vega, B. Environmental pollutants as risk factors for neurodegenerative disorders: Alzheimer and Parkinson diseases. *Front. Cell. Neurosci.* **2015**, *9*, 124. [CrossRef] [PubMed]
37. Sasaguri, H.; Nilsson, P.; Hashimoto, S.; Nagata, K.; Saito, T.; De Strooper, B.; Hardy, J.; Vassar, R.; Winblad, B.; Saido, T.C. APP mouse models for Alzheimer's disease preclinical studies. *EMBO J.* **2017**, *36*, 2473–2487. [CrossRef] [PubMed]
38. Cheignon, C.; Tomas, M.; Bonnefont-Rousselot, D.; Faller, P.; Hureau, C.; Collin, F. Oxidative stress and the amyloid beta peptide in Alzheimer's disease. *Redox Biol.* **2018**, *14*, 450–464. [CrossRef]
39. Chandravanshi, L.P.; Gupta, R.; Shukla, R.K. Developmental Neurotoxicity of Arsenic: Involvement of Oxidative Stress and Mitochondrial Functions. *Biol. Trace Elem. Res.* **2018**, *186*, 185–198. [CrossRef]
40. Hannan, M.A.; Dash, R.; Sohag, A.A.M.; Haque, M.N.; Moon, I.S. Neuroprotection against Oxidative Stress: Phytochemicals Targeting TrkB Signaling and the Nrf2-ARE Antioxidant System. *Front. Mol. Neurosci.* **2020**, *13*, 116. [CrossRef]
41. Dwivedi, N.; Flora, S.J. Concomitant exposure to arsenic and organophosphates on tissue oxidative stress in rats. *Food Chem. Toxicol.* **2011**, *49*, 1152–1159. [CrossRef]
42. Roy, N.K.; Murphy, A.; Costa, M. Arsenic Methyltransferase and Methylation of Inorganic Arsenic. *Biomolecules* **2020**, *10*, 1351. [CrossRef]
43. Dash, R.; Mitra, S.; Ali, M.C.; Oktaviani, D.F.; Hannan, M.A.; Choi, S.M.; Moon, I.S. Phytosterols: Targeting neuroinflammation in neurodegeneration. *Curr. Pharm. Des.* **2021**, *27*, 383–401. [CrossRef]
44. Medda, N.; Patra, R.; Ghosh, T.K.; Maiti, S. Neurotoxic Mechanism of Arsenic: Synergistic Effect of Mitochondrial Instability, Oxidative Stress, and Hormonal-Neurotransmitter Impairment. *Biol. Trace Elem. Res.* **2020**, *198*, 8–15. [CrossRef]
45. Firdaus, F.; Zafeer, M.F.; Anis, E.; Ahmad, M.; Afzal, M. Ellagic acid attenuates arsenic induced neuro-inflammation and mitochondrial dysfunction associated apoptosis. *Toxicol. Rep.* **2018**, *5*, 411–417. [CrossRef] [PubMed]
46. He, Q.; Chen, B.; Chen, S.; Zhang, M.; Duan, L.; Feng, X.; Chen, J.; Zhou, L.; Chen, L.; Duan, Y. MBP-activated autoimmunity plays a role in arsenic-induced peripheral neuropathy and the potential protective effect of mecobalamin. *Environ. Toxicol.* **2021**, *36*, 1243–1253. [CrossRef]
47. Rahman, M.A.; Rahman, M.H.; Biswas, P.; Hossain, M.S.; Islam, R.; Hannan, M.A.; Uddin, M.J.; Rhim, H. Potential therapeutic role of phytochemicals to mitigate mitochondrial dysfunctions in Alzheimer's disease. *Antioxidants* **2021**, *10*, 23. [CrossRef]
48. Prakash, C.; Soni, M.; Kumar, V. Mitochondrial oxidative stress and dysfunction in arsenic neurotoxicity: A review. *J. Appl. Toxicol.* **2016**, *36*, 179–188. [CrossRef]
49. Prakash, C.; Soni, M.; Kumar, V. Biochemical and Molecular Alterations Following Arsenic-Induced Oxidative Stress and Mitochondrial Dysfunction in Rat Brain. *Biol. Trace Elem. Res.* **2015**, *167*, 121–129. [CrossRef] [PubMed]
50. King, A.P.; Wilson, J.J. Endoplasmic reticulum stress: An arising target for metal-based anticancer agents. *Chem. Soc. Rev.* **2020**, *49*, 8113–8136. [CrossRef]
51. Liu, X.; Gao, Y.; Liu, Y.; Zhang, W.; Yang, Y.; Fu, X.; Sun, D.; Wang, J. Neuroglobin alleviates arsenic-induced neuronal damage. *Environ. Toxicol. Pharmacol.* **2021**, *84*, 103604. [CrossRef] [PubMed]
52. Delaney, P.; Nair, A.R.; Palmer, C.; Khan, N.; Sadler, K.C. Arsenic induced redox imbalance triggers the unfolded protein response in the liver of zebrafish. *Toxicol. Appl. Pharmacol.* **2020**, *409*, 115307. [CrossRef]
53. Rana, S.V.S. Endoplasmic Reticulum Stress Induced by Toxic Elements-a Review of Recent Developments. *Biol. Trace Elem. Res.* **2020**, *196*, 10–19. [CrossRef] [PubMed]
54. Wei, S.; Qiu, T.; Yao, X.; Wang, N.; Jiang, L.; Jia, X.; Tao, Y.; Wang, Z.; Pei, P.; Zhang, J.; et al. Arsenic induces pancreatic dysfunction and ferroptosis via mitochondrial ROS-autophagy-lysosomal pathway. *J. Hazard. Mater.* **2020**, *384*, 121390. [CrossRef]
55. Zhang, W.; Cui, X.; Gao, Y.; Sun, L.; Wang, J.; Yang, Y.; Liu, X.; Li, Y.; Guo, X.; Sun, D. Role of pigment epithelium-derived factor (PEDF) on arsenic-induced neuronal apoptosis. *Chemosphere* **2019**, *215*, 925–931. [CrossRef] [PubMed]
56. Sinha, K.; Das, J.; Pal, P.B.; Sil, P.C. Oxidative stress: The mitochondria-dependent and mitochondria-independent pathways of apoptosis. *Arch. Toxicol.* **2013**, *87*, 1157–1180. [CrossRef]

57. Fu, S.C.; Lin, J.W.; Liu, J.M.; Liu, S.H.; Fang, K.M.; Su, C.C.; Hsu, R.J.; Wu, C.C.; Huang, C.F.; Lee, K.I.; et al. Arsenic induces autophagy-dependent apoptosis via Akt inactivation and AMPK activation signaling pathways leading to neuronal cell death. *Neurotoxicology* **2021**, *85*, 133–144. [CrossRef]
58. Dash, R.; Jahan, I.; Ali, M.C.; Mitra, S.; Munni, Y.A.; Timalsina, B.; Hannan, M.A.; Moon, I.S. Potential roles of natural products in the targeting of proteinopathic neurodegenerative diseases. *Neurochem. Int.* **2021**, *145*, 105011. [CrossRef]
59. Dash, R.; Ali, M.C.; Jahan, I.; Munni, Y.A.; Mitra, S.; Hannan, M.A.; Timalsina, B.; Oktaviani, D.F.; Choi, H.J.; Moon, I.S. Emerging potential of cannabidiol in reversing proteinopathies. *Ageing Res. Rev.* **2021**, *65*, 101209. [CrossRef]
60. Tam, L.M.; Wang, Y. Arsenic Exposure and Compromised Protein Quality Control. *Chem. Res. Toxicol.* **2020**, *33*, 1594–1604. [CrossRef]
61. Andersson, S.; Romero, A.; Rodrigues, J.I.; Hua, S.S.; Hao, X.X.; Jacobson, T.; Karl, V.; Becker, N.; Ashouri, A.; Rauch, S.; et al. Genome-wide imaging screen uncovers molecular determinants of arsenite-induced protein aggregation and toxicity. *J. Cell. Sci.* **2021**, *134*, jcs258338. [CrossRef]
62. Priya Wadgaonkar, F.C. Connections between endoplasmic reticulum stress-associated unfolded protein response, mitochondria, and autophagy in arsenic-induced carcinogenesis. *Semin. Cancer Biol.* **2021**, in press. [CrossRef]
63. Dodson, M.; Liu, P.F.; Jiang, T.; Ambrose, A.J.; Luo, G.; de la Vega, M.R.; Cholanians, A.B.; Wong, P.K.; Chapman, E.; Zhang, D.D. Increased O-GlcNAcylation of SNAP29 Drives Arsenic-Induced Autophagic Dysfunction. *Mol. Cell. Biol.* **2018**, *38*. [CrossRef] [PubMed]
64. Dodson, M.; de la Vega, M.R.; Harder, B.; Castro-Portuguez, R.; Rodrigues, S.D.; Wong, P.K.; Chapman, E.; Zhang, D.D. Low-level arsenic causes proteotoxic stress and not oxidative stress. *Toxicol. Appl. Pharmacol.* **2018**, *143*, 106–113. [CrossRef] [PubMed]
65. Bolt, A.M.; Zhao, F.; Pacheco, S.; Klimecki, W.T. Arsenite-induced autophagy is associated with proteotoxicity in human lymphoblastoid cells. *Toxicol. Appl. Pharmacol.* **2012**, *264*, 255–261. [CrossRef] [PubMed]
66. Weidling, I.; Swerdlow, R.H. Mitochondrial Dysfunction and Stress Responses in Alzheimer's Disease. *Biology* **2019**, *8*, 39. [CrossRef]
67. Wang, Y.; Zhao, H.J.; Liu, Y.C.; Guo, M.H.; Tian, Y.; Huang, P.Y.; Xing, M.W. Arsenite induce neurotoxicity of common carp: Involvement of blood brain barrier, apoptosis and autophagy, and subsequently relieved by zinc (II) supplementation. *Aquat. Toxicol.* **2021**, *232*, 105765. [CrossRef]
68. Zhao, H.; Wang, Y.; Liu, J.; Guo, M.; Fei, D.; Yu, H.; Xing, M. The cardiotoxicity of the common carp (*Cyprinus carpio*) exposed to environmentally relevant concentrations of arsenic and subsequently relieved by zinc supplementation. *Environ. Pollut.* **2019**, *253*, 741–748. [CrossRef] [PubMed]
69. Wang, Y.; Zhao, H.; Liu, Y.; Nie, X.; Xing, M. Zinc exerts its renal protection effect on arsenic-exposed common carp: A signaling network comprising Nrf2, NF-κB and MAPK pathways. *Fish Shellfish. Immunol.* **2020**, *104*, 383–390. [CrossRef]
70. Wang, Y.; Zhao, H.; Nie, X.; Guo, M.; Jiang, G.; Xing, M. Zinc application alleviates the adverse renal effects of arsenic stress in a protein quality control way in common carp. *Environ. Res.* **2020**, *191*, 110063. [CrossRef]
71. Zhao, H.; Wang, Y.; Guo, M.; Fei, D.; Mu, M.; Yu, H.; Xing, M. Hepatoprotective effects of zinc (II) via cytochrome P-450/reactive oxygen species and canonical apoptosis pathways after arsenite waterborne exposure in common carp. *Chemosphere* **2019**, *236*, 124869. [CrossRef]
72. Wang, Y.; Zhao, H.; Mu, M.; Guo, M.; Xing, M. Zinc offers splenic protection through suppressing PERK/IRE1-driven apoptosis pathway in common carp (Cyprinus carpio) under arsenic stress. *Ecotoxicol. Environ. Saf.* **2021**, *208*, 111473. [CrossRef] [PubMed]
73. Cao, A.L.; Beaver, L.M.; Wong, C.P.; Hudson, L.G.; Ho, E. Zinc deficiency alters the susceptibility of pancreatic beta cells (INS-1) to arsenic exposure. *Biometals* **2019**, *32*, 845–859. [CrossRef]
74. Garla, R.; Sharma, N.; Shamli; Kaushal, N.; Garg, M.L. Effect of Zinc on Hepatic and Renal Tissues of Chronically Arsenic Exposed Rats: A Biochemical and Histopathological Study. *Biol. Trace Elem. Res.* **2021**. [CrossRef] [PubMed]
75. Milton, A.G.; Zalewski, P.D.; Ratnaike, R.N. Zinc protects against arsenic-induced apoptosis in a neuronal cell line, measured by DEVD-caspase activity. *Biometals* **2004**, *17*, 707–713. [CrossRef]
76. Wei, Y.Y.; Huang, H.; Xia, Y.K.; Wei, L.M.; Chen, X.; Zhang, R.Y.; Duan, W.W.; Su, L.; Rahman, M.L.; Rahman, M.; et al. Antagonistic effect of early stage zinc on arsenic toxicity induced preterm birth during pregnancy: Evidence from a rural Bangladesh birth cohort. *Chin. Med. J.* **2021**, *134*, 619–621. [CrossRef]
77. Adedara, I.A.; Fabunmi, A.T.; Ayenitaju, F.C.; Atanda, O.E.; Adebowale, A.A.; Ajayi, B.O.; Owoeye, O.; Rocha, J.B.T.; Farombi, E.O. Neuroprotective mechanisms of selenium against arsenic-induced behavioral impairments in rats. *Neurotoxicology* **2020**, *76*, 99–110. [CrossRef] [PubMed]
78. Samad, N.; Rao, T.; Rehman, M.H.u.; Bhatti, S.A.; Imran, I. Inhibitory Effects of Selenium on Arsenic-Induced Anxiety-/Depression-Like Behavior and Memory Impairment. *Biol. Trace Elem. Res.* **2021**. [CrossRef] [PubMed]
79. Ren, Z.; Deng, H.; Wu, Q.; Jia, G.; Wen, N.; Deng, Y.; Zhu, L.; Zuo, Z.; Deng, J. Effect of Selenium on Brain Injury in Chickens with Subacute Arsenic Poisoning. *Biol. Trace Elem. Res.* **2021**. [CrossRef]
80. Ren, Z.; Deng, H.; Deng, Y.; Tang, W.; Wu, Q.; Zuo, Z.; Cui, H.; Hu, Y.; Yu, S.; Xu, S.Y.; et al. Effects of Selenium on Arsenic-Induced Liver Lesions in Broilers. *Biol. Trace Elem. Res.* **2021**, *199*, 1080–1089. [CrossRef] [PubMed]
81. Adedara, I.A.; Adebowale, A.A.; Atanda, O.E.; Fabunmi, A.T.; Ayenitaju, A.C.; Rocha, J.B.T.; Farombi, E.O. Selenium abates reproductive dysfunction via attenuation of biometal accumulation, oxido-inflammatory stress and caspase-3 activation in male rats exposed to arsenic. *Environ. Pollut.* **2019**, *254 Pt B*, 113079. [CrossRef]

82. Zwolak, I. The Role of Selenium in Arsenic and Cadmium Toxicity: An Updated Review of Scientific Literature. *Biol. Trace Elem. Res.* **2020**, *193*, 44–63. [CrossRef]
83. Sarkar, N.; Das, B.; Bishayee, A.; Sinha, D. Arsenal of Phytochemicals to Combat Against Arsenic-Induced Mitochondrial Stress and Cancer. *Antioxid. Redox Signal.* **2020**, *33*, 1230–1256. [CrossRef]
84. Zhang, Q.Y.; Wang, F.X.; Jia, K.K.; Kong, L.D. Natural Product Interventions for Chemotherapy and Radiotherapy-Induced Side Effects. *Front. Pharmacol.* **2018**, *9*, 1253. [CrossRef] [PubMed]
85. Salehi, B.; Machin, L.; Monzote, L.; Sharifi-Rad, J.; Ezzat, S.M.; Salem, M.A.; Merghany, R.M.; El Mahdy, N.M.; Kılıç, C.S.; Sytar, O.; et al. Therapeutic Potential of Quercetin: New Insights and Perspectives for Human Health. *ACS Omega* **2020**, *5*, 11849–11872. [CrossRef]
86. Sarkozi, K.; Papp, A.; Horváth, E.; Máté, Z.; Ferencz, A.; Hermesz, E.; Krisch, J.; Paulik, E.; Szabó, A. Green tea and vitamin C ameliorate some neuro-functional and biochemical signs of arsenic toxicity in rats. *Nutr. Neurosci.* **2016**, *19*, 102–109. [CrossRef]
87. Vazhappilly, C.G.; Devanga, R.N.K.; Palamadai, K.S.; Rangasamy, A.K. In Vitro Protective Potentials of Annona muricata Leaf Extracts Against Sodium Arsenite-induced Toxicity. *Curr. Drug Discov. Technol.* **2015**, *12*, 59–63.
88. Jomova, K.; Jenisova, Z.; Feszterova, M.; Baros, S.; Liska, J.; Hudecova, D.; Rhodes, C.J.; Valko, M. Arsenic: Toxicity, oxidative stress and human disease. *J. Appl. Toxicol.* **2011**, *31*, 95–107. [CrossRef] [PubMed]
89. Das, B.; Mandal, S.; Chaudhuri, K. Role of arginine, a component of aqueous garlic extract, in remediation of sodium arsenite induced toxicity in A375 cells. *Toxicol. Res.* **2014**, *3*, 191–196. [CrossRef]
90. Das, T.; Roychoudhury, A.; Sharma, A.; Talukder, G. Modification of clastogenicity of three known clastogens by garlic extract in mice in vivo. *Environ. Mol. Mutagen.* **1993**, *21*, 383–388. [CrossRef] [PubMed]
91. Aslani, M.R.; Najarnezhad, V.; Mohri, M. Individual and combined effect of meso-2,3-dimercaptosuccinic acid and allicin on blood and tissue lead content in mice. *Planta Med.* **2010**, *76*, 241–244. [CrossRef] [PubMed]
92. Amagase, H.; Petesch, B.L.; Matsuura, H.; Kasuga, S.; Itakura, Y. Intake of garlic and its bioactive components. *J. Nutr.* **2001**, *131*, 955S–962S. [CrossRef] [PubMed]
93. Gupta, R.; Flora, S.J. Therapeutic value of Hippophae rhamnoides L. against subchronic arsenic toxicity in mice. *J. Med. Food* **2005**, *8*, 353–361. [CrossRef] [PubMed]
94. Gupta, R.; Flora, S.J. Effect of Centella asiatica on arsenic induced oxidative stress and metal distribution in rats. *J. Appl. Toxicol.* **2006**, *26*, 213–222. [CrossRef]
95. Tiwari, H.; Rao, M.V. Curcumin supplementation protects from genotoxic effects of arsenic and fluoride. *Food Chem. Toxicol.* **2010**, *48*, 1234–1238. [CrossRef] [PubMed]
96. Mishra, D.; Flora, S.J. Quercetin administration during chelation therapy protects arsenic-induced oxidative stress in mice. *Biol. Trace Elem. Res.* **2008**, *122*, 137–147. [CrossRef]
97. Ghosh, A.; Mandal, A.K.; Sarkar, S.; Panda, S.; Das, N. Nanoencapsulation of quercetin enhances its dietary efficacy in combating arsenic-induced oxidative damage in liver and brain of rats. *Life Sci.* **2009**, *84*, 75–80. [CrossRef]
98. Bjorklund, G.; Oliinyk, P.; Lysiuk, R.; Rahaman, M.S.; Antonyak, H.; Lozynska, I.; Lenchyk, L.; Peana, M. Arsenic intoxication: General aspects and chelating agents. *Arch. Toxicol.* **2020**, *94*, 1879–1897. [CrossRef]
99. Aziz, S.N.; Boyle, K.J.; Rahman, M. Knowledge of arsenic in drinking-water: Risks and avoidance in Matlab, Bangladesh. *J. Health Popul. Nutr.* **2006**, *24*, 327–335.

toxics

MDPI

Review

Commonalities between Copper Neurotoxicity and Alzheimer's Disease

Roshni Patel and **Michael Aschner** *

Department of Molecular Pharmacology, Albert Einstein College of Medicine, 1300 Morris Park Avenue, Bronx, NY 10461, USA; roshnipatel6511@gmail.com
* Correspondence: michael.aschner@einsteinmed.org

Abstract: Alzheimer's disease, a highly prevalent form of dementia, targets neuron function beginning from the hippocampal region and expanding outwards. Alzheimer's disease is caused by elevated levels of heavy metals, such as lead, zinc, and copper. Copper is found in many areas of daily life, raising a concern as to how this metal and Alzheimer's disease are related. Previous studies have not identified the common pathways between excess copper and Alzheimer's disease etiology. Our review corroborates that both copper and Alzheimer's disease target the hippocampus, cerebral cortex, cerebellum, and brainstem, affecting motor skills and critical thinking. Additionally, Aβ plaque formation was analyzed beginning from synthesis at the APP parent protein site until Aβ plaque formation was completed. Structural changes were also noted. Further analysis revealed a relationship between amyloid-beta plaques and copper ion concentration. As copper ion levels increased, it bound to the Aβ monomer, expediting the plaque formation process, and furthering neurodegeneration. These conclusions can be utilized in the medical community to further research on the etiology of Alzheimer's disease and its relationships to copper and other metal-induced neurotoxicity.

Keywords: Alzheimer's disease; copper

Citation: Patel, R.; Aschner, M. Commonalities between Copper Neurotoxicity and Alzheimer's Disease. *Toxics* **2021**, *9*, 4. https://doi.org/10.3390/toxics9010004

Received: 12 December 2020
Accepted: 5 January 2021
Published: 7 January 2021

1. Introduction

1.1. Alzheimer's Disease

Alzheimer's disease is a progressive neurodegenerative disorder that targets neuron communication and can result in loss of cell function or cell death [1–3]. This is due to the buildup of abnormally structured proteins called amyloid plaques and neurofibrillary tangles between neurons, essentially blocking communication [3]. Nerve cell death consequently results in loss of brain tissue. Alzheimer's disease initially disrupts bodily communication, metabolism, and repair processes [3]. As its effects advance, the patient may experience memory loss as well as changes in language, reasoning, and social behavior [3]. Alzheimer's disease is also a fatal form of dementia, and, according to a 2020 study, is the sixth-leading cause of death in the United States [4].

It is estimated that one in 10 Americans age 65 or older, about 5.8 million people, suffer from Alzheimer's disease [5]. Women show a greater prevalence of Alzheimer's disease with 3.5 million cases, about two-thirds of Americans with Alzheimer's disease, while men show a prevalence of 2.1 million [6]. Between 2000 and 2017, the death rate due to Alzheimer's disease increased by 145% with 121,404 cases reporting Alzheimer's disease as the cause of death in 2017 [6]. This number is expected to rise in the coming years as it is estimated that the prevalence of Alzheimer's disease in the United States will continue to grow to 13.8 million [6].

Alzheimer's disease proves to be a major social and financial burden on the families of those affected. It is estimated that 83 percent of unpaid aid, referred to as informal care, provided to those with Alzheimer's comes from family members, friends, or unpaid caregivers [7]. One study estimated that in 2018, over 18.5 billion hours of informal care were provided to those with Alzheimer's disease and like dementias, equating to $233.9 billion [6]. Lifetime costs of someone with Alzheimer's disease are estimated at $350,174 where 70 percent of these costs are associated with family care [8]. This poses a significant problem as 41 percent of informal caregivers have a household income of less than $50,000, putting an outstanding strain on family members, especially since many of those placed in these situations often believe they had no choice in taking on such a role [6].

1.2. Copper

Copper, an essential trace element found in the brain, liver, and kidneys, enables the body to form red blood cells, maintain bone health, and can help prevent cardiovascular disease and osteoporosis. Copper is also a key element in maintaining lung function as it plays a vital role in metabolic processes such as cellular respiration [9]. Copper stored in the human body can be used for protein and energy production [10,11]. For adults, healthy copper levels range between 50 and 80 mg; levels exceeding this range are considered toxic and can lead to a buildup of copper in the kidneys, brain, and eyes. This causes a burden on the body as it may result in possible liver cell death, permanent nerve damage, oxidative stress, and reduced cell proliferation [10,12]. A lethal dose of copper ranges from 10 to 20 g [12]. Common signs of copper toxicity include, but are not limited to, headaches, bloody vomit, diarrhea, and black stools.

In the bloodstream, copper exists in two forms: 85 to 90 percent of copper found in the blood is bound to ceruloplasmin, a protein that plays a role in iron metabolism; the remaining 10 to 15 percent is free-floating copper, sometimes loosely bonded to other molecules [12,13]. Copper toxicity may occur as a result of various exposures. Common means of excess copper exposure occur from consumption of acidic foods cooked in uncoated copper cookware, exposure to excess copper in drinking water, breathing air or dust containing copper, as well as other environmental sources [12,14]. Other instances include copper salt topical creams for burn treatment, as well as in farming as a pesticide, and the leather industry [12].

Several genetic disorders are associated with copper related diseases. Examples of such illnesses include Wilson's disease and Menkes disease. Wilson's disease, a genetic disorder where the body is unable to filter out excess copper and therefore builds up in body tissue, occurs due to mutations in the ATP7B gene [13]. ATP7B codes for the protein ATPase 2, a copper-transporter found in the liver and brain [13,15]. This mutation can result in hepatic toxicity as well as adverse effects on the central nervous system, disrupting homeostatic bodily functions [13]. Similarly, Menkes disease, an X-linked recessive disorder, is a result of mutation in the ATP7A gene; the purpose of ATP7A is to code for copper regulation and copper absorption from food [15]. Symptoms of Menke's induced copper deficiency include hair loss, slow growth and development, and neurological effects [15].

2. Methodology

This study aimed to clarify the relationship between copper toxicity and Alzheimer's disease, more specifically the effect of copper on neurological pathways and amyloid-beta plaque formation. Such an association has been previously studied by various researchers; therefore, this study utilized a systematic review approach of published literature to synthesize the known data and information regarding this relationship (See Figure 1). Data sets such as PubMed, Science Direct, and Google Scholar were utilized to collect journals, publications, and review journals relating to the topic. Journals were chosen using keywords: copper, Alzheimer's disease, neurotoxicity, and amyloid beta plaques. These data sets were sourced using Zotero, and online reference manager. These journals were reviewed and trends and patterns regarding copper toxicity and its relationship with Alzheimer's disease were analyzed.

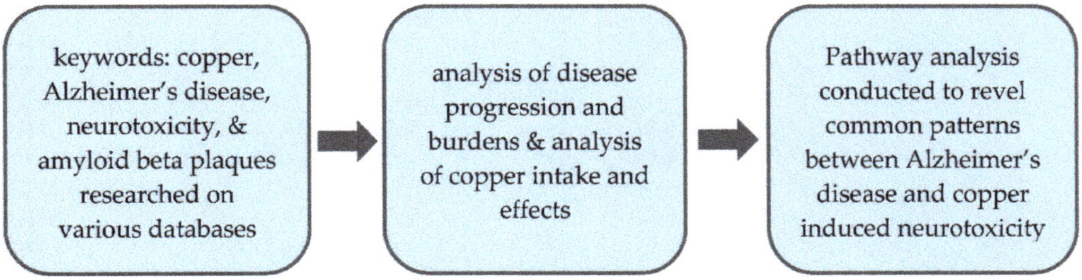

Figure 1. Methodology overview.

To begin the data collection process, preliminary research was conducted to gain an overarching understanding of the topic. Such research included analyzing the progression and development of Alzheimer's disease and its effects on a patient and those around them. Social and financial burdens were also analyzed to develop the societal impacts of Alzheimer's disease. Similarly, journals pertaining to copper toxicity and its effects on the brain were analyzed to understand the common causes of copper toxicity, the prevalence of copper in everyday life, and common forms of entry into the human body.

Once comprehensive background knowledge of the individual research subjects was obtained, trends and patterns relating the two were sought and noted. Firstly, patterns in the pathways of both copper and Alzheimer's disease effects were noted in order to analyze if a correlation could be derived between the two. Patterns were analyzed in terms of the route copper and Alzheimer's disease took in the brain; pathways were named in terms of the section of the brain that was affected, such as cerebral cortex, cerebellum, and brainstem. As more trends arose, they were subsequently categorized into their respective clusters, such as common pathways of copper and Alzheimer's disease, as well as the relationship between Aβ plaques and copper binding sites. After the organization of the detailed clusters was analyzed and categorized respectively, sections were created to study the beta-amyloid plaque and copper relationship. This area included the analysis of plaque formation from protein strand to tangle as well as how shape affects the purpose of the plaque. Then, copper's relationship to the development of these plaques was introduced and the two were analyzed in concurrence to reveal significant findings such as coppers role in plaque development as well as how different copper concentrations affect plaque growth.

3. Copper Pathways to the Brain

3.1. Role of Copper in the Human Body

Copper is an element essential to the homeostasis of the human body due to its role in energy production, iron metabolism, neuropeptide activation, and connective tissue and neurotransmitter synthesis [16]. Copper's most crucial property however is its role in cuproenzyme ceruloplasmin [17]. Ceruloplasmin is the major copper-carrying protein in the blood, accounting for over 95% of copper transport in human plasma [16,17]. Studies have also found that copper plays a crucial role in the formation of red blood cells, immune system function, brain development, gene expression, and other physiologic processes [16]. Through the use of ceruloplasmin transport, copper is able to reach the brain, which uses a significant portion of copper found in the body for brain development and regulation of the nervous system [13,16,17]. Copper is also significantly used in the liver to convert iron into its ferric form as well as absorb iron into the gut [13].

3.2. Common Forms of Copper Intake

Copper toxicity is an associated result of exposure to copper contaminated sources. Common exposure sites include water supplies, copper pipes, uncoated copper cookware, birth control, dietary supplements, food, and fungicides. Water supplies are often contaminated by farm operations and other industrial waste practices that become runoff and enter reservoirs and public wells [18]. This waster is then transported to nearby populations for consumption and other daily activities. Such an instance was accounted for in a 2016 journal highlighting the widespread water contamination of heavy metals, specifically copper and lead, in New South Wales, Australia [18]. This study analyzed tap water from 212 different homes in the region; of these samples, almost 100 percent tested positive for copper, with 5 percent testing positive for excess copper [18]. For the houses testing within the normal copper range, it was also found that drinking water contributed to 6 to 13 percent of the average daily intake of copper [18]. A high copper positive percentage was expected since many of these homes utilize copper pipes and fittings for water supply [18]. Subsequently, copper pipes are a great cause of concern regarding copper contamination as acidic water may cause erosion of the pipe or fitting, releasing copper particles directly into the water [18]. Therefore, this study concluded there is a great concern for copper levels due to their frequent occurrence at high concentrations, as this water is then used for drinking, washing of cookware, as well as for bathing, and personal hygiene, creating various opportunities for direct ingestion of copper contaminated water, increasing the prevalence of copper toxicity in the region [18]. Other sources of copper contamination include improperly coated copper cookware, such as pots and pans. If not coated with another non-reactive metal, copper from the cookware can enter food and therefore the human body. As copper levels in the body increase, excess copper begins to pool in tissues, leading to copper toxicity. Similar to copper pipes, unprotected copper cookware can corrode in the presence of acidic foods, resulting in another means of entry into the human body [12]. Copper toxicity can also be caused through the use of birth control, as it can raise copper levels, which also destroys Vitamin C, another essential nutrient for the human body's optimal function [19]. Dietary supplements and other foods can also be a source of excess copper. The recommended daily dose of copper for those aged 19 and older is 900 mcg, with the upper limit at 10,000 mcg [13]. Overconsumption of copper supplements or foods high in copper content, such as legumes, mushrooms, chocolate, liver, and nuts and seeds, can result in copper toxicity when not consumed in moderation [13]. Copper sulfate, an inorganic copper and sulfur compound used as a fungicide and algaecide in swimming pools, creates a toxic environment for microorganisms that kill small plants and animals by shocking roots and causing copper toxicity, respectively [20]. Due to its multiple points of entry into the human body, copper contamination is a growing concern, especially considering the large quantities in which it is found in the activities of daily life.

3.3. Alzheimer's Disease and Copper Pathways

The adult human brain contains approximately 100 billion neurons, directly correlating with brain mass. As healthy aging occurs, neurogenesis, the process of regeneration by neural plasticity, slows down, and neuron proliferation rate decreases [21]. Dementias, such as Alzheimer's, can speed up neurodegenerative processes, causing memory loss, progressing to the point where the patient is no longer capable of independently performing activities of daily living. Referred to as cerebral atrophy, the loss of neurons decreases brain mass overall or in specific areas [22]. Alzheimer's disease begins by targeting the hippocampus and entorhinal cortex located in the temporal lobe [3,23]. The temporal lobe is responsible for the connection of the network responsible for memory, navigation, and the perception of time [23]. The targeting of this region decreases the number of transmitters in the temporal lobe, causing neuron death and therefore size reduction, which can be seen on MRI scans of the brain [23]. A 2020 study found that the temporal lobe of those with Alzheimer's disease decreases at a rate of 15.1 percent per year whereas the normal neurodegeneration rate is a significantly lower 1.5 percent [24]. The study concluded that cognitive decline and subsequent memory loss are a direct result of temporal lobe atrophy [24]. As the disease progresses, the cerebral cortex also becomes impaired [3,25]. The cerebral cortex is the outer layer of neural tissue located at the front of the head, covering the outer portion of the cerebrum [26]. The cerebrum controls language, reasoning, social behaviors, emotion, muscle movement, hearing, vision, and other sensory controls [3,26]. Alzheimer's disease can also be characterized by neural mass loss and astrocytosis in the cerebral cortex [3,25]. A 2018 study found that about 25 percent of individuals who die by the age of 75 presented with substantial cerebral lesions resembling those of Alzheimer's disease [25]. This study concluded that the identification of cellular brain structure is essential to understanding neurodegenerative disease progression [25]. The order of progression for Alzheimer's disease begins in the hippocampus and expands outward to the temporal lobes, frontal lobes and prefrontal cortex, parietal lobes, occipital lobes, cerebellum, and finally the brainstem [27]. Once the disease reaches the brain stem, autonomous functions of the body will cease, ultimately proving fatal [27].

Copper toxicity results in the pooling of copper in different tissues of the body [10,12,28]. Prominent areas of copper pooling include the liver, brain, and eyes [12]. A 2013 study of male Wistar rats conducted by Pal et al. found copper toxicity effects on the rat brain include swelling and increased number of astrocytes, star-shaped glial cells, and copper deposition in the choroid plexus, which is located in close contact with the cerebral cortex [29,30]. The study concluded that copper toxicity in male Wistar rats causes impaired spatial memory and neuromuscular coordination, swelling of astrocytes, copper deposition in the choroid plexus, neuronal degeneration, and augmented levels of copper in the hippocampus [29]. This is a significant finding due to the implications of these symptoms; impaired spatial memory and neuromuscular coordination can result in difficulty walking and navigating, swelling of astrocytes can result in brain edema and fulminant hepatic failure, impairment of the choroid plexus can cause variations in the development of cerebrospinal fluid possibly leading to overproduction causing pressure in the brain, neuronal degeneration can lead to neuron death, and augmented levels on copper in the hippocampus can lead to various dementias [31–34]. Dementias may be a result of the various plaques copper can induce, the most prevalent of which are amyloid-beta plaques associated with Alzheimer's disease [35]. Another 2018 study, conducted by Kardos et al., found copper levels to be significantly higher in the cerebellum, choroid plexus, ventricle system, and substantia nigra, a region of the midbrain [36]. This study also found that free excess copper often pools in the soma of cerebellar granule and cortical pyramidal neurons, as well as the hippocampus and spinal cord [36]. One 2012 study analyzed possible reasons for high copper levels in the brain and theorized that a high-fat diet caused an increase in copper levels, which they theorized was also correlated to an increased risk of Alzheimer's disease [37]. The study found that the highest 20 percent of those with copper intake lost cognition at six times the rate of groups with lower copper consumption if they also had a high-fat diet [37]. The study hypothesized that the ingestion of free-floating copper from sources such as drinking water and copper supplements played a major factor in the onset of Alzheimer's as it caused high levels of copper pooling in the brain [37]. This study concluded that there is a strong correlation between copper levels and the prevalence of Alzheimer's disease [37]. It is believed this correlation exists due to the commonalities between affected areas by Alzheimer's disease and copper toxicity [36,37]. Such common areas include the hippocampus, cerebral cortex, cerebellum, and brainstem, which affect memory, information processing, motor skills, and regulation of autonomous functions, respectively [27,36,37]. This supports the theory that copper may play a role in the onset of Alzheimer's disease as these regions are also the pathway of the general progression of the disease [27].

4. Beta-Amyloid and Copper Relationship

4.1. Beta-Amyloid Plaques

Alois Alzheimer characterized Alzheimer's disease as an elevated number of plaques in the brain in the early twentieth century [38]. Then, in the 1980s, it was found that these plaques consist of amyloid-beta (Aβ) peptides, protein fragments from the amyloid precursor protein (APP) known to play an important role in synaptic physiology by regulating synaptic vesicle release and scaling [38,39]. This was an important discovery as the amyloid precursor protein had been known to regulate cell growth, motility, neurite outgrowth, and cell survival [39]. Aβ peptides are also naturally produced in the brain and normally exist in healthy levels in the cerebrospinal fluid and serum [40]. As research in the field progressed, John Hardy and Gerald Higgins proposed the Amyloid Cascade hypothesis in 1992 [41].

The Amyloid Cascade hypothesis postulates that as proteostasis changes due to aging, the amyloid precursor protein is broken down to form Aβ, which results in an abnormal accumulation of amyloid-beta plaques in various regions of the brain [35,42]. This hypothesis was significant since the formation of the Aβ monomers could cause the deposition of extracellular fibrils, resulting in neuron death and the formation of senile plaques [35]. A 2014 study on the Amyloid Cascade hypothesis, however, found that low doses of Aβ are stimulatory to the brain whereas high doses were inhibitory [38]. Since advanced medicine has been unable to find a cure for Alzheimer's disease based on this hypothesis, new avenues of approach are being investigated, leading to the development of new lines of reasoning [43]. One such new hypothesis that challenges the widely accepted Amyloid Cascade hypothesis is the Beta-Amyloid Dysfunction (BAD) hypothesis proposed by Heinz Hillen in 2019 [43]. The BAD hypothesis differs from the Amyloid Cascade hypothesis by building on the homeostasis of the Aβ peptide in the synaptic vesicle cycle [43]. Further, the BAD hypothesis accounts for the physiological Aβ monomer, the pathophysiology of Aβ deposits, and reduced Aβ cerebrospinal fluid levels in Alzheimer's disease whereas the Amyloid Cascade hypothesis does not [43]. Other hypotheses that were proposed but are not widely accepted include the infection hypothesis and vascular hypothesis [35].

The amyloid-beta buildup can cause multiple health concerns and detrimentally impact patient health. When Aβ plaques accumulate outside nerve cells, neural synapses are blocked and action potentials, or electrical impulses in the brain, are blocked, hindering communication and resulting in neurodegeneration [44–46]. This can cause a variety of illnesses, such as Alzheimer's disease, Parkinson's disease, Huntington's disease, type 2 diabetes, amyotrophic lateral sclerosis, secondary amyloidosis, and prion diseases, among others [47]. In Alzheimer's disease specifically, these Aβ plaques can also result in neurofibrillary tangles, further hindering neuron communication [35]. Mutations in the previously described APP protein cause the development of amyloid-beta plaques due to improper folding of the peptide [48,49]. A 2017 study by Han et al. examined how Aβ-42 induces neural apoptosis; the study concluded that Aβ-42 targets the mitochondria of the neuron, causing neurodegeneration [50].

The human body has some protection against the buildup of Aβ monomers in the form of brain macrophages which can engulf and destroy the peptide, called microglia [35]. However, the aging process results in a decreased amount of these macrophages, allowing for unchecked Aβ buildup [35]. Unregulated buildup induces an inflammatory response, causing neuroinflammation and neuron death [35]. This response works in conjunction with Tau proteins which are also stimulated by the inflammatory response, creating neurofibrillary tangles that restrict neuron function and lead to neurodegeneration [35].

Amyloid-beta peptides are created through the incorrect folding of proteins, resulting in amyloidosis [51]. Protease, an enzyme that breaks down large proteins, recognizes these abnormalities and discards them [51]. However, sometimes the amyloid to protease ratio is imbalanced, with more amyloids than the enzymes can handle, resulting in a buildup [51]. Other instances resulting in buildup are caused by the formation of a rigid structure from the amyloid mass, which exceeds the protease's capabilities [51]. When this mass is excreted out of the cell, it folds to create a beta-sheet, which can go on to wreak damage on the area, as seen in the case of Alzheimer's disease [49]. The Aβ peptide consists of 37 to 49 amino acids that stem from the previously defined APP protein [48]. These monomers can aggregate into various structures, including oligomers, protofibrils, and amyloid fibrils, which are usually larger than the latter two and insoluble [48]. Due to their larger size and insolubility, these fibrils can rearrange themselves into clumps, forming Aβ plaques [48]. Therefore, the APP's amyloid fibril form is said to be a primary component of amyloid plaques found in the brain during Alzheimer's disease [48,49]. These plaques are most commonly found in the neocortex, a part of the cerebral cortex associated with sight and hearing [48]. Figure 2 derived from the findings of Chen et al. provides more information into the process by which Aβ fibrils are created.

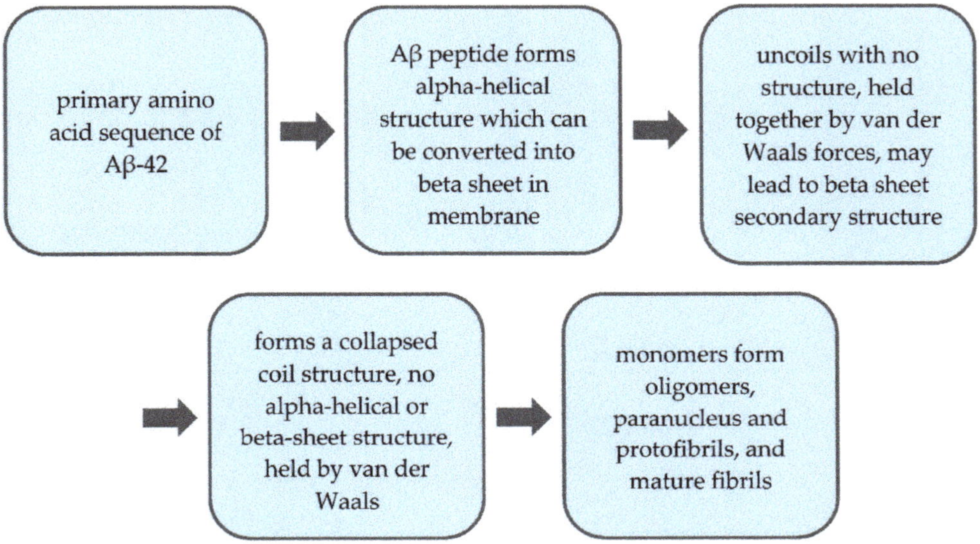

Figure 2. Structural process of Aβ-42 monomer to Aβ fibrils.

It is important to note that Aβ is thought to be intrinsically structured, and therefore is unable to be crystallized [48]. Therefore, scientists aim to optimize conditions to stabilize the Aβ peptide in order to study its properties [48]. The primary Aβ-42 amino acid sequence was discovered in 1984 from extracellular deposits and amyloid plaques and was found to range from 37 to 49 amino acids [48,49]. Early models of the Aβ peptide found it to fold into α-helical and β-sheet structures in membranous environments [48]. Portions of the peptide structure then uncoil and either form various or no structures, depending on their placement in the peptide chain; a helix to coil transformational transaction is also promoted [48]. The Aβ peptide collapses into nonstructural loops, strands, and turns when placed in water [48,49]. Although this chain no longer has a set structure, it is held together by van der Waals forces [48]. Beta secondary structures may be formed through fibrillization, and two helical regions connected by a β-turn are formed [48]. It is believed that this aggregation of Aβ peptides into Aβ fibrils is what causes the onset of neurotoxicity and dementias such as Alzheimer's disease [48].

4.2. Copper Toxicity and Aβ Levels

Alzheimer's disease is characterized by neurodegeneration, elevated heavy metal concentrations in the brain, and amyloid-beta plaque buildup [40]. Some heavy metals that are believed to have a relation to the onset of Alzheimer's disease include zinc, iron, and copper [40]. Of the three, copper is believed to have a more significant impact on the development of Alzheimer's disease due to the amyloid-beta plaques two copper-binding sites, which can be used to promote reactive oxygen species (ROS) generations [1,40]. When bound to plaques, copper cycles between the +I and +II oxidative states, boosting ROS generation; ROS may damage lipids, RNA, DNA, or proteins, effectively advancing aging and possibly Alzheimer's disease [52,53]. One 2018 study conducted by *Bagheri* et al. found that copper ions have a high affinity to the Aβ peptide copper-binding site, increasing the proportions of β-sheet and α-helix structures in Aβ aggregations, forming plaques [1]. This study also found that increased copper ion concentrations result in greater fibril formation, once again increasing Aβ plaque buildup [1]. This happens due to the β-amyloid's ability to transform into β-sheet structures when bound to two copper ions, which not only stimulates the aforementioned β-sheet conformity but also aggregation and toxicity, as depicted in greater detail in Figure 3 [1]. In vitro studies performed in the early 2000s were able to confirm the formation of these structures in copper toxic Alzheimer's patients [1].

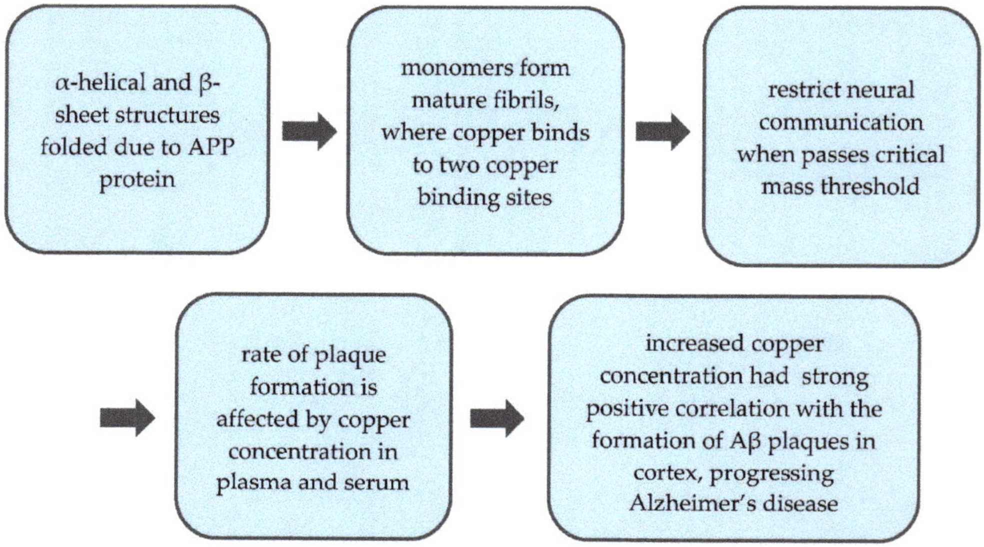

Figure 3. Process of APP protein to Aβ plaque formation in accordance with the effect of copper.

Several factors are believed to play a role in the copper's ability to bind to these Aβ aggregates, inducing a fibril formation process [54]. Such factors include pH of 7.0, increased concentrations of free copper in plasma, and serum copper concentrations, which can also cause the onset of a rare type of Alzheimer's disease [1,54,55]. Furthermore, the study conducted by Bagheri et al. found a 1:1 ratio of free-floating copper in serum to Aβ aggregates, a significant advancement in the development in the comprehensive understanding of Alzheimer's disease [1]. Over the course of its six meta-analyses, this study aimed to evaluate copper concentrations in AD in different biological matrices, this study concluded that free copper/copper ion, concentration in serum was significantly greater in patients who were diagnosed with Alzheimer's disease than those who were considered healthy [1]. This conclusion helps develop past findings, such as the fact that Aβ plaques exhibit high copper concentrations among other trace metals, to understand the reasoning behind why such phenomena occur [1].

As previously described, this increase in copper leads to an increase in Aβ plaques through the bonding of free copper and β structures. This bonding causes the shape of the β sheet to change again, creating fibrils that can aggregate. These masses can then hinder the brain's ability to successfully communicate information and allow the body to function. However, another 2016 study by Kitazawa et al. establishes an alternate line of reasoning for this process [56]. In this study, Kitazawa et al. propose that high copper ion concentrations in serum can trigger an inflammatory response in the brain [56]. This inflammatory response in turn obstructs the brain from breaking down Aβ aggregates, causing a buildup as seen in Alzheimer's disease [56]. Essentially, this study focused on excess copper concentration and its effects on microglia, which were previously defined as the body's natural way to break down and dispose of excess Aβ plaques before they pose a threat to the patient's homeostatic processes [35,56]. This study found that excess copper concentrations correlated with a decrease in microglial activity as well as brain inflammation [35,56].

When operating in conjunction, these processes accelerate neurodegeneration and the formation of Aβ plaques, detrimentally affecting the patient's brain twofold [35,56]. This study, like others cited herein, concluded that copper toxicity may trigger the progression of Alzheimer's disease by targeting homeostatic processes [1,35,40,54,56].

5. Conclusions

This study sought to investigate the relationship between Alzheimer's disease and copper toxicity. Initial research into the individual factors found Alzheimer's disease to be one of the most prevalent forms of dementia, accounting for 5.8 million cases in the United States alone [5]. As there is no cure for this illness to date, the disease proves fatal [4,27]. Care for those who suffer from Alzheimer's disease is extremely costly; those from lower-income demographics diagnosed with this illness face a significant financial strain, further establishing why a cure for this illness is needed [7]. Alzheimer's disease targets neuron function through the formation of amyloid-beta plaques, which hinder the neuron's ability to successfully transmit nerve impulses, causing neurodegeneration including cell death [3,57]. In turn, this loss of brain mass results in decreased cognitive and motor skills [24]. This study sought to understand the role of copper in the etiology of Alzheimer's disease.

Alzheimer's disease manifests in the inner portions of the brain, such as the hippocampus and temporal lobes, and extends outwards with the following order of progression: temporal lobe, frontal lobe, prefrontal cortex, parietal lobe, occipital lobe, cerebellum, and brainstem [27,36,37]. When analyzed alongside the regions targeted by Alzheimer's disease, commonalities with copper toxicity, such as accumulation in hippocampus, cerebral cortex, cerebellum, and brainstem were found, concomitant with adverse effects on memory, information processing, motor skills, and regulation of autonomous functions [1,24].

We further reviewed the relationship between amyloid-beta plaques and copper. Firstly, we reviewed the structure and formation of Aβ beginning from its parent protein, APP [38,39]. From the APP protein, structures such as α-helical and β-sheets were addressed to determine the process by which such structures are folded [48]. Finally, the process from monomer to plaque, including aggregation, was addressed [48]. The effect of copper ions on these plaques was then reviewed, suggesting that excess copper, given its high affinity for Aβ plaques, can bind to the plaques at two distinct copper-binding sites [1,40,54]. When these plaques reach a critical mass and cluster between neurons, they restrict neuronal communication [1,54,56]. Factors that affect the rate of plaque formation include copper concentrations in plasma and serum [1,54,55]. Increased copper concentration had a strong positive correlation with the formation of Aβ plaques in the cortex, advancing the progression of Alzheimer's disease [1,35,40,54–56].

These findings can be used in the medical field to derive treatment options and solutions to this ongoing problem. An additional timely study is to target the copper-binding site in the Aβ monomer to investigate how this change affects Aβ plaque generation. Another study should investigate if free-floating copper also plays a role in the rate at which Aβ monomers aggregate into Aβ plaques, as this would provide insight into how to directly combat the formation of plaques.

Author Contributions: Validation, R.P. and M.A.; formal analysis, R.P.; investigation, R.P.; data curation, R.P.; writing—original draft preparation, R.P.; writing—review and editing, M.A.; supervision, M.A. All authors have read and agreed to the published version of the manuscript.

Funding: This research was funded in part by National Institute of Environmental Health Sciences (NIEHS), grant numbers R01ES07331 and R01ES10563.

Institutional Review Board Statement: Not applicable.

Informed Consent Statement: Not applicable.

Data Availability Statement: Not applicable.

Acknowledgments: M.A. was supported in part by grants from the National Institute of Environmental Health Sciences (NIEHS) R01ES10563 and R01ES07331.

Conflicts of Interest: The authors declare no conflict of interest.

References

1. Bagheri, S.; Squitti, R.; Haertlé, T.; Siotto, M.; Saboury, A.A. Role of Copper in the Onset of Alzheimer's Disease Compared to Other Metals. *Front. Aging Neurosci.* **2018**, *9*. [CrossRef]
2. DeTure, M.A.; Dickson, D.W. The Neuropathological Diagnosis of Alzheimer's Disease. *Mol. Neurodegener.* **2019**, *14*, 32. [CrossRef]
3. National Institute on Aging What Happens to the Brain in Alzheimer's Disease? | National Institute on Aging. Available online: https://www.nia.nih.gov/health/what-happens-brain-alzheimers-disease (accessed on 14 July 2020).
4. Derby, C.A. Trends in the Public Health Significance, Definitions of Disease, and Implications for Prevention of Alzheimer's Disease. *Curr. Epidemiol. Rep.* **2020**, *7*, 68–76. [CrossRef]
5. Hebert, L.E.; Weuve, J.; Scherr, P.A.; Evans, D.A. Alzheimer Disease in the United States (2010–2050) Estimated Using the 2010 Census. *Neurology* **2013**, *80*, 1778–1783. [CrossRef]
6. Alzheimer's Association. 2019 Alzheimer's Disease Facts and Figures. *Alzheimer's Dement.* **2019**, *15*, 321–387. [CrossRef]
7. Friedman, E.M.; Shih, R.A.; Langa, K.M.; Hurd, M.D. US Prevalence And Predictors Of Informal Caregiving For Dementia. *Health Aff. Proj. Hope* **2015**, *34*, 1637–1641. [CrossRef]
8. Jutkowitz, E.; Kane, R.L.; Gaugler, J.E.; MacLehose, R.F.; Dowd, B.; Kuntz, K.M. Societal and Family Lifetime Cost of Dementia: Implications for Policy. *J. Am. Geriatr. Soc.* **2017**, *65*, 2169–2175. [CrossRef]
9. Ghazizadeh, H.; Yaghooti-Khorasani, M.; Darroudi, S.; Esmaily, H.; Sharifan, P.; Tayefi, M.; reza Seyedi, S.M.; Mohammadi-Bajgiran, M.; Ghaffarian-Zirak, R.; Tavallaie, S.; et al. Evaluation of the Association between the Healthy Eating Index and the Level of Serum and Dietary Intake of Copper and Zinc. *Obes. Med.* **2020**, *19*, 100277. [CrossRef]
10. Freeborn, D.; Haldeman-Englert, C. Total Copper (Blood)—Health Encyclopedia—University of Rochester Medical Center. Available online: https://www.urmc.rochester.edu/encyclopedia/content.aspx?contenttypeid=167&contentid=total_copper_blood (accessed on 6 October 2020).
11. Roberts, E.A.; Sarkar, B. Liver as a Key Organ in the Supply, Storage, and Excretion of Copper. *Am. J. Clin. Nutr.* **2008**, *88*, 851S–854S. [CrossRef]
12. Royer, A.; Sharman, T. Copper Toxicity. In *StatPearls*; StatPearls Publishing: Treasure Island, FL, USA, 2020.
13. Collins, J.F.; Klevay, L.M. Copper12. *Adv. Nutr.* **2011**, *2*, 520–522. [CrossRef]
14. Pohanka, M. Copper and Copper Nanoparticles Toxicity and Their Impact on Basic Functions in the Body. *Bratisl. Med. J.* **2019**, *120*, 397–409. [CrossRef]
15. De Bie, P.; Muller, P.; Wijmenga, C.; Klomp, L.W.J. Molecular Pathogenesis of Wilson and Menkes Disease: Correlation of Mutations with Molecular Defects and Disease Phenotypes. *J. Med. Genet.* **2007**, *44*, 673–688. [CrossRef]
16. Ross, A.C.; Caballero, B.H.; Cousins, R.J.; Tucker, K.L.; Ziegler, T.R. *Modern Nutrition in Health and Disease: Eleventh Edition*; Wolters Kluwer Health Adis (ESP): Alphen aan den Rijn, The Netherlands, 2012; ISBN 978-1-60547-461-8.
17. Ramos, D.; Mar, D.; Ishida, M.; Vargas, R.; Gaite, M.; Montgomery, A.; Linder, M.C. Mechanism of Copper Uptake from Blood Plasma Ceruloplasmin by Mammalian Cells. *PLoS ONE* **2016**, *11*. [CrossRef]
18. Harvey, P.J.; Handley, H.K.; Taylor, M.P. Widespread Copper and Lead Contamination of Household Drinking Water, New South Wales, Australia. *Environ. Res.* **2016**, *151*, 275–285. [CrossRef]
19. Babi, E.; Tariba, B.; Kovai, J.; Piznet, A.; Varnai, V.; Macan, J. Relevance of Serum Copper Elevation Induced by Oral Contraceptives: A Meta-Analysis—Contraception. *Contraception* **2012**, *87*, 790–800. [CrossRef]
20. Jalali, K.; Nouairi, I.; Kallala, N.; M'Sehli, W.; Zribi, K.; Mhadhbi, H. Germination, Seedling Growth, and Antioxidant Activity in Four Legume (Fabaceae) Species under Copper Sulphate Fungcide Treatment. *Pak. J. Bot.* **2018**, *50*, 1599–1606.
21. Snyder, J.S. Recalibrating the Relevance of Adult Neurogenesis. *Trends Neurosci.* **2019**, *42*, 164–178. [CrossRef] [PubMed]
22. Coupé, P.; Manjón, J.V.; Lanuza, E.; Catheline, G. Lifespan Changes of the Human Brain In Alzheimer's Disease. *Sci. Rep.* **2019**, *9*. [CrossRef]
23. Graham, N.; Warner, J. *Alzheimer's Disease and Other Dementias*; Family Doctor Publications Ltd.: Poole, UK, 2009; ISBN 978-1-903474-61-7.
24. Brandes, R. The Current Neuroscientific Understanding of Alzheimer's Disease. *Purs. J. Undergrad. Res. Univ. Tenn.* **2020**, *10*, 2.
25. Li, Z.; Del-Aguila, J.L.; Dube, U.; Budde, J.; Martinez, R.; Black, K.; Xiao, Q.; Cairns, N.J.; Dougherty, J.D.; Lee, J.-M.; et al. Genetic Variants Associated with Alzheimer's Disease Confer Different Cerebral Cortex Cell-Type Population Structure. *Genome Med.* **2018**, *10*, 43. [CrossRef]
26. Hu, X.; Hu, Z.-L.; Li, Z.; Ruan, C.-S.; Qiu, W.-Y.; Pan, A.; Li, C.-Q.; Cai, Y.; Shen, L.; Chu, Y.; et al. Sortilin Fragments Deposit at Senile Plaques in Human Cerebrum. *Front. Neuroanat.* **2017**, *11*. [CrossRef] [PubMed]
27. Patel, H.; Dobson, R.J.B.; Newhouse, S.J. A Meta-Analysis of Alzheimer's Disease Brain Transcriptomic Data. *J. Alzheimer's Dis.* **2019**, *68*, 1635–1656. [CrossRef] [PubMed]
28. Yarris, L. Copper on the Brain. Available online: https://newscenter.lbl.gov/2013/05/24/copper-on-the-brain/ (accessed on 14 July 2020).

29. Pal, A.; Badyal, R.K.; Vasishta, R.K.; Attri, S.V.; Thapa, B.R.; Prasad, R. Biochemical, Histological, and Memory Impairment Effects of Chronic Copper Toxicity: A Model for Non-Wilsonian Brain Copper Toxicosis in Wistar Rat. *Biol. Trace Elem. Res.* **2013**, *153*, 257–268. [CrossRef] [PubMed]
30. Yadav, S.; Younger, S.H.; Zhang, L.; Thompson-Peer, K.L.; Li, T.; Jan, L.Y.; Jan, Y.N. Glial Ensheathment of the Somatodendritic Compartment Regulates Sensory Neuron Structure and Activity. *Proc. Natl. Acad. Sci. USA* **2019**, *116*, 5126–5134. [CrossRef]
31. Nazer, H.; Nazer, D.; Windle, M.; Cuffari, C.; Deodhar, J. Pediatric Fulminant Hepatic Failure: Background, Pathophysiology, Epidemiology. 2020. Available online: https://emedicine.medscape.com/article/929028-overview (accessed on 6 January 2021).
32. Nehring, S.M.; Tadi, P.; Tenny, S. Cerebral Edema. In *StatPearls*; StatPearls Publishing: Treasure Island, FL, USA, 2020.
33. Pal, A.; Prasad, R. Regional Distribution of Copper, Zinc and Iron in Brain of Wistar Rat Model for Non-Wilsonian Brain Copper Toxicosis. *Indian J. Clin. Biochem.* **2016**, *31*, 93–98. [CrossRef] [PubMed]
34. Stokum, J.A.; Kurland, D.B.; Gerzanich, V.; Simard, J.M. Mechanisms of Astrocyte-Mediated Cerebral Edema. *Neurochem. Res.* **2015**, *40*, 317–328. [CrossRef]
35. Fulop, T.; Witkowski, J.M.; Bourgade, K.; Khalil, A.; Zerif, E.; Larbi, A.; Hirokawa, K.; Pawelec, G.; Bocti, C.; Lacombe, G.; et al. Can an Infection Hypothesis Explain the Beta Amyloid Hypothesis of Alzheimer's Disease? *Front. Aging Neurosci.* **2018**, *10*. [CrossRef]
36. Kardos, J.; Héja, L.; Simon, Á.; Jablonkai, I.; Kovács, R.; Jemnitz, K. Copper Signalling: Causes and Consequences. *Cell Commun. Signal.* **2018**, *16*. [CrossRef]
37. Brewer, G.J. Copper Toxicity in Alzheimer's Disease: Cognitive Loss from Ingestion of Inorganic Copper. *J. Trace Elem. Med. Biol.* **2012**, *26*, 89–92. [CrossRef]
38. Morley, J.E.; Farr, S.A. The Role of Amyloid-Beta in the Regulation of Memory. *Biochem. Pharmacol.* **2014**, *88*, 479–485. [CrossRef]
39. O'Brien, R.J.; Wong, P.C. Amyloid Precursor Protein Processing and Alzheimer's Disease. *Annu. Rev. Neurosci.* **2011**, *34*, 185–204. [CrossRef] [PubMed]
40. Gaetke, L.M.; Chow-Johnson, H.S.; Chow, C.K. Copper: Toxicological Relevance and Mechanisms. *Arch. Toxicol.* **2014**, *88*, 1929–1938. [CrossRef] [PubMed]
41. Hardy, J.A.; Higgins, G.A. Alzheimer's Disease: The Amyloid Cascade Hypothesis. *Science* **1992**, *256*, 184–186. [CrossRef] [PubMed]
42. Barage, S.H.; Sonawane, K.D. Amyloid Cascade Hypothesis: Pathogenesis and Therapeutic Strategies in Alzheimer's Disease. *Neuropeptides* **2015**, *52*, 1–18. [CrossRef]
43. Hillen, H. The Beta Amyloid Dysfunction (BAD) Hypothesis for Alzheimer's Disease. *Front. Neurosci.* **2019**, *13*. [CrossRef] [PubMed]
44. Lodish, H.; Berk, A.; Zipursky, S.L.; Matsudaira, P.; Baltimore, D.; Darnell, J. Overview of Neuron Structure and Function. In *Molecular Cell Biology*, 4th ed.; W.H. Freeman and Company: New York, NY, USA, 2000.
45. Rudenko, G. Neurexins—Versatile Molecular Platforms in the Synaptic Cleft. *Curr. Opin. Struct. Biol.* **2019**, *54*, 112–121. [CrossRef]
46. Tricaud, N. Myelinating Schwann Cell Polarity and Mechanically-Driven Myelin Sheath Elongation. Available online: https://www.frontiersin.org/articles/10.3389/fncel.2017.00414/full (accessed on 10 October 2020).
47. Mukherjee, A.; Soto, C. Prion-Like Protein Aggregates and Type 2 Diabetes. *Cold Spring Harb. Perspect. Med.* **2017**, *7*. [CrossRef]
48. Chen, G.; Xu, T.; Yan, Y.; Zhou, Y.; Jiang, Y.; Melcher, K.; Xu, H.E. Amyloid Beta: Structure, Biology and Structure-Based Therapeutic Development. *Acta Pharmacol. Sin.* **2017**, *38*, 1205–1235. [CrossRef]
49. Greenwald, J.; Riek, R. Biology of Amyloid: Structure, Function, and Regulation. *Structure* **2010**, *18*, 1244–1260. [CrossRef]
50. Han, X.-J.; Hu, Y.-Y.; Yang, Z.-J.; Jiang, L.-P.; Shi, S.-L.; Li, Y.-R.; Guo, M.-Y.; Wu, H.-L.; Wan, Y.-Y. Amyloid β-42 Induces Neuronal Apoptosis by Targeting Mitochondria. *Mol. Med. Rep.* **2017**, *16*, 4521–4528. [CrossRef]
51. Bendlin, B.B.; Zetterberg, H. Screening with a High-Precision Blood-Based Assay for Alzheimer Disease. *Neurology* **2019**, *93*, 737–738. [CrossRef] [PubMed]
52. Cheignon, C.; Tomas, M.; Bonnefont-Rousselot, D.; Faller, P.; Hureau, C.; Collin, F. Oxidative Stress and the Amyloid Beta Peptide in Alzheimer's Disease. *Redox Biol.* **2017**, *14*, 450–464. [CrossRef] [PubMed]
53. Schieber, M.; Chandel, N.S. ROS Function in Redox Signaling and Oxidative Stress. *Curr. Biol.* **2014**, *24*, R453–R462. [CrossRef] [PubMed]
54. Talwar, P.; Grover, S.; Sinha, J.; Chandna, P.; Agarwal, R.; Kushwaha, S.; Kukreti, R. Multifactorial Analysis of a Biomarker Pool for Alzheimer Disease Risk in a North Indian Population. *Dement. Geriatr. Cogn. Disord.* **2017**, *44*, 25–34. [CrossRef]
55. González, C.; Martín, T.; Cacho, J.; Breñas, M.T.; Arroyo, T.; García-Berrocal, B.; Navajo, J.A.; González-Buitrago, J.M. Serum Zinc, Copper, Insulin and Lipids in Alzheimer's Disease Epsilon 4 Apolipoprotein E Allele Carriers. *Eur. J. Clin. Investig.* **1999**, *29*, 637–642. [CrossRef]
56. Kitazawa, M.; Hsu, H.-W.; Medeiros, R. Copper Exposure Perturbs Brain Inflammatory Responses and Impairs Clearance of Amyloid-Beta. *Toxicol. Sci.* **2016**, *152*, 194–204. [CrossRef]
57. Segal, M. Dendritic Spines: Morphological Building Blocks of Memory. *Neurobiol. Learn. Mem.* **2017**, *138*, 3–9. [CrossRef]

toxics

Review

Direct and Indirect Neurotoxic Potential of Metal/Metalloids in Plants and Fungi Used for Food, Dietary Supplements, and Herbal Medicine

Peter S. Spencer [1,2,*] and Valerie S. Palmer [1]

1 Department of Neurology, Oregon Health & Science University, Portland, OR 97239-3098, USA; palmerv@ohsu.edu
2 Oregon Institute of Occupational Health Sciences, Oregon Health & Science University, Portland, OR 97239-3098, USA
* Correspondence: spencer@ohsu.edu

Abstract: Plants and mushrooms bioconcentrate metals/metalloids from soil and water such that high levels of potentially neurotoxic elements can occur in cultivated and wild species used for food. While the health effects of excessive exposure to metals/metalloids with neurotoxic potential are well established, overt neurological disease from prolonged ingestion of contaminated botanicals has not been recognized. However, the presence of metal elements may affect levels of botanical neurotoxins in certain plants and mushrooms that are established causes of acute and chronic neurological disease.

Keywords: soil and water pollution; heavy metals; morels; grasspea; cassava; neurodegeneration

Citation: Spencer, P.S.; Palmer, V.S. Direct and Indirect Neurotoxic Potential of Metal/Metalloids in Plants and Fungi Used for Food, Dietary Supplements, and Herbal Medicine. *Toxics* **2021**, *9*, 57. https://doi.org/10.3390/toxics9030057

Received: 23 February 2021
Accepted: 12 March 2021
Published: 16 March 2021

1. Introduction

Plants and mushrooms that are/are not used for food can sequester metals/metalloids drawn from soil and water that have the potential for human neurotoxic effects. Although this property of botanicals is exploited for the bioremediation of metal-contaminated soils, the possibility of toxic/neurotoxic effects in the consumer of metal-concentrating plants and mushrooms has been rarely addressed. This requires a synthesis of mycology, botany, and toxicology, not only of the human and mammal but also of the plant and mushroom.

Three scenarios are considered here: certain metals contained in plants and mushrooms used for food pose: (a) a direct threat to human neurological function and/or (b) an indirect threat because they can modulate the concentration of botanical compounds (hydrazine, dencichine, cyanogens) that by themselves pose a neurotoxic hazard to the consumer. Examples of the latter include certain plants (grasspea, cassava) and mushrooms (false morel), food use of which can precipitate neurotoxic disease.

2. Botanical Uptake of Metal Elements Required for Biological Function

The major elements in soil (>100 mg/kg), and minerals derived therefrom, include aluminum, calcium, carbon, iron, magnesium, nitrogen, oxygen, potassium, silicon, sodium, sulfur, and titanium. Other soil elements include barium, chlorine, manganese, phosphorus, strontium, and zirconium [1,2]. Common trace elements in soil (<100 mg/kg) comprise arsenic, chromium, cobalt, copper, lead, lithium, nickel, selenium, and zinc [1]. Whereas Ca, Fe, K, Mn, and P are required for normal human physiological function, prolonged excessive exposure to some of the listed elements (Al, As, Cu, Fe, Li, Pb), among others (Cd, Hg, Tl, Zn), can result in neurologic dysfunction or overt disease.

Soil goes through various stages of aging that result in changes in its chemistry and that of associated water. The resulting chemical environment is reflected in the organisms that feed on soil nutrients [1,2]. Fungi play a vital role in the soil environment by acting as a bridge between soil microbes and plants, thereby facilitating nutrient cycling and plant health and disease control [3]. Fungi are particularly effective in changing the solubility of metals by employing three major reaction types that change the speciation of metal complexes: reduction, methylation, and dealkylation [4]. Fungal hyphae connect with the root systems of plants, making the constant exchange of nutrients possible [5]. Certain metals are biologically active in mushrooms, including Ca, Co, Cu, Fe, K, Mg, Mn, Na, Ni, and Zn [6]. Fungi secrete a chelator that binds iron, solubilizing the element and allowing it to enter cells [7].

Twenty elements are considered essential for plant growth, including those that may derive from air (carbon, oxygen, hydrogen), air and soil (nitrogen), or soil alone, such as boron, calcium, chorine, copper, iron, magnesium, manganese, molybdenum, and zinc [8]. Certain fungi and plants can form a symbiotic relationship in which the mushrooms help plants acquire trace nutrients in exchange for carbon. Fungi specialize in concentrating elements and passing them along to plants. About 80% of plant roots participate in this type of relationship, meaning that almost all nutrients taken up by plants first transit through fungi [9]. Thus, the function of fungi is to accumulate essential elements (notably K, Na, and Zn) not only for mushroom growth but also for that of plants and, indirectly, for animals and humans that consume components as food [2].

3. Fungal Uptake of Metals/Metalloids with Neurotoxic Potential

While fungi absorb metals in soil required for their normal growth and pigmentation, they also take up and bioaccumulate soil metals that are present in concentrations that would be harmful to plants [4]. In turn, certain metals found in soils, often due to anthropogenic activities, can harm mushrooms by competing for binding with elements that fungi need for metabolism and reproduction [2,10]. In general, fungi adapt to the soil content of metals [11], as illustrated by arsenic and lead, but high levels of certain elements can perturb fungal growth and development [12]. The ability of mushrooms to accumulate heavy metals is relevant to their use by humans for food, nutrient supplementation, and mycomedicine.

Numerous studies have measured the concentration of various elements in mushrooms, including metals/metalloids with potential for human neurological disease. Arsenic concentrations in the fruit bodies of 37 common edible mushrooms ranged from >0.05 mg/kg to 146.9 mg/kg [13]. The ability to accumulate arsenic was found in mushrooms with saprotrophic nutrition (feeding on nonliving organic matter), including the Basidiomycetes families of Agaricaceae, Tricholomataceae, and Gasteromycetaceae. By contrast, arsenic accumulation was not detectable in xylophagous (rotting wood-feeding) or mycorrhizal (plant-fungus symbiotic) species of edible mushrooms. Forms of arsenic found in mushrooms include arsenobetaine, arsenate, arsenocholine, and unidentified compounds containing the trimethylarsonium ion [14,15]. Fresh fruit bodies reportedly contain about a tenfold lower arsenic level than dried specimens [13]. Studies from northwest Spain and Dhaka, Bangladesh, reported low mean concentrations of arsenic (0.27 and 0.51 mg/kg dry weight, respectively) in wild and cultivated mushrooms [16,17].

As with arsenic, widely divergent levels of lead have been found in fungi. Some reports do not clarify whether content refers to fresh fungal tissue or dry weight (dw). A study from Tuscany, Italy, reported a range of lead concentrations from 0.4 to 15.5 mg/kg in fungi and 22 to 51 mg/kg in soil [18]. Lead content ranged from 1.9 to 10.8 mg/kg in mushrooms collected from three sites in China [19] and between 0.4 and 36 mg/kg in Sweden [20]. A similar concentration range (0.60–11.4 mg/kg) was found in wild mushrooms collected in Turkey [21], while lower levels (up to 2.4 mg/kg) were found in species collected in parts of Germany, Macedonia, Greece, and Turkey [22–25]. High levels of lead were measured in wild mushrooms (76.00 $\pm$ 9.78 mg/kg) and edible species (6.46–27.33 mg/kg) sold in Nigeria [26,27].

Several other studies have examined the concentration of metals in fungi, including elements with neurotoxic potential. Reports focused on tissue mercury content [28,29] found high levels (4.9–22 mg kg dw) in edible *Boletus* species in the mercuriferous belt of southwestern China [30] and lower levels (2.28 mg/kg) in *B. edulis* (edible Penny Bun mushroom) collected in Croatia [31]. Cadmium and silver are also taken up by *B. edulis* [32]. Analysis of 14 wild edible mushrooms collected in Yunnan, China, identified high concentrations of manganese (13.5–113 mg/kg) and iron (67.5–843 mg/kg) [33]. Lead, cadmium, mercury, and selenium were found in 60 species of common edible mushrooms collected mainly in the province of Reggio Emilia, Italy [34]. Species of *Agaricus, Macrolepiota, Lepista,* and *Calocybe* accumulate a high content of cadmium and mercury, even in unpolluted areas, but the concentration of these metals increases considerably in heavily polluted sites, such as in the vicinity of both working and abandoned metal smelters or inside cities [35]. Blanching and pickling edible mushrooms reduce their metal content [35].

A recent article analyzed 200 European publications (published between 2001 and 2016) that describe the concentration of selected elements in mushrooms [36]. Papers dealing with elements such as Cd, Cu, Fe, Pb, and Zn originated primarily from Turkey, Poland, Spain, and the Czech Republic. Many studies underlined the need to assess the risk to human health arising from the consumption of mushrooms taken from various contaminated habitats because polluted soils and water directly impact the concentration of elements in mushrooms. Those with a high lead content were collected from soils impacted by former metallurgical and mining activities. For example, in Příbram, Czech Republic, the upper soil layer had a lead concentration of 36,234 mg/kg, while the stipe of *B. edulis* growing in this area contained 165 mg/kg dw [37]. A copper concentration of 427 mg/kg dw was measured in samples of *B. edulis* collected near a copper smelter in Norway [38], while high concentrations of lead (11,460 mg/kg dw), manganese, and copper were measured in *Lepista (Clitocybe) nuda* (edible Wood Blewit mushroom) collected from the Eskişehir forest area of Turkey [39]. Many years of traffic pollution were blamed for the very high levels of iron (9685 mg/kg dw) in *Omphalotus olearius* (Jack-O'-Lantern mushroom), a poisonous xylophagous fungal species taken from the forest along the Balıkesir-Manisa highway in Turkey [40]. Other studies linked vehicular pollution to the lead content of certain fungal species collected near heavily trafficked roads [41]. The ability of mushrooms, such as *Pleurotus* species, to biosorb heavy metals has important applications for remediation of polluted soil and water of industrial origin [12]. Uptake of heavy metals by the mycelia of *P. ostreatus* (Oyster mushroom) increased proportionally to their concentration in the medium on which the fungus was grown [42].

Botanicals contain polyvalent phytic acid, which can bind bi- and trivalent cations of various elements. At neutral pH, metal binding to phytic acid corresponds to Cu > Zn > Ni > Co > Mn > Fe > Ca [43]. The cap of mushrooms produces stress-related factors (metallothionein) that govern the uptake of metal ions [44]. Cysteine-rich oligopeptides (phytochelatin family) bind a large fraction of cadmium in the caps of *B. edulis* when the edible mushroom is exposed to excess metals [45]. Fungi also bisorb and sequester heavy metals via melanin, a negatively charged hydrophobic pigment formed by the polymerization of indolic and phenolic compounds. Experiments with melanin extracted from *Armillaria cepistipes* (Ringless Honey Mushroom) revealed a differential metal affinity, namely $Pb^{2+} > Cr^{3+} > Ni^{2+} > Cd^{2+} > Zn^{2+} > Ca^{2+}$—with an extreme preference for Pb^{2+} (80% removal) over the essential metals (0% and 12% removal for Ca^{2+} and Zn^{2+}, respectively) [46]. Fungal melanin production can be both constitutive and facultative, production increasing according to environmental stressors, such as UV radiation, drying, high concentrations of salts, heavy metals, and radionuclides [47]. Melanized fungi are thus candidates for soil and water bio(myco)remediation [48].

4. Heavy Metals in Mushrooms with Potential Neurotoxicity

The True Morel *Morchella esculenta*, a facultative mycorrhizal mushroom widely prized by gourmets, illustrates the potential human health threat of consuming mushrooms that bioaccumulate metals from contaminated soil. In the USA, lead arsenate (PbHAsO4) was widely used as a moth pesticide from the late 1800s, replaced by DDT in the 1950s, and banned from use in fruit orchards in 1988. However, the potentially neurotoxic elements (Pb, As) persist in the topsoils on which morels grow. A study of 29 abandoned apple orchards in the northeastern USA revealed a range of lead and arsenic in soil (19.20–2450 and 3.08–244.00 mg/kg, respectively) and in the fruit bodies of *M. esculenta* (0.5–13.00 and 0.15–2.85 mg/kg, respectively) growing on polluted soils. The respective concentrations were statistically associated (r = 0.81) for lead and arsenic in soil, and for soil and morel lead content (r = 0.94, r = 0.57, respectively). Almost all (94%) of the arsenic stored in morel tissues was in the inorganic form, and the levels of the two elements in morel fruitbodies were considered to pose a human health risk [49]. A mycologist who consumed *M. esculenta* collected from apple orchards had elevated levels of urinary arsenic and lead; he complained of symptoms consistent with sensory (arsenic) neuropathy that resolved following chelation therapy [50].

False Morel mushrooms, such as *Gyromitra esculenta*, a poisonous species that is nevertheless eaten by some, contain gyromitrin (acetaldehyde *N*-methyl-*N*-formyl-hydrazone) (Figure 1, center). While the function of fungal gyromitrin is unknown, hydrazones can form complexes with metals, such as Ni^{2+}, Cu^{2+}, Zn^{2+}, and Cd^{2+} [51]. The principal metabolite of gyromitrin, monomethylhydrazine (MMH) (Figure 1, left), is an acutely neurotoxic compound that interferes with pyridoxine utilization by both glutamic acid decarboxylase and γ-aminobutyric acid (GABA) transaminase, leading to decreased concentrations of the inhibitory neurotransmitter GABA in the brain and consequent induction of seizures [52]. Levels of MMH in *G. esculenta* vary and can be reduced by prolonged desiccation [53]. Consumption of the False Morel *G. gigas* collected from soils near a closed lead mine in the French Alps has recently been linked to a focus of amyotrophic lateral sclerosis; while the authors did not measure lead levels in local False Morels, they attributed induction of neurodegenerative disease to the genotoxic properties of MMH [54].

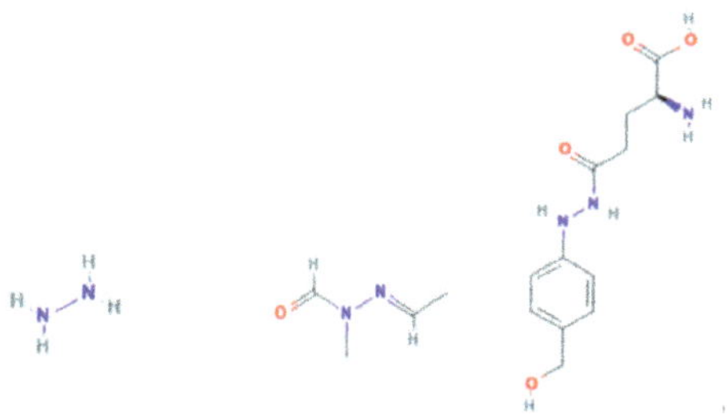

Figure 1. Monomethylhydrazine (**left**), gyromitrin (**center**), and agaritine (**right**). Source: PubChem, National Institutes of Health National Library of Medicine https://pubchem.ncbi.nlm.nih.gov/, accessed on 14 March 2021.

Hydrazine is a reducing agent that is used industrially to reduce metal salts and oxides to pure metals. Phenylhydrazines, notably agaritine (*N*2-(γ-L-glutamyl)-4-hydroxymethylph enylhydrazine) (Figure 1, right), are found in *Agaricus* mushrooms. Concentrations range from 200 to 500 mg agaritine/kg fresh weight in cultivated species of the universally eaten *A. biporus* (Button Mushroom). Wild samples of *A. elvensis* have been reported to contain up to 10,000 mg agaritine/kg fresh weight [55,56]. Heavy metals (mg/kg) in the fruiting bodies of *A. bisporus* growing wild in Poland included: Cd, 0.68–6.14; Cr, 0.38–6.93; Cu, 1.90–101.71; Fe, 33.01–432.24; Mn, 2.86–387.43; Ni, 0.20–3.09; Pb, 0.98–42.83, and Zn, 31.87–124.84 [57]. Samples collected near Kermanshah City, Iran, contained levels of arsenic (mean 65.23 $\pm$ 13.57 mg/kg) and zinc (mean 66.23 $\pm$ 2.80 mg/kg) that exceeded the maximum permissible limit [58].

5. Plant Uptake of Metals/Metalloids with Neurotoxic Potential

The presence of heavy metals in plants used for food and their associated human health threats have been comprehensively reviewed by Rai and colleagues [59]. The root causes of this problem are attributed to the rapid pace of urbanization, changes in land use, and industrialization, especially in countries with high populations, such as India and China. Whereas in parts of Asia and Africa, inadequately treated wastewater, effluent, and sludge used for irrigation are the main sources of plant contamination, sources of heavy metals in food plants grown in America, Europe, and Oceania arise predominantly from particulate matter (from industrial and transport sectors) and agricultural practices. Sludge from distilleries and the chemical, electroplating, textile, and leather industries is often found to contain significantly high concentrations of heavy metals, such as Cr, Cu, Fe, Mn, Ni and Pb. Gold mining is a leading source of heavy metal contamination (especially Cd, Hg, Pb) in soil and food crops. Plants used in Asian herbal medicines acquire metal contaminants (e.g., As, Cd, Cr, Cu, Fe, and Pb) when grown near smelting or other industrial areas [60–62].

Certain plant species can immobilize metals in soil and groundwater through absorption and accumulation by roots, adsorption onto roots, or precipitation within the root zone, a process known as phytostabilization. Other species absorb and hyperaccumulate metals/metalloid contaminants through a phytoextraction process. Several plants, such as *Brassica* spp., can accumulate lead in concentrations >50 mg/g dw. The roots of *B. juncea* (Oriental Mustard) concentrated mercury 100-270X (dw) above initial solution concentrations, but only 0.7–2% was translocated to the shoots [63]. Numerous metals have been measured in plants used for cereals, fruits, and vegetables; for example, samples of rice (*Oryza sativa*) contained chromium (15–465 mg/kg), manganese (61–356 mg/kg), and lead (16–16,500 mg/kg), and high levels of arsenic have been found in rice and lettuce (*Lactuca sativa*) [59].

6. Heavy Metals in Plants with Potential Neurotoxicity

6.1. Grasspea

The legume *Lathyrus sativus* (grasspea) (Figure 2), prolonged heavy consumption of which triggers the irreversible central motorsystem disease lathyrism [64,65], is a strong metal accumulator of lead and cadmium in all parts of the plants [66–68]. Root tissues of lead-exposed grasspea showed a six-fold, two-fold, and three-and-a-half-fold reduction in calcium, zinc, and copper contents, respectively, which indicated the plant tolerates a deficiency in essential nutrients while storing large amounts of lead [69]. The uptake of cadmium and copper in grasspea shoots was exponential over the range of concentrations tested [70]. The distribution of lead in *L. sativus* has a selective character: leaves > roots > stems > seeds. While lead is also found in the seed, soil pollution with heavy metals does not affect seed quality, such that the high nutrient (K, O, Cu, Fe, Mn, Zn) and protein content (23.18–29.54%) is preserved [71]. Grasspea can also bioaccumulate arsenic in roots > shoots [72]. These properties are important because grasspea is considered an ideal crop for resource-poor farmers, is widely eaten on the Indian subcontinent, and serves as both a regular and famine food in the northern Ethiopian highlands [73,74].

The neurotoxic property of grasspea is attributable to its content of the excitotoxic nonprotein amino acid β-*N*-oxalylamino-L-alanine (L-BOAA, *syn.* dencichine) (Figure 3), also known as β-oxalyldiaminopropionic acid (β-ODAP) and, perhaps, also to its low content of methionine and cysteine [68]. L-BOAA/β-ODAP is found in all parts of the plant, with concentrations of 0.5–2.5% in traditional varieties of grasspea [75]. Zinc deficiency and oversupply of iron to the roots of *L. sativus* induce increases in the L-BOAA/β-ODAP content in ripe seed [76]. The biosynthesis of L-BOAA/β-ODAP and its genetic and environmental regulation are under intensive study because grasspea is tolerant of drought and waterlogging, flourishes without inputs, and the seed has significant protein content, such that grasspea varieties with low-L-BOAA/β-ODAP content may potentially serve as a safe and valuable foodstuff [68,77,78]. However, environmental factors such as drought, zinc deficiency, iron oversupply, and the presence of heavy metals in the soil reportedly can considerably increase the level of L-BOAA/ß-ODAP in the seed of grasspea [79].

Figure 2. *Lathyrus sativus* (grasspea), an annual legume cultivated in arid and semiarid areas, has attractive flowers (**A**,**B**) and yields nutritious seed (**C**). The seeds are a rich source of protein and starch (**D**). Reproduced with permission from Xu et al. [68], doi:10.3390/ijms18030526, accessed on 23 February 2021.

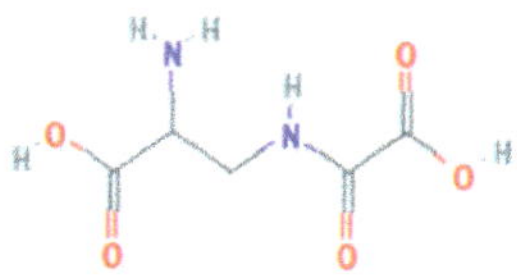

Figure 3. Structure of β-*N*-oxalylamino-L-alanine. Source: PubChem, National Institutes of Health National Library of Medicine https://pubchem.ncbi.nlm.nih.gov/, accessed on 14 March 2021.

The addition of arbuscular mycorrhizal fungi (AMF) to grasspea seedlings promoted plant growth of grasspea under sulfate stress [80]. AMF are associated with most terrestrial plants: they penetrate the roots and form symbiotic relationships that protect the plant from diverse stressors, including resistance to temperature extremes, drought, waterlogging, salinity, and heavy metals [81]. *L. sativus* is highly tolerant of such environmental stressors, which suggests AMF are normally associated with the plant and thus able to provide some protection against heavy metals. The subject is of importance to the health of people with dietary reliance on grasspea in countries such as Bangladesh, Ethiopia, India, and Nepal [77].

6.2. Cassava

The tubers and leaves of *Manihot esculenta* (cassava) provide a source of food for huge numbers of people in Asia, Africa, and South America (Figure 4). For example, cassava tubers supply ~70% of the daily calorie input for >50 million people in Nigeria, the world's largest cassava producer (49 million metric tons/year). Fresh cassava root contains >30% carbohydrate, >2% protein, 0.1% fat, and >75% moisture content [82,83]. However, *M. esculenta* also synthesizes cyanogenic glucosides (linamarin and its methylated relative lotaustralin; 97:3), the concentration of which must be substantially reduced before the tuber is used for human consumption. The method of processing cassava roots, the duration of tuber storage, and the type of meal preparation determine the amount of consumer exposure to cassava cyanogens and heavy metals [84]. Failure to remove the cyanogenic glucosides and their products before human ingestion can precipitate acute cyanide (HCN) poisoning or, with continued daily consumption, the subacute onset of an irreversible motorsystem disease (cassavism), which is known as *konzo* in the Democratic Republic of Congo and *mantakassa* in Mozambique [85,86]. Whether the daily consumption of cassava-derived *gari* in Nigeria is also responsible for endemic peripheral neuropathy with optic atrophy and sensorineural deafness has been questioned.

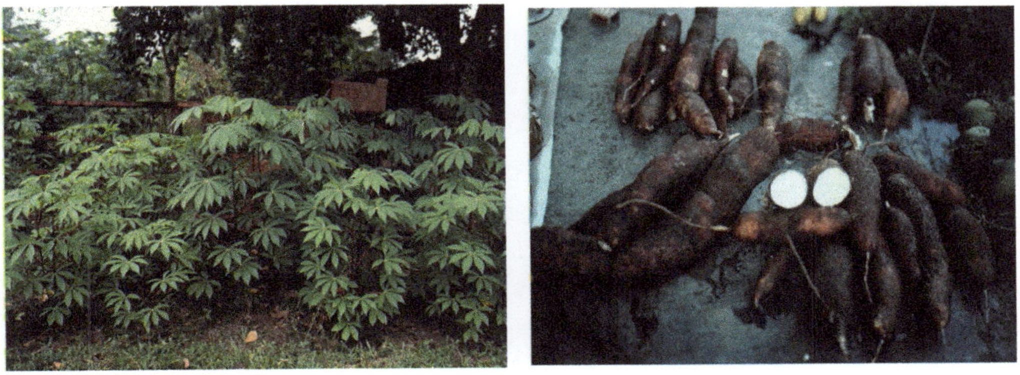

Figure 4. *Manihot esculenta* Crantz. leaves (**left**) and tubers (**right**).

Acute toxicity from cassava processing and/or ingestion results from cyanide ion (CN^-) binding to iron and copper sites in cytochrome *c* oxidase, which is required for electron transport and energy generation. Cassavism is associated with elevated levels of urinary thiocyanate (SCN^-) produced in the consumer by the addition of endogenous methionine/cystine sulfur to cyanide by thiosulfate sulfurtransferase (rhodanese) [87], the enzyme activity of which is inhibited by certain metal ions (Au, Ni, Pt, and Zn) [88]. Laboratory animals given a sulfur-free diet also eliminate CN as cyanate (OCN^-) [89], prolonged exposure to which (in the form of NaOCN) can precipitate peripheral neuropathy [90]. All three anions (CN^-, OCN^-, and SCN^-) form metal complexes, at least in the chemistry laboratory [91–93]. Cyanide also forms stable complexes with cobalt, gold, iron, and mercury, a property used commercially to leach the metal from gold-bearing ore [94]. Linamarin may be involved in the uptake of gold and mercury into cassava roots [95].

Several studies have shown heavy metals in cassava and the processed flour derived from its root. Cassava in communities in Nigeria's River State contained metal elements in both the leaves (Fe > Cu > Cr > Ni > Pb > Cd) and tubers (Fe > Cu > Cr > Ni > Pb > Cd) [96]. Metal concentrations in the tubers (Cd, Ni, and Pb) and leaves (Cd, Cr, and Pb) of cassava plants grown on soils around a lead sulfide (galena) mining area in Nigeria's Ebonyi State exceeded WHO standards [97]. Exceedances for Al, As, Ca, Cr, Cu, Fe, K, Mn, Pb, and Zn were found in cassava tubers growing around two cement factories in Ogun State, Nigeria [98]. Cassava grown on soil contaminated with crude oil in the Niger Delta Region contained heavy metals (Fe > Zn > Ni > Pb > Cd > Cr) in their leaves and tubers [99].

Automobile emissions added to the metal content of soil samples and cassava plants collected from farmlands along a major expressway in Nigeria's Delta State [100]. Mean levels (mg/kg) of heavy metals (Fe> Zn> Ni> Cr > Pb) in soil samples were 142.93 ± 42.16 for Fe, 59.34 ± 25.21 for Zn, 24.98 ± 15.57 for Ni, 14.27 ± 5.39 for Cr, and 13.63 ± 5.41 for Pb. Corresponding mean metal concentrations (mg/kg) in cassava leaves and tubers were, respectively, 21.70 ± 3.45 and 9.62 ± 3.53 for Fe, 4.15 ± 1.01 and 1.15 ± 0.44 for Zn, 5.12 ± 2.75 and 0.37 ± 0.63 for Cr, and for leaves only, 3.46 ± 1.58 for Pb. Plant concentration factor values corresponded to Cr > Pb > Fe > Ni. Cassava harvested from farmlands along highways in Owerri, Nigeria, contained leaf concentrations of Zn > Cu >Pb> Ni > Cd [101].

Metal contamination also occurs from the common practice of sun-drying cassava flour by the roadside [102]. Significantly elevated concentrations of Pb > Fe > Cu > Cr > Mn > Zn > Ni > Cd > Co were found in cassava flour processed by roadside drying along a Nigerian highway [103]. Samples of cassava flour purchased from the capital city of Nigeria's Osun State contained lead concentrations (up to 0.34 mg/kg) that far exceeded the Nigerian Industrial Standard Permissible Level (0.1 mg/kg). However, flour levels (mg/kg) of CN (0.03–0.09), cadmium (0.01–0.0), copper (0.35–0.62), iron (0.1–0.6), nickel (0.20–0.49), selenium (3.46–5.43), and zinc (0.182–0.3) were within permissible limits [104].

While several studies have revealed the potential for heavy metal contamination of cassava roots and leaves used for food in sub-Saharan Africa, whether and which metals can influence the content, speciation, or neurotoxic potential of the plant's cyanogens and their metabolites are unknown. Metal neurotoxicity has not been reported or looked for in those who use cassava as a food staple.

7. Key Conclusions

Botanicals used for food, nutritional supplements, and medicinal purposes can bioaccumulate metals/metalloids with neurotoxic potential in concentrations that, in some circumstances, exceed permissible limits for human consumption. While certain heavy metals/metalloids, such as arsenic, lead, manganese, and mercury [105–108], have established human neurotoxic potential, especially during brain development, their effects from ingestion of contaminated botanicals have rarely if ever resulted in diagnosed neurological illness. Associations have been reported between heavy metals and certain neurodegenerative diseases, including amyotrophic lateral sclerosis [109,110], Parkinson's disease [111,112], and Alzheimer's disease [113,114], but evidence of cause–effect relationships is generally lacking. Chemical principles in some mushrooms (hydrazone/hydrazines) and plants (dencichine, cyanogens) have established human neurotoxicity [115,116], their concentration varying in response to the uptake of metal elements from soil and water. Such botanical species normally pose serious health hazards to humans because of their potential for acute, chronic, and possibly long-latency neurological disease, the risk for such effects varying with the amount and duration of consumption. Witness the extraordinarily high prevalence (5.5% in 2019) of spastic paraparesis (lathyrism) in northeast Ethiopia, with somewhat higher rates (6.2%) among those who use traditional iron-containing clay pots to cook grasspea [117]. The metallic content of botanicals used for food, thus has the potential for direct and indirect adverse effects on the human nervous system. On the other hand, the ability of certain botanicals to bioaccumulate certain metals/metalloids is exploited to protect human health by reducing soil and water pollution via phytoremediation and mycoremediation [118,119].

Author Contributions: Both authors searched and retrieved published literature. Spencer wrote the manuscript, and Palmer reviewed the content and recommended additions. Both authors have read and agreed to the published version of the manuscript.

Funding: This work received no external funding.

Institutional Review Board Statement: This work did not require institutional review.

Informed Consent Statement: This work did not require informed consent.

Data Availability Statement: All data were drawn from published works as cited.

Acknowledgments: Third World Medical Research Foundation assisted with the collection of information on food use and neurological effects of grasspea and cassava.

Conflicts of Interest: The authors declare no conflict of interest.

References

1. Sposito, G. *The Chemistry of Soils*; Oxford University Press: Oxford, UK; New York, NY, USA, 2008.
2. Briggs, E.D.; Mobilization of Metals by Fungi in Historic Cemeteries. Bachelor's Thesis, Bates College, Lewiston, MI, USA, 2015. Available online: https://scarab.bates.edu/geology_theses/25 (accessed on 14 March 2021).
3. Went, F.W.; Stark, N. The biological and mechanical role of soil fungi. *Science* **1968**, *160*, 444. [CrossRef]
4. Morley, G.F.; Sayer, J.A.; Wilkinson, S.C.; Ghareib, M.M.; Gadd, G.M. Fungal sequestration, mobilization, and transformation of metals and metalloids. In *Fungi and Environmental Change*; Frankland, J.C., Magan, N., Gadd, G.M., Eds.; Cambridge University Press: New York, NY, USA, 1996; pp. 235–256.
5. Colpaert, J.V.; Van Tichelen, K.K. Mycorrhizas and environmental stress. In *Fungi and Environmental Change*; Frankland, J., Magan, J., Gadd, G.M., Eds.; British Mycological Society Symposia (20); Cambridge University Press: Cambridge, UK, 1996. [CrossRef]
6. Tobin, J.M.; Cooper, D.G.; Neufeld, R.J. Investigation of the mechanism of metal uptake by denatured *Rhizopus arrhizus* biomass. *Enzy Microb Technol.* **1990**, *12*, 591–595 101016/0141. [CrossRef]
7. Garraway, M.O.; Evans, R.C. *Fungal Nutrition and Physiology*; Wiley: New York, NY, USA, 1984.
8. OSU. National Forage and Grassland Curriculum. Available online: https://forages.oregonstate.edu/nfgc/eo/onlineforagecurriculum/instructormaterials/availabletopics/fertilization/elements (accessed on 14 March 2021).
9. Martino, E.; Perotto, S. Mineral transformations by mycorrhizal fungi. *Geomicrobiol. J.* **2010**, *27*, 609–623. [CrossRef]
10. Gadd, G.M. Interactions of Fungi with Toxic Metals. In *The Genus Aspergillus*; Powell, K.A., Renwick, A., Peberdy, J.F., Eds.; Springer: Boston, MA, USA, 1994; Volume 69. [CrossRef]
11. Giller, K.E.; Witter, E.; McGrath, S.P. Heavy metals and soil microbes. *Soil Biol. Biochem.* **2009**, *41*, 2031–2037. [CrossRef]
12. Kapahi, M.; Sachdeva, S. Mycoremediation potential of *Pleurotus* species for heavy metals: A review. *Bioresour. Bioprocess.* **2017**, *4*, 32. [CrossRef]
13. Vetter, J. Arsenic content of some edible mushroom species. *Eur. Food Res. Technol.* **2004**, *219*, 71–74. [CrossRef]
14. Kuehnelt, D.; Goessler, W.; Irogolic, K.J. Arsenic compounds in terrestrial organisms II: Arsenocholine in the mushroom *Amanita muscaria*. *Appl. Organomet. Chem.* **1997**, *11*, 459–470. [CrossRef]
15. Slejkovec, Z.; Byrne, A.R.; Smodis, B.; Rossbach, M. Preliminary studies on arsenic species in some environmental samples. *Anal. Bioanal. Chem.* **1996**, *354*, 592–595. [CrossRef] [PubMed]
16. Melgar, M.J.; Alonso, J.; García, M.A. Total contents of arsenic and associated health risks in edible mushrooms, mushroom supplements and growth substrates from Galicia (NW Spain). *Food Chem. Toxicol.* **2014**, *73*, 44–50. [CrossRef]
17. Rashid, M.H.; Rahman, M.M.; Correll, R.; Naidu, R. Arsenic and other elemental concentrations in mushrooms from Bangladesh: Health risks. *Int. J. Environ. Res. Publ. Health* **2018**, *15*, 919. [CrossRef] [PubMed]
18. Giannaccini, G.; Betti, L.; Palego, L.; Mascia, G.; Schmid, L.; Lanza, M.; Mela, A.; Fabbrini, L.; Biondi, L.; Lucacchini, A. The trace element content of top-soil and wild edible mushroom samples collected in Tuscany, Italy. *Environ. Monit. Assess.* **2012**, *184*, 7579–7595. [CrossRef] [PubMed]
19. Chen, X.H.; Zhou, H.B.; Qiu, G.Z. Analysis of several heavy metals in wild edible mushrooms from regions of China. *Bull. Environ. Contam. Toxicol.* **2009**, *83*, 280–285. [CrossRef] [PubMed]
20. Tyler, G. Metals in sporophores of basidiomycetes. *Trans. Br. Mycol. Soc.* **1980**, *74*, 41–49. [CrossRef]
21. Mendil, D.; Uluözlü, Ö.D.; Tüzen, M.; Hasdemir, E.; Sarı, H. Trace metal levels in mushroom samples from Ordu, Turkey. *Food Chem.* **2005**, *91*, 463–467. [CrossRef]
22. Stijve, T.; Andre, D.; Lucchini, G.; Goessler, W. Lanthanides and other less common metals in mushrooms. *Dtsch. Lebensmittel-Rundschau.* **2002**, *98*, 82–87.
23. Soylak, M.; Saraçoğlu, S.; Tüzen, M.; Mendil, D. Determination of trace metals in mushroom samples from Kayseri, Turkey. *Food Chem.* **2005**, *92*, 649–652. [CrossRef]
24. Sesli, E.; Tuzen, M.; Soylak, M. Evaluation of trace metal contents of some wild edible mushrooms from Black sea region, Turkey. *J. Hazard Mater.* **2008**, *160*, 462–467. [CrossRef]
25. Ouzouni, P.K.; Petridis, D.; Koller, W.-D.; Riganakos, K.A. Nutritional value and metal content of wild edible mushrooms collected from West Macedonia and Epirus, Greece. *Food Chem.* **2009**, *115*, 1575–1580. [CrossRef]
26. Udochukwu, U.; Nekpen, B.O.; Udinyiwe, O.C.; Omeje, F.I. Bioaccumulation of heavy metals and pollutants by edible mushroom collected from Iselu market Benin-city. *Int. Curr. Microbiol. Appl. Sci.* **2014**, *3*, 52–57.
27. Ndimele, C.C.; Ndimele, P.E.; Chukwuka, K.S. Accumulation of heavy metals by wild mushrooms in Ibadan, Nigeria. *J. Health Pollut.* **2017**, *7*, 26–30. [CrossRef] [PubMed]
28. Melgar, M.J.; Alonso, J.; García, M.A. Mercury in edible mushrooms and underlying soil: Bioconcentration factors and toxicological risk. *Sci. Total Environ.* **2009**, *407*, 5328–5334. [CrossRef] [PubMed]
29. Falandysz, J.; Gucia, M.; Brzostowski, A.; Kawano, M.; Bielawski, L.; Frankowska, A.; Wyrzykowska, B. Content and bioconcentration of mercury in mushrooms from northern Poland. *Food Addit Contam.* **2003**, *20*, 247–253. [CrossRef]
30. Falandysz, J.; Zhang, J.; Wang, Y.Z.; Saba, M.; Krasińska, G.; Wiejak, A.; Li, T. Evaluation of mercury contamination in fungi *Boletus* species from latosols, lateritic red earths, and red and yellow earths in the circum-Pacific mercuriferous belt of southwestern China. *PLoS ONE* **2015**, *10*, e0143608. [CrossRef] [PubMed]

31. Širić, I.; Kos, I.; Kasap, A.; Kaić, A.; Držaić, V.; Rakić, L. Mercury bioaccumulation by wild edible mushrooms. In Proceedings of the 52st Croatian and 12th International Symposium on Agriculture, Dubrovnik, Croatia, 12–17 February 2017. Available online: https://www.cabdirect.org/globalhealth/abstract/20173178523 (accessed on 23 February 2021).
32. Koroleva, Y.; Vakhranyova, O.; Okhrimenko, M. [Accumulation of trace elements by wild mushrooms in West part of Russia (South-Eastern Baltic)]. *Pollut. Atmosph.* **2015**, *226*, 2–10. [CrossRef]
33. Zhu, F.; Qu, L.; Fan, W.; Qiao, M.; Hao, H.; Wang, X. Assessment of heavy metals in some wild edible mushrooms collected from Yunnan Province, China. *Environ. Monit. Assess.* **2011**, *179*, 191–199. [CrossRef]
34. Cocchi, L.; Vescovi, L.; Petrin, L.E.; Petrini, O. Heavy metals in edible mushrooms in Italy. *Food Chem.* **2006**, *98*, 277–284. [CrossRef]
35. Kalač, P.; Svoboda, L.; Havlíčková, B. Contents of cadmium and mercury in edible mushrooms. *J. Appl. Biomed.* **2004**, *2*, 15–20. [CrossRef]
36. Świsłowski, P.; Dołhańczuk-Śródka, A.; Rajfur, M. Bibliometric analysis of European publications between 2001 and 2016 on concentrations of selected elements in mushrooms. *Environ. Sci. Pollut. Res. Int.* **2020**, *27*, 22235–22250. [CrossRef] [PubMed]
37. Komárek, M.; Chrastný, V.; Štíchová, J. Metal/metalloid contamination and isotopic composition of lead in edible mushrooms and forest soils originating from a smelting area. *Environ. Int.* **2007**, *33*, 677–684. [CrossRef] [PubMed]
38. Collin-Hansen, C.; Yttri, K.E.; Andersen, R.A.; Berthelsen, B.O.; Steinnes, E. Mushrooms from two metal-contaminated areas in Norway: Occurrence of metals and metallothionein-like proteins. *Geochem. Explor. Environ. Anal.* **2002**, *2*, 121–130. [CrossRef]
39. Yamaç, M.; Yildiz, D.; Sarikürkcü, C.; Çelikkollu, M.; Solak, M.H. Heavy metals in some edible mushrooms from the Central Anatolia, Turkey. *Food Chem.* **2007**, *103*, 263–267. [CrossRef]
40. Yilmaz, F.; Isiloglu, M.; Merdivan, M. Heavy metal levels in some macrofungi. *Turk. J. Bot.* **2003**, *27*, 45–56.
41. Jorhem, L.; Sundström, B. Levels of some trace elements in edible fungi. *Z. Lebensm Unters Forsch.* **1995**, *201*, 311–316. [CrossRef]
42. Sanglimsuwan, S.; Yoshida, N.; Morinaga, T.; Murooka, Y. Resistance to and uptake of heavy metals in mushrooms. *J. Ferment. Bioeng.* **1993**, *5*, 112–114. [CrossRef]
43. Kalac, P. *Mineral Composition and Radioactivity of Edible Mushrooms*; Academic Press: New York, NY, USA, 2019.
44. Damodaran, D.; Balakrishnan, R.M.; Shetty, V.K. The uptake mechanism of Cd(II), Cr(VI), Cu(II), Pb(II), and Zn(II) by mycelia and fruiting bodies of *Galerina vittiformis*. *BioMed Res. Internat.* **2013**, 149120. [CrossRef] [PubMed]
45. Collin-Hansen, C.; Pedersen, S.A.; Andersen, R.A.; Steinnes, E. First report of phytochelatins in a mushroom: Induction of phytochelatins by metal exposure in *Boletus edulis*. *Mycologia* **2007**, *99*, 161–174. [CrossRef]
46. Tran-Ly, A.N.; Ribera, J.; Schwarze, F.W.M.R.; Brunelli, M.; Fortunato, G. Fungal melanin-based electrospun membranes for heavy metal detoxification of water. *Sustain. Mat. Technol.* **2020**, *23*, e00146. [CrossRef]
47. Gessler, N.N.; Egorova, A.S.; Belozerskaya, T.A. Melanin pigments of fungi under extreme environmental conditions. *Appl. Biochem. Microbiol.* **2014**, *50*, 105–113. [CrossRef]
48. Eisenman, H.C.; Casadevall, A. Synthesis and assembly of fungal melanin. *Appl. Microbiol. Biotechnol.* **2012**, *93*, 931–940. [CrossRef]
49. Shavit, E.; Shavit, E. Lead and arsenic in *Morchella esculenta* fruitbodies collected in lead arsenate contaminated apple orchards in the northeastern United States: A preliminary study. *Fungi Mag. Spring* **2010**, *3*, 11–18.
50. Shavit, E. Arsenic in morels. Morels collected in New Jersey apple orchards blamed for arsenic poisoning. *Fungi* **2014**, *1*, 2–10.
51. Kandil, F.; Chebani, M.K.; Al Zoubi, W. Synthesis of macrocyclic bis-hydrazone and their use in metal cations extraction. *Int. Sch. Res. Not.* **2012**, 208284. [CrossRef] [PubMed]
52. Horowitz, K.M.; Erwin, L.; Kong, E.L.; Horowitz, B.Z. Gyromitra mushroom toxicity. In *StatPearls [Internet] Last update 21 August 2020*. Available online: https://www.ncbi.nlm.nih.gov/books/NBK470580/ (accessed on 14 March 2021).
53. Andary, C.; Privat, G.; Bourrier, M.-J. Variations of monomethylhydrazine content in Gyromitra esculenta. *Mycologia* **1985**, *77*, 259–264. [CrossRef]
54. Lagrange, E.; Vernoux, J.P.; Reis, J.; Palmer, V.; Camu, W.; Spencer, P.S. An amyotrophic lateral sclerosis hot spot in the French Alps associated with genotoxic fungi. *J. Neurol. Sci.* Under review. Available online: https://hal.archives-ouvertes.fr/hal-01658408/ (accessed on 14 March 2021).
55. Andersson, H.C.; Gry, J. *Phenylhydrazines in the Cultivated Mushroom (Agaricus bisporus)—Occurrence, Biological Properties, Risk Assessment and Recommendations*; TemaNord: Copenhagen, Denmark, 2004; Volume 558, p. 123.
56. Schulzova, V.; Hajslova, J.; Peroutka, R.; Hlavasek, J.; Gry, J.; Andersson, H.C. Agaritine content of 53 *Agaricus* species collected from nature. *Food Add. Contam.* **2009**, *26*, 82–93. [CrossRef] [PubMed]
57. Bosiacki, M.; Krzebietke, S.J.; Siwulski, M.; Sobieralka, K. The content of selected heavy metals in fruiting bodies of *Agaricus bisporus* (Lange) Imbach. wild growing in Poland. *J. Elementol.* **2018**, *23*. [CrossRef]
58. Ardakani, S.S.; Jahangard, A. Toxicological assessment of inorganic arsenic and zinc content in button mushrooms. *J. Adv. Environ. Health Res.* **2017**, *5*, 246–251. [CrossRef]
59. Rai, P.K.; Lee, S.S.; Zhang, M.; Tsang, Y.F.; Kim, K.-H. Heavy metals in food crops: Health risks, fate, mechanisms, and management. *Environ. Int.* **2019**, *125*, 365–385. [CrossRef]
60. Ernst, E. Toxic heavy metals and undeclared drugs in Asian herbal medicines. *Trends Pharmacol. Sci.* **2002**, *23*, 136–139. [CrossRef]
61. Kulhari, A.; Sheorayan, A.; Bajar, S.; Sarkar, S.; Chaudhury, A.; Kalia, R.K. Investigation of heavy metals in frequently utilized medicinal plants collected from environmentally diverse locations of north western India. *Springerplus* **2013**, *2*, 676. [CrossRef]

62. Luo, L.; Wang, B.; Jiang, J.; Fitzgerald, M.; Huang, Q.; Yu, Z.; Li, H.; Zhang, J.; Wei, J.; Yang, C.; et al. Heavy metal contaminations in herbal medicines: Determination, comprehensive risk assessments, and solutions. *Front. Pharmacol.* **2021**. [CrossRef]
63. Tangahu, B.V.; Abdullah, S.R.S.; Basri, H.; Idris, M.; Anuar, N.; Mukhlisin, M. A review on heavy metals (As, Pb, and Hg) uptake by plants through phytoremediation. *Int. J. Chem. Engin.* **2011**, 939161. [CrossRef]
64. Ludolph, A.C.; Hugon, J.; Dwivedi, M.P.; Schaumburg, H.H.; Spencer, P.S. Studies on the aetiology and pathogenesis of motor neuron diseases. 1. Lathyrism: Clinical findings in established cases. *Brain* **1987**, *110*, 149–165. [CrossRef]
65. Haimanot, R.T.; Kidane, Y.; Wuhib, E.; Kalissa, A.; Alemu, T.; Zein, Z.A.; Spencer, P.S. Lathyrism in rural northwestern Ethiopia: A highly prevalent neurotoxic disorder. *Int. J. Epidemiol.* **1990**, *19*, 664–672. [CrossRef] [PubMed]
66. Abdelkrim, S.; Jebara, S.H.; Saadani, O.; Chiboub, M.; Abid, G.; Mannai, K.; Jebara, M. Heavy metal accumulation in *Lathyrus sativus* growing in contaminated soils and identification of symbiotic resistant bacteria. *Arch. Microbiol.* **2019**, *201*, 107–121. [CrossRef] [PubMed]
67. Abdelkrim, S.; Jebara, S.H.; Saadani, O.; Abid, G.; Taamalli, W.; Zemni, H.; Mannai, K.; Louati, F.; Jebara, M. In situ effects of *Lathyrus sativus*—PGPR to remediate and restore quality and fertility of Pb and Cd polluted soils. *Ecotoxicol. Environ. Saf.* **2020**, *192*. [CrossRef]
68. Xu, Q.; Liu, F.; Chen, P.; Jez, J.M.; Krishnan, H.B. β-*N*-Oxalyl-l-α,β-diaminopropionic ccid (β-ODAP) content in *Lathyrus sativus*: The integration of nitrogen and sulfur metabolism through β-cyanoalanine synthase. *Int. J. Mol. Sci.* **2017**, *18*, 526. [CrossRef] [PubMed]
69. Brunet, J.; Repellin, A.; Varrault, G.; Terry, N.; Zuily-Fodil, Y. Lead accumulation in the roots of grass pea (*Lathyrus sativus* L.): A novel plant for phytoremediation systems? *C. R. Biol.* **2008**, *331*, 859–864. [CrossRef] [PubMed]
70. Nagati, V.B.; Koyyati, R.; Marx, P.; Chinnapaka, V.D.; Padigya, P.R.M. Effect of heavy metals on seed germination and plant growth. *Int. J. Pharm. Tech. Res.* **2015**, *7*, 528–534.
71. Angelova, V.R. Assessment of soil contamination on the content of macro and microelements and the quality of grass pea seeds (*Lathyrus sativus* L.). *Int. J. Environ. Ecol. Engin.* **2019**, *13*, 651–658. Available online: https://zenodo.org/record/3593104#.YE5nDlknYq8 (accessed on 14 March 2021).
72. Talukdar, D. Effect of arsenic-induced toxicity on morphological traits of *Trigonella foenum-graecum* L. and *Lathyrus sativus* L. during germination and early seedling growth. *Curr. Res. J. Biol. Sci.* **2011**, *3*, 116–123. Available online: https://www.semanticscholar.org/paper/Effect-of-Arsenic-induced-Toxicity-on-Morphological-Talukdar/a185693915e14d2b8b50bf8b079f3ac1af54fa9b (accessed on 14 March 2021).
73. Hillocks, R.J.; Maruthi, M.N. Grass pea (Lathyrus sativus): Is there a case for further crop improvement? *Euphytica* **2012**, *186*, 647–654. [CrossRef]
74. Khandare, A.L.; Babu, J.J.; Ankulu, M.; Aparna, N.; Shirfule, A.; Rao, G.S. Grass pea consumption & present scenario of neurolathyrism in Maharashtra State of India. *Indian J. Med. Res.* **2014**, *140*, 96–101.
75. Kumar, S.; Bejiga, G.; Ahmed, S.; Nakkoul, H.; Sarker, A. Genetic improvement of grass pea for low neurotoxin (β-ODAP) content. *Food Chem. Toxicol.* **2011**, *49*, 589–600. [CrossRef]
76. Lambein, F.; Haque, R.; Khan, J.K.; Kebede, N.; Kuo, Y.H. From soil to brain: Zinc deficiency increases the neurotoxicity of Lathyrus sativus and may affect the susceptibility for the motorneurone disease neurolathyrism. *Toxicon* **1994**, *32*, 461–466. [CrossRef]
77. Spencer, P.S. (Ed.) The Grass Pea: Threat and Promise. In *Proceedings of the International Network for the Improvement of Lathyrus sativus and the Eradication of Lathyrism, London, UK*; Third World Medical Research Foundation: New York, NY, USA, 1989; 244p.
78. Yan, Z.Y.; Spencer, P.S.; Li, Z.X.; Liang, Y.M.; Wang, Y.F.; Wang, C.Y.; Li, F.M. *Lathyrus sativus* (grass pea) and its neurotoxin ODAP. *Phytochemistry* **2006**, *67*, 107–121. [CrossRef] [PubMed]
79. Jiao, C.J.; Jiang, J.L.; Ke, L.M.; Cheng, W.; Li, F.M.; Li, Z.X.; Wang, C.Y. Factors affecting β-ODAP content in *Lathyrus sativus* and their possible physiological mechanisms. *Food Chem. Toxicol.* **2011**, *49*, 543–549. [CrossRef]
80. Jin, L.; Sun, X.; Wang, X.; Shen, Y.; Hou, F.; Chang, S.; Wang, C. Synergistic interactions of arbuscular mycorrhizal fungi and rhizobia promoted the growth of *Lathyrus sativus* under sulphate salt stress. *Symbiosis* **2010**, *50*, 157–164. [CrossRef]
81. Diagne, N.; Ngom, M.; Djighaly, P.I.; Fall, D.; rie Hocher, V.; Svistoonoff, S. Roles of arbuscular mycorrhizal fungi on plant growth and performance: Importance in biotic and abiotic stressed regulation. *Diversity* **2020**, *12*, 370. [CrossRef]
82. Chikezie, P.C.; Ojiako, O.A. Cyanide and aflatoxin loads of processed cassava (*Manihot esculenta*) tubers (garri) in Njaba, Imo State, Nigeria. *Toxicol. Internat.* **2013**, *20*, 261–267. [CrossRef] [PubMed]
83. Adeolu, A.T.; Adewoye, S.O. Efficacy of cassava peel extracts for the removal of heavy metals from hospital sewage sludge in Nigeria. *J. Health Pollut.* **2019**, *9*, 190908. [CrossRef]
84. Ezeh, E.; Okeke, O.; Aburu, C.M.; Anya, O.U. Comparative evaluation of the cyanide and heavy metal levels in traditionally processed cassava meal products sold within Enugu Metropolis. *Int. J. Environ. Sci. Nat. Res.* **2018**, *12*, 47–52. [CrossRef]
85. Tshala-Katumbay, D.D.; Spencer, P.S. Toxic disorders of the upper motor neuron system. In *Handbook of Clinical Neurology*; Eisen, A., Shaw, P., Eds.; Motor Neuron Disorders and Related Diseases; Elsevier: Edinburgh, UK, 2007; Volume 82, pp. 353–372.
86. Oluwole, O.; Onabolu, A.; Cotgreave, I.A.; Rosling, H. Incidence of ataxic polyneuropathy and its relationship to exposure to cyanide in a Nigerian community. *J. Neurol. Neurosurg. Psychiat.* **2003**, *74*, 1417–1422. [CrossRef]

87. Paula Cardoso, A.; Ernesto, M.; Nicala, D.; Mirione, E.; Chavane, L.; N'zwalo, H.; Chikumba, S.; Cliff, J.; Paulo Mabota, A.; Rezaul Haque, M.; et al. Combination of cassava flour cyanide and urinary thiocyanate measurements of school children in Mozambique. *Int. J. Food Sci. Nutr.* **2004**, *55*, 183–190. [CrossRef]
88. Lijk, L.J.; Kalk, K.H.; Brandenburg, N.P.; Hol, W.G. Binding of metal cyanide complexes to bovine liver rhodanese in the crystalline state. *Biochemistry* **1983**, *22*, 2952–2957. [CrossRef] [PubMed]
89. Tor-Agbidye, J.; Palmer, V.S.; Sabri, M.I.; Craig, A.M.; Blythe, L.L.; Spencer, P.S. Dietary deficiency of cystine and methionine in rats alters thiol homeostasis required for cyanide detoxification. *J. Toxicol. Environ. Health Part A* **1998**, *55*, 583–595. [CrossRef]
90. Ohnishi, A.; Peterson, C.M.; Dyck, P.J. Axonal degeneration in sodium cyanate-induced neuropathy. *Arch. Neurol.* **1975**, *32*, 530–534. [CrossRef] [PubMed]
91. Tuan, D.F.-T.; Reed, J.W.; Hoffman, R. Studies of the linkage and bonding of triatomics intransition metal complexes. Part 2. NCS^- complexes. *J. Mol. Struct. (Theochem)* **1991**, *232*, 111–121. [CrossRef]
92. Matveichuk, Y.V.; Rakhman'ko, E.M.; Yasinetskii, V.V. Thiocyanate complexes of *d* metals: Study of aqueous solutions by UV, visible, and IR spectrometry. *Russian J. Inorg. Chem.* **2015**, *60*, 100–104. [CrossRef]
93. DeGroot, H.P.; Hanusa, T.P. Cyanide complexes of the transition metals. In *Encyclopedia of Inorganic and Bioinorganic Chemistry*; Scott, R.A., Ed.; John Wiley: New York, NY, USA, 2020. [CrossRef]
94. International Cyanide Management Code Gold Mining Industry. Cyanide Chemistry. Available online: https://www.cyanidecode.org/cyanide-facts/cyanide-chemistry (accessed on 14 March 2021).
95. Alcantara, H.J.P.; Jativa, F.; Doronila, A.I.; Anderson, C.W.N.; Siegele, R.; Spassov, T.G.; Sanchez-Palaciois Boughton, B.A.; Kolev, S.D. Localization of mercury and gold in cassava (*Manihot esculenta* Crantz). *Environ. Sci. Pollut. Res.* **2020**, *27*, 18498–18509. [CrossRef] [PubMed]
96. Nwineewii, J.D. Heavy metal concentrations in cassava leaves and tubers harvested from some communities in Gokana, Rivers State, Nigeria. *Int. J. Res. Sci. Innov.* **2020**, *7*, 167–172. Available online: https://www.rsisinternational.org/virtual-library/papers/heavy-metal-concentrations-in-cassava-leaves-and-tubers-harvested-from-some-communities-in-gokana-rivers-state-nigeria/ (accessed on 14 March 2021).
97. Ajiwe, V.I.E.; Chukwujindu, A.; Chukwujindu, C.N. Heavy metals concentration in cassava tubers and leaves from a galena mining area in Ishiagu, IVO L.G.A of Ebonyi State Nigeria. *IOSR J. Appl. Chem.* **2018**, *11*, 54–58 doi 109790/5736.
98. Moses, M.O. Evaluation of heavy metals in cassava tubers grown around two major cement factories in Ogun State, Nigeria. *Int. J. Res. Stud. Biosci.* **2016**, *4*, 26–29. [CrossRef]
99. Harrison, U.E.; Osu, S.R.; Ekanem, J. Heavy metals accumulation in leaves and tubers of cassava (*Manihot Esculenta* Crantz) grown in crude oil contaminated soil at Ikot Ada Udo, Nigeria. *J. Appl. Sci. Environ. Manag.* **2018**, *22*, 845–851. [CrossRef]
100. Osakwe, S.A.; Okolie, L.P. Physicochemical characteristics and heavy metals contents in soils and cassava plants from farmlands along a major highway in Delta State, Nigeria. *J. Appl. Sci. Environ. Manag.* **2015**, *19*, 695–704. [CrossRef]
101. Okerke, J.N.; Nduka, J.N.; Adanma, U.A.; Odangowei, O. Heavy metals in cassava (*Manihot esculenta* Crantz) harvested from farmlands along highways in Owerri, Nigeria. *Turk. J. Agric. Food Sci. Tchnol.* **2020**, *8*, 800–806. [CrossRef]
102. Obanijesu, E.O.; Olajide, J.O. Trace metal pollution study on cassava flour's roadside drying technique in Nigeria. In *Appropriate Technologies for Environmental Protection in the Developing World*; Yanful, E.K., Ed.; Springer: Dordrecht, The Netherlands, 2009. [CrossRef]
103. Audu, A.A.; M Waziri, M.; Olasinde, T.T. Proximate analysis and levels of some heavy metals in cassava flour processed by roadside drying along Abuja-Lokoja highway, Nigeria. *Ind. J. Fund Appl. Life Sci.* **2012**, *2*, 55–58.
104. Sawyerr, O.H.; Odipe, O.E.; Raimi, M.O.; Ogungbemi, O.H. Assessment of cyanide and some heavy metals concentration in consumable cassava flour 'lafun' across Osogbo Metropolis, Nigeria. *MOJ Eco. Environ. Sci.* **2018**, *3*, 369–372. Available online: https://papers.ssrn.com/sol3/papers.cfm?abstract_id=3294152 (accessed on 23 February 2021).
105. Vahidnia, A.; van der Voet, G.B.; de Wolff, F.A. Arsenic neurotoxicity—A review. *Hum. Exp. Toxicol.* **2007**, *26*, 823–832. [CrossRef] [PubMed]
106. Feldman, R.G.; White, R.F. Lead neurotoxicity and disorders of learning. *J. Child. Neurol.* **1992**, *7*, 354–359. [CrossRef] [PubMed]
107. Dobson, A.W.; Erikson, K.M.; Aschner, M. Manganese neurotoxicity. *Ann. N. Y. Acad. Sci.* **2004**, *1012*, 115–128. [CrossRef]
108. Farina, M.; Rocha, J.B.; Aschner, M. Mechanisms of methylmercury-induced neurotoxicity: Evidence from experimental studies. *Life Sci.* **2011**, *89*, 555–563. [CrossRef] [PubMed]
109. Bozzoni, V.; Pansarasa, O.; Diamanti, L.; Nosari, G.; Cereda, C.; Ceroni, M. Amyotrophic lateral sclerosis and environmental factors. *Funct. Neurol.* **2016**, *31*, 7–19. [CrossRef] [PubMed]
110. Ash, P.E.A.; Dhawan, U.; Boudeau, S.; Lei, S.; Carlomagno, Y.; Knobel, M.; Al Mohanna, L.F.A.; Boomhower, S.R.; Newland, M.C.; Sherr, D.H.; et al. Heavy metal neurotoxicants induce ALS-linked TDP-43 pathology. *Toxicol. Sci.* **2019**, *167*, 105–115. [CrossRef] [PubMed]
111. Ball, N.; Teo, W.P.; Chandra, S.; Chapman, J. Parkinson's Disease and the environment. *Front. Neurol.* **2019**, *10*, 218. [CrossRef]
112. Bjorklund, G.; Stejskal, V.; Urbina, M.A.; Dadar, M.; Chirumbolo, S.; Mutter, J. Metals and Parkinson's disease: Mechanisms and biochemical processes. *Curr. Med. Chem.* **2018**, *25*, 2198–2214. [CrossRef]
113. Huat, T.J.; Camats-Perna, J.; Newcombe, E.A.; Valmas, N.; Kitazawa, M.; Medeiros, R. Metal toxicity links to Alzheimer's Disease and neuroinflammation. *J. Mol. Biol.* **2019**, *431*, 1843–1868. [CrossRef]

114. Bakulski, K.M.; Seo, Y.A.; Hickman, R.C.; Brandt, D.; Vadari, H.S.; Hu, H.; Park, S.K. Heavy metals exposure and Alzheimer's Disease and related dementias. *J. Alzheimer's Dis.* **2020**, *76*, 1215–1242. [CrossRef] [PubMed]
115. Spencer, P.S.; Kisby, G.E. Role of hydrazine-related chemicals in cancer and neurodegenerative disease. *Chem. Res. Toxicol.* **2021**, in press.
116. Spencer, P.S.; Palmer, V.S. Food plant chemicals linked with neurological and neurodegenerative disease. *Adv. Neurotoxicol.* **2017**, *1*, 247–278.
117. Hussien, F.M.; Dagnaw, M.M.; Ahmed, A.Y.; Hassen, H.Y. Lathyrism and socioeconomic disparities: A neglected public health problem in northeast Ethiopia. *Am. J. Trop. Med. Hyg.* **2021**. [CrossRef] [PubMed]
118. Li, C.; Ji, X.; Luo, X. Phytoremediation of heavy metal pollution: A bibliometric and scientometric analysis from 1989 to 2018. *Int. J. Environ. Res. Public Health* **2019**, *16*, 4755. [CrossRef] [PubMed]
119. Nahid, A.; Amin-Ul Mannan, M. Mycoremediation: Expunging environmental pollutants. *Biotchnol. Rep. Amst.* **2020**, *26*, e00452. [CrossRef]

toxics

MDPI

Review

Mechanisms of Metal-Induced Mitochondrial Dysfunction in Neurological Disorders

Hong Cheng [1], Bobo Yang [2], Tao Ke [2], Shaojun Li [3], Xiaobo Yang [1,4], Michael Aschner [2,*] and Pan Chen [2,*]

1 Department of Occupational Health and Environmental Health, School of Public Health, Guangxi Medical University, Nanning 530021, China; chenghong@stu.gxmu.edu.cn (H.C.); yangx@gxmu.edu.cn (X.Y.)
2 Department of Molecular Pharmacology, Albert Einstein College of Medicine, Bronx, NY 10461, USA; bobo.yang@einsteinmed.org (B.Y.); tao.ke@einsteinmed.org (T.K.)
3 Department of Toxicology, School of Public Health, Guangxi Medical University, Nanning 530021, China; lishaojun0613@163.com
4 Department of Public Health, School of Medicine, Guangxi University of Science and Technology, Liuzhou 545006, China
* Correspondence: michael.aschner@einsteinmed.org (M.A.); pan.chen@einsteinmed.org (P.C.); Tel.: +1-718-430-2317 (M.A.); +1-718-430-4047 (P.C.)

Abstract: Metals are actively involved in multiple catalytic physiological activities. However, metal overload may result in neurotoxicity as it increases formation of reactive oxygen species (ROS) and elevates oxidative stress in the nervous system. Mitochondria are a key target of metal-induced toxicity, given their role in energy production. As the brain consumes a large amount of energy, mitochondrial dysfunction and the subsequent decrease in levels of ATP may significantly disrupt brain function, resulting in neuronal cell death and ensuing neurological disorders. Here, we address contemporary studies on metal-induced mitochondrial dysfunction and its impact on the nervous system.

Keywords: mitochondrial dysfunction; neurological disorders; metals; neurotoxicity

Citation: Cheng, H.; Yang, B.; Ke, T.; Li, S.; Yang, X.; Aschner, M.; Chen, P. Mechanisms of Metal-Induced Mitochondrial Dysfunction in Neurological Disorders. *Toxics* **2021**, *9*, 142. https://doi.org/10.3390/toxics9060142

Academic Editors: Richard Ortega, Asuncion Carmona and Pamela Lein

Received: 7 May 2021
Accepted: 14 June 2021
Published: 17 June 2021

1. Introduction

Mitochondria play a key role in many cellular physiological and pathological processes, including energy metabolism, calcium homeostasis, lipid biosynthesis, and apoptosis [1]. One of their main functions is to produce adenosine triphosphate (ATP) by coupling the electron transport chain (ETC) with phosphorylation. The ETC consists of four major protein–metal complexes (I–V) which primarily serve to generate a proton gradient to drive the production of ATP [2]. Superoxide anion, a byproduct of the ETC's operation, is extremely unstable and rapidly converted into hydrogen peroxide (H_2O_2) and ROS in the cytoplasm [3]. However, excessive production of ROS may cause oxidative stress, ETC dysfuction, mitochondrial structural damage [4,5], and oxidative damage to proteins, DNA, and lipids [6].

Neurons are highly polarized cells, heavily dependent on the energy generated by mitochondria, and the brain consumes about 20% of the body's resting ATP, while it accounts for only about 2% of the body's mass [7,8]. In addition, mitochondria are necessary calcium-buffering organelles in neurons as they regulate local calcium dynamics to control neurotransmitter release [9]. Mitochondrial dysfunction has been implicated in a variety of diseases, and is a causative factor in several neurodegenerative diseases, including Alzheimer's disease (AD), Parkinson's disease (PD), Huntington's disease (HD), autism, and amyotrophic lateral sclerosis (ALS) [10–12].

Among the chemical elements that humans are exposed to, metals play an important role in both health and disease. Metals are natural components of the Earth's crust and enter the biosphere through a variety of human activities [13]. They are generally classified into two groups: essential and non-essential metals. The main routes of human exposure include ingestion, inhalation, and dermal contact [14]. The brain is able to regulate these metals effectively under physiological conditions. However, excessive exposure to metals, such as arsenic (As), aluminum (Al), cadmium (Cd), lead (Pb), copper (Cu), and manganese (Mn) may lead to their accumulation, and ensuing neurodegeneration [15]. Mitochondrial impairment and metal dyshomeostasis have been linked to some neurodegenerative disorders including AD, PD, HD, and ALS [12]. Metals can cause neurodegeneration by disrupting mitochondrial function, and thereby deplete ATP, induce ROS production, and ultimately lead to cell death through apoptotic and/or necrotic mechanisms [16]. There has been a growing interest in understanding the metabolism of neurotoxic metals and their role in the etiology of various neurodegenerative diseases, and a great deal of research has been done for this purpose. However, the effects of various metals on different neurodegenerative diseases are not identical, and their specific mechanisms of damage have yet to be fully clarified. Therefore, in this review, we summarize the latest reports on the mechanism of mitochondrial dysfunction in neurodegenerative diseases caused by metal exposure.

2. Neurological Disorders with Mitochondrial Dysfunction and Oxidative Stress

2.1. Alzheimer's Disease (AD)

AD is a well-known age-related neurodegenerative disorder characterized by progressive decline in cognitive function and pathological features of increased neuronal cell death [17]. The etiological hypotheses for AD mainly include genetics [18], decreased acetylcholine synthesis [19], accumulation of neurotoxic protein plaques of amyloid-β (Aβ) peptide [20], fibrous tangles with high phosphorylation of tau protein (P-tau) [21], or irregular mitochondrial function and dynamics [2]. However, the pathogenesis of AD remains unclear. It has been demonstrated that mitochondrial dysfunction is an early event in AD pathogenesis, characterized by decreased metabolism, disruption of Ca^{2+} homeostasis, elevated ROS levels, lipid peroxidation, and apoptosis [22]. An increased association between mitochondria-associated membranes (MAM) and mitochondria has also been linked to the pathogenesis of AD [23,24]. Moreover, variations in mtDNA have also been found to be related to the pathogenesis of AD, such as mutations in the heteroplasmic somatic mtDNA control region [25] and mitochondrial point/missense mutations in genes encoding cytochrome c oxidase subunits I, II, and III [26].

Although aging is a major risk factor for AD, extensive epidemiological evidence suggests that exposure to environmental toxins, particularly pesticides, metals, and solvents, may increase the risk of developing neurodegenerative diseases [27]. Neurotoxic metals, such as Pb [28], Hg [29], Al [30], Cd [31], and As [32] have been implicated in AD due to their ability to increase Aβ peptide and P-tau phosphorylation, leading to senile/amyloid plaques and neurofibrillary tangles (NFTs). Synergistic exposure to Pb, As, and Cd has been shown to further enhance the expression of amyloid precursor protein (APP) and BACE1, which in turn maximizes the induction of Aβ production [32]. A recent in vivo study showed that chronic inorganic arsenic (iAs) exposure aggravated AD-like pathology in 3xTgAD mouse brain, including reduced ATP content and complex I levels, as well as increased ROS formation in the hippocamus. In addition, higher immunopositive responses to amyloid isoforms and phosphorylated tau were observed in the frontal cortex and hippocampus [33]. It has been proved that Al in the brain can regulate the expression, distribution, and accumulation of APP and induce the maladjustment of iron-modulated signaling pathways through its interaction with the IRE mRNA regions, thus stimulating Fe^{2+}-induced membrane lipid peroxidation and causing oxidative damage [30,34,35].

A large body of evidence suggests a role for essential metal ion dysregulation in the etiology of AD, in particular the accumulation of Cu, zinc (Zn), and iron (Fe) in the amyloid plaques [36–38]. It is well established that an increase in loosely bound Cu and Fe in human AD brains can promote oxidative stress [39]. It is worth noting that although the total copper content in the AD brain is lower, the proportion of redox-active exchangeable Cu is higher, which is positively correlated with increased oxidative damage and AD neuropathology [40]. Similarly, tau displays redox activity when it binds to Cu, leading to further oxidative damage in the brain [41]. Mitochondrial ferritin deficiency aggravates the neurotoxicity induced by β-amyloid in mice, which may be related to the increase in intracellular iron accumulation and oxidative stress levels [42]. In vitro studies have demonstrated that binding of iron to Aβ peptide can promote Aβ aggregation and further increase the neurotoxicity of Aβ [43–45] by regulating the redox potential to the level at which iron's redox cycling occurs, which not only leads to the production of oxidative species, but also consumes essential oxygen and biological reductants [46]. ROS or exogenous oxidants can promote the release of harmful zinc from metallothionin, which in turn leads to mitochondrial dysfunction and further oxidative stress [47–49], and affects protein aggregation [50–52]. Studies have shown that zinc content is particularly high in AD neurons expressing mutant APP, PSEN1, and tau [53,54]. A systematic review and meta-analysis has shown that patients with AD had lower serum Mn levels, suggesting that Mn deficiency may be a risk factor for AD [55]. However, the link between Mn and AD remains very limited. The expression level of MnSOD in the hippocampal CA1–CA4 region of AD patients was 3–11 fold higher than that of the control group, suggesting the normal compensatory mechanism of Mn-dependent antioxidant enzymes may not be sufficient to protect the hippocampus from free radical oxidative damage [56].

2.2. Parkinson's Disease (PD)

PD is the second most prevalent and incidental neurodegenerative disease, affecting more than 2% of the population older than 65 years old [57]. Typical symptoms of PD include rigidity, bradykinesia, and test tremor [58]. The main pathological features of PD include selective loss of dopaminergic neurons in the substantia nigra (SN) region of the brain and more widespread aggregation of protein α-synuclein in Lewy bodies (LB) [59]. PD is accociated with mitochondrial dysfunction and calcium and dopamine (DA) dyshomeostasis, as well as abnormal autophagy and proteostasis [60].

Numerous studies have posited that mitochondrial dysfunction plays a key role in the pathogenesis of PD. The first line of evidence was documented in 1989 by Schapira and co-workers as they found a decrease in complex I of the ETC in the SN pars compacta (SNpc) of PD patients [61], which has been further confirmed [62]. The expression of mitochondrial proteins were changed, such as the molecular chaperones [63], the protease HtrA2 [64], a and b hemoglobins [65], or the outer mitochondrial membrane VDAC1 [66]. Recent studies showed that a vicious cycle between α-synuclein aggregation and mitochondrial impairment may exist in DA neurons [67]. Alterations in the PD-related genes DJ-1, PINK1, parkin, alpha-synuclein, and LRRK2 can directly or indirectly lead to mitochondrial dysfunction, resulting in increased ROS production and susceptibility to oxidative stress [68,69]. Oxidative stress plays an important role in the degeneration of dopaminergic neurons in PD [70]. Accumulating evidence has shown that oxidative stress is elevated in the brains of PD patients of both genetic and sporadic cases, and oxidative stress markers can be found in the SNpc DA neurons and their striatal axons [69,71]. DA metabolism, mitochondrial dysfunction, and neuroinflammatory processes are the main contributors to oxidative stress augmentation in PD [72].

More than 90% of PD cases are sporadic, and the etiology is associated with the complex interaction between genetic susceptibility and environmental stimuli [73]. The role of metals in the pathogenesis of PD has been the focus in medical chemistry and neurotoxicology [74,75]. Epidemiological studies have reported positive correlations between PD and long-term exposure to metals, such as Mn, Hg, Cu, Pb, Zn, Fe, and Al [76,77]. Metal exposure has been associated with key factors in the pathogenesis of PD, such as mitochondrial dysfunction, alterations in metal homeostasis, and aggregation of a-synuclein [30,78–81]. Epidemiological studies have reported a significant dose–response relationship between PD patients and blood Hg levels [82]. Hg has been found to cause neuron loss and cognitive and motor impairments in animal models, and further in vitro studies have shown that mercury exposure can cause apoptosis and oxidative stress [83–85]. SN neurons contain neuromelanin that can bind Fe and generate free radicals, causing cell death and lipid peroxidation [86]. Fe can also induce dopamine oxidation in SN neurons, which leads to the release of additional free radicals [87]. The binding of Cu to α-synuclein can induce oxidative damage of the protein and the oxidation of some C-terminal residues can promote protein aggregation [88–91]. High concentrations of free Zn in the PD anterior olfactory nucleus was detected and the colocalization of free zinc and alpha-synuclein suggested the role of zinc in the pathogenesis of PD [92]. In a mouse model of PD, Mn exposure can enhance mitochondrial dysfunction to aggravate neurodegeneration and progressive motor deficits [93].

2.3. Huntington's Disease (HD)

HD is an inherited autosomal dominant neurodegenerative disorder caused by a CAG amplification of the huntingtin (Htt) gene. The typical clinical manifestations are chorea-like involuntary movements, dementia, and psychiatric symptoms [94], and the pathological features are selective loss of striatal neurons and aggregation of the mutant Htt protein [95,96]. Disruptions of mitochondrial energy metabolism were found in the brains of patients with advanced HD, including reduced activities of mitochondrial respiratory complexes II–IV and aconitase [97,98]. In addition, in vitro studies with samples from presymptomatic and pathological Grade 1 HD patients showed no changes in striatal or cortical complexes I–IV activity, suggesting that mitochondrial energy impairment is a late event in the progression of the disease, rather than a cause [99,100]. HD patients also exhibit weight loss, which may be due to the mitochondrial ATP synthesis disorder [101]. The pathogenesis of HD is related to mitochondrial dysfunction, which is manifested by reduced ATP/ADP ratio, decreased O_2 consumption, increased mitochondrial ROS and fragmentation, abnormal lactate/pyruvate levels, and decreased mitochondrial membrane potential [102]. These apparent mitochondrial dysfunctions may be related to interaction with mutant Htt [103]. Mutant Htt can interfere with mitochondrial function by binding to Drp1 to disrupt the balance of mitochondrial fission–fusion dynamics, reducing anterograde and retrograde axonal mitochondrial transport, and binding to peroxisome proliferator-activated receptor coactivator-1α (PGC-1α) protein which is involved in mitochondrial biogenesis and antioxidant defenses [104,105]. The ultimate result of these mitochondrial injury is a reduction in ATP production, with ensuing neuronal dysfunction followed by death [106].

Although evidence of metal involvement in HD pathogenesis is limited, histological and MRI studies demonstrate elevated basal ganglia iron levels in HD patients [107]. Agrawal and Fox found that mitochondrial iron accumulated in a mouse model HD brain, and neonatal iron supplementation could increase the accumulation of mitochondrial iron in the brains and enhanced markers of mitochondrial dysfunction [108]. Cu has been shown to promote aggregation of huntingtin protein [109]; however, it is not clear whether the abnormal distribution of Cu interacts with mitochondria [59]. The HD-associated mitochondrial inhibitor 3-nitropropionic acid (3-NPA) causes Zn accumulation in vitro or in vivo [110]. Studies have found that Mn deficiency is related to HD [111–116], and exogenous Mn supplementation can promote the clearance of mutant HTT protein aggregates in striatum cells [117,118]. A recent study reported that Mn-induced mitochondrial dysfunction in HD cells could only be detected at an exposure dose above the acute toxicity threshold [119]. One study found that Cd exposure increased oxidative stress, caused apoptosis, and altered metal transport in heterozygous HTT striatum cells [120].

2.4. Autism

Autism spectrum disorder (ASD) is a complex neurodevelopmental disorder characterized by impairments in reciprocal social interaction and communication, as well as restricted and stereotyped patterns of interests and behaviors [121]. The pathogenesis of ASD is unclear, and its clinical manifestations are varied. The etiology of ASD may involve a variety of genetic and environmental factors [122]. Some studies suggested that the idiopathic risk factors may include obstetric complications, fetal hypoxia, maternal or paternal age, gestational bleeding, gestational diabetes, prenatal diet, and medication [123,124]. A growing amount of evidence has indicated that mitochondrial dysfunction plays an important role in the development of ASD. ASD is also associated with redox abnormality and oxidative stress [122]. A few studies have suggested that children with autism have limited availability of thiol and reduced glutathione (GSH) storage capacity, resulting in decreased detoxification, increased oxidative stress and DNA damage, and chronic inflammatory responses [125–128].

Some studies support a significant relationship between ASD and metal exposure [11]. Arora et al. measured the tooth-matrix biomarkers from twin samples, and found that the absorption of Mn and Zn decreased while Pb increased in the ASD patients. In addition, Mn and Pb were also associated with the severity and characteristics of ASD [129]. A systematic review and meta-analysis concluded that early life iAS exposure is positively associated with ASD, and the relationship between lead exposure and autism risk is controversial [130]. Another recent systematic review and meta-analysis indicated that existing evidence supports significant associations between ASD and Al, Cd, and Hg, respectively [131]. Some data demonstrate that the neurotoxic mechanisms of which metals trigger or accelerate the onset of ASD include oxidative stress, endoplasmic reticulum (ER) stress, and destruction of essential metalloproteins, which further lead to or promote neuroinflammation, excitatory toxicity, and apoptosis [11]. In a mouse model, perinatal Pb exposure significantly reduced the activities of SOD, glutathione peroxidase (GPx), and glutathione-disulfide reductase (GSR) in the hypothalamus, corpora quadrigemina, and corpus striatum [132]. A recent cross-sectional study found that zinc levels in hair were inversely associated with the severity of autism symptoms [133]. Fe deficiency was more common in children with ASD compared to the control group [134], and low serum Fe and ferritin levels may be associated with attention deficit hyperactivity disorder [135].

2.5. Amyotrophic Lateral Sclerosis (ALS)

ALS is a devastating motor neuron disorder that typically affects men and women between the ages of 50 and 60, and is characterized by progressive muscle weakness, paralysis, and death within a few years of onset [94]. The majority of cases are sporadic, but about 10% are inherited [17]. Disruption of mitochondrial structure, dynamics, bioenergetics, and calcium buffering has been considered to be directly involved in the pathogenesis of ALS [1,17]. Many identified ALS genes play a role in mitochondrial-related functions; for example, superoxide dismutase 1 (SOD1), ALS2, fused in sarcoma/translocated in sarcoma (FUS), VAMP-associated protein type B and C (VAPB), and open reading frame 72 on chromosome 9 (C9orf72). Evidence gathered from patient studies as well as in vitro and in vivo studies strongly reveals that mitochondrial dysfunction is a core event in ALS [1]. Indeed, increased levels of ALS-associated mutant mitochondrial SOD1 may lead to mitochondrial aberrations in ALS [59]. SOD1 mutations are the most common mutation found in ALS, present in about 20% of familial cases and about 2% of overall cases [136]. Mutant SOD1 has been reported to be involved in pathogenesis of ALS through oxidative stress, ER stress, glutamate toxicity, mitochondrial dysfunction, axonal transport disruption, extracellular toxicity, and amyloid aggregation [137]. Under normal circumstances, mitochondria convert 1–3% of oxygen molecules into superoxide radicals, which are later eliminated by SOD1. Thus, in the absence of SOD1, the slowed dismutation process will lead to oxidative stress [138]. Expansion of GGGGCC (G4C2) repeats in the C9orf72 is the most common genetic cause of ALS with frontotemporal dementia (C9-ALS/FTD). Increased ROS and mitochondria hyperpolarization have been reported in the fibroblasts of C9-ALS/FTD patients [139]. A recent genetic study releaved that poly (GR), a dipeptide translated from G4C2 repeat transcript, could be inhibited by yeast mitochondrial escape 1-like ATPase (YME1L) and mitochondria-associated noncanonical Notch signaling [140].

The role of heavy metal exposure (such as Pb, Se, Hg, Cd, and Fe) as a risk factor for ALS has been studied [141]. A recent systematic review and meta-analysis indicated that environmental/occupational Pb exposure was positively proportional to the risk of ALS [142].

In vivo studies have shown that Hg accumulates in the nervous system and damages the axons of motor neurons, consistent with the typical pathological changes of neuron degeneration in ALS [143]. Pb and methyl-mercury (MeHg) can induce ALS-linked TAR DNA-binding protein 43 (TDP-43) accumulation in neurons [144]. Beqollari et al. found that exposure to low doses of MeHg could accelerate the onset of ALS in a SOD1-G93A mouse model probably through glutamate-mediated excitotoxicity [145]. High concentrations of Cd have been detected in blood, cerebrospinal fluid (CSF), and gray and white matter in ALS patients [146–148]. Interestingly, a case-report study of ALS showed that Cd disrupted the blood–brain barrier (BBB), decreased SOD1 levels in brain, and enhanced the glutamate excitability in glial cells [138]. In addition, higher Mn contents in the CSF of ALS patients have also been reported, suggesting that the regulation of Mn distribution in human body might play a role in the etiology of ALS [148]. Peters et al. found blood Se and Zn concentrations were negatively correlated with ALS, while blood Cu content was positively correlated with ALS [149]. Se has a protective effect on ALS, which may be related to the protective antioxidant mechanism [150]. Besides, Cu and Zn may play a more direct role in the pathogenesis of ALS, because both are cofactors for cytosolic SOD1. Most polymorphisms lead to misfolding of the SOD1 monomer, reducing its affinity for Zn and exposing the Cu binding site, and this conformational change leads the enzyme to generate rather than detoxify ROS [151,152]. Free Fe level was also higher in the CSF of ALS patients compared to controls [153,154], which may increase iron redox activity and ROS production [155].

3. Molecular Mechanisms of Metal-Induced Mitochondrial Dysfunction

3.1. Arsenic (As)

As, a widely distributed toxic metalloid, is a risk for about 200 million people in more than 24 countries around the world [156,157]. It can be absorbed through skin, digestive tract, and inhalation. After absorption, As can be distributed to various organs, including kidney, lung, liver, and spleen in the animal and human bodies [158,159]. More seriously, As can enter the central nervous system (CNS) through the BBB and accumulate in different brain regions [160–162]. In vivo studies showed that excessive exposure to As induced neuronal apoptosis, which interrupted the neurodevelopment and cognitive functions of rats [163–165]. Epidemiological studies in rural-dwelling adults and elders also show that As (3–15 μg/L) levels in water negatively correlated with the scores of cognitive performance and memory, indicating that As is a neurotoxic metalloid [166], which also acts as a risk factor for AD [33,167–169]. However, the mechanisms of As-induced neurotoxicity remain unclear.

To date, As-induced neurotoxicity has been related to Aβ overproduction [32,170], inflammatory responses [171,172], thiamine deficiency [173], oxidative stress, disruption of neurotransmitters [163,171], cytoskeletal gene expression, mitochondrial dysfunction, and disruption of acetyl cholinesterase activity [166,167,174]. Among them, mitochondrial dysfunction has been demonstrated to play a key role in As-induced neurotoxicity. Several in vitro studies have shown that As may induce adverse effects on mitochondrial functions. For example, Haga et al. [175] suggested that aggregated mitochondria were found in A172 cells after 50 μM arsenic trioxide (As_2O_3) treatment for 8 h. Subsequently, other investigators also suggested that sodium arsenite ($NaAsO_2$) or As_2O_3 treatment induced mitochondrial dysfunction via increasing intracellular Ca^{2+} levels, mitochondrial membrane potential (MMP), or calpain 1 levels in N_2A cells [176], SHSY-5Y cells [177], and primary astrocytes [178], as well as rats' primary neuronal cells [179]. Moreover, in vivo studies have also verified the critical roles of oxidative stress and mitochondrial dysfunctions in As-induced neurotoxicity [180,181].

It is well known that the mitochondrion is the main source of ROS formation, as well as a major target of ROS [182]. Oxidative stress is closely related to mitochondrial dysfunctions induced by As. Yadav et al. [183] showed that the activities of oxidative stress marker enzymes MnSOD and CAT were decreased by As in the mitochondrial fraction of different brain regions (including striatum, hippocampus, and frontal cortex) of rats via increasing ROS, and lipid peroxidation after exposure to $NaAsO_2$ for 28 days [181,183]. Similar results were found in sub-chronic As exposure studies done by other investigators which indicated that MnSOD, CAT, Gpx, GR, and GST activity were decreased in the mitochondrial fraction of rat brain [184,185]. Moreover, various studies suggested that As directly impaired the mitochondrial respiratory system via oxidative stress. Dwivedi et al. [180] indicated that As caused oxidative stress which in turn inhibited the activities of complexes I, II, and IV in the mitochondria of rat brain. These results have been corroborated by other labs [181,185]. Furthermore, excessive As exposure disrupted oxidative phosphorylation, and thus interrupted the ATP synthesis and mitochondrial respiration in the mitochondria of the brain [180,186]. Consistent with these results, sub-chronic exposure to low levels of As has been shown to decrease gene expression of the mitochondrial complexes II, IV, and V in mice brains [187,188]. All of the above-mentioned studies suggested that the mechanisms of oxidative stress involved in As-induced mitochondrial dysfunctions play a pivotal role in As-induced neurotoxicity.

In summary, these studies suggest that the mitochondrial dysfunction in the CNS is the most important mechanism of As-induced neurotoxicity. It includes impairments of Ca^{2+} homeostasis [177,189], abnormal mitochondrial dynamics [190,191], and changes in membrane potential and permeability [174,192], which induces neuronal injuries via the mediating mitochondria-dependent pathway.

3.2. *Aluminum (Al)*

Al is a ubiquitously distributed metal on the earth, and it can be easily absorbed via skin contact, inhalation, and ingestion. Al sulfate has been ubiquitously used for water purifying, food processing, and the medicine and pharmaceutical industry, which ensure its presence in human bodies [193]. An increasing number of studies have shown that Al could accumulate in various mammalian organs, including bone, kidney, lung, liver, spleen, and brain [194–196]. Growing evidence has also suggested that Al accumulations in various brain regions may cause neurotoxic symptoms and learning impairment [196,197]. Studies in rodents indicated that chronic Al exposure led to Al accumulation in the hippocampus and caused neurobehavioral impairment [198–200]. Other studies also reported that Al caused neurofibrillary degeneration [197]. Altmann et al. showed that the impairment in cerebral function may be related to the concentrations of Al in the contaminated water [201]. Additionally, epidemiological studies suggested that Al has been considered as a potential risk factor in the development of neurodegenerative diseases, such as AD [196,202], PD [203,204], and ALS, etc. [205–207].

Several studies have proposed that mitochondrial dysfunction may play a critical role in the toxic effects of Al, including neurotoxicity [197,208]. Rao et al. [209] have shown that the ROS formation and mitochondrial respiratory activity, as well as glutathione depletion, were increased in the glial cells after being treated with Al for 24 h. Other groups have also depicted that Al exposure increased ROS formation and impaired the cytochrome c oxidase, which impaired mitochondrial functions in various neuronal cell types, including PC12 [210–212], SH-SY5Y neuroblastoma cells [213,214], and rat and cerebellar granule neuronal cells [42,215]. Mitochondrial dysfunction was also observed in in vivo studies [216,217]. Acute exposure to 50 μM Al maltonate via intracisternal injection caused the release of cytochrome c (cyt-c), accompanied by decreased Bcl-2, upregulated Bax, p53, and caspase-3, and DNA fragmentation in the mitochondria of rabbit brain [218]. Subsequently, Kumar et al. also reported that sub-chronic Al exposure for 12 weeks resulted in elevated ROS generation, and decreased ATP synthesis and cytochrome levels in a rat's brain, which implied disruption of mitochondrial function [219]. In addition, their other study also suggested that Al exposure decreased MnSOD and aconitase activities in different regions of the rat brain [220]. Additionally, transmission electron microscope results showed that Al exposure caused mitochondrial swelling and vacuolization structures, and thus increased the diameter of mitochondria in the hippocampus nerve cells of mice and rats [208,219]. Finally, Al exposure upregulated the autophagy-related proteins LC3-II and Beclin-1, while downregulating p62 expression, suggesting that Al-induced learning and memory impairments may be related to mitophagy [208].

Recently, oxidative stress and mitochondrial disorders have been suggested as major targets for Al-induced neurotoxicity. For example, quercetin has shown protective effects on Al-induced mitochondrial swelling and chromatin condensation in rat hippocampus [221]. Naringin also has protective effects on memory impairment of sub-chronic Al-exposed rats via preventing the activations of mitochondrial oxidative damage in the brain [222]. Subsequently, *Centella asiatica*, which has antioxidant properties, was shown to ameliorate memory impairment and the activation of oxidative stress and decrease mitochondrial enzyme activity in the hippocampus and cerebral cortex induced by Al [223]. In addition, some other natural compounds also have been shown to have neuroprotective effects on Al-induced neurotoxicity, such as crocin, curcumin, and polyphenols [197,224,225]. These studies indicate that inhibition of oxidative stress and mitochondrial dysfunction may be a therapeutic strategy to prevent the neuronal injuries induced by Al.

3.3. Copper (Cu)

Cu is an essential trace metal for human health. Cu takes part in many cellular enzymatic activities, including energy production, redox balance, and neurotransmitter biosynthesis [226]. An adequate amount of copper is critical for the maintenance of redox balance in the mitochondria [227]. The mitochondria are both a regulatory hub for Cu homeostasis and a target of Cu toxicity [228]. For example, Cu is required for metallation of the catalytic core of cytochrome c oxidase, a mitochondrial metalloenzyme in the respiratory complex chain [229]. However, overload of mitochondrial Cu is detrimental to the function of respiratory complexes, leading to elevation of ROS and mitochondria dysfunction. Wilson's disease is a genetic disorder caused by excessive mitochondrial copper in the liver [227].

Brain mitochondria are particularly sensitive to the detrimental effects of Cu [230]. Compared to the mitochondria in the liver, kidney, and heart, brain mitochondria are susceptible to elevated levels of Cu, which attacks free thiols in large molecules that are indispensable for maintaining neuronal cell function [230]. The membrane potential, efficiency in ATP production, and structural integrity of brain mitochondria were prone to damage caused by excessive Cu [230]. Chronic Cu exposure led to spatial memory impairment that was associated with mitochondrial damage in the hippocampus [231]. Specifically, beta-amyloid-induced memory deficit in rats is exacerbated by Cu exposure. Meanwhile, analysis of isolated mitochondria from rat hippocampus following Cu exposure demonstrated a significant decline in mitochondria health, including increased lipid peroxidation and glutathione oxidation [231]. Mishandling of Cu in the mitochondria has been linked to age-related neurodegenerative disorders [232–234]. In a mice model of AD, a proteomics study showed that low levels of Cu exposure (0.13 ppm, 2 months) induced deficits in mitochondrial dynamics, leading to increased H_2O_2 production and reduced cytochrome oxidase activity [232]. Common biochemical characteristics of PD include accumulation of iron and diminished Cu content in degenerated brain regions. The disruption of Cu metabolism was believed to be involved in the pathological process in loss of catecholamine neurons [233]. Additionally, in a 6-hydroxydopamine (6-OHDA)-induced-PD model, Cu exposure increased oxidation of 6-OHDA, resulting in an increase in the rate of p-quinone formation and H_2O_2 accumulation. In the same model, the 6-OHDA-induced lipid peroxidation and protein oxidation were potentiated by Cu exposure [234].

Mitochondrial dysfunction following chronic Cu exposure involves oxidative stress, collapse in mitochondrial membrane potential, depletion of GSH, comprised function of respiratory complexes, reduction in APT production, and structural damage to the mitochondria [230,231]. Experimental evidence showed that free protein thiols in the mitochondria are potential toxic targets of Cu [230]. GSH supplementation attenuated Cu-induced lipid peroxidation but failed to protect oxidized thiols [234]. In addition, the induction of the mitochondrial permeability transition (MPT) was associated with Cu-induced astrocytic injury [235]. Furthermore, mitochondrial health in the hippocampus is a potential in vivo target of Cu. A recent study showed that mitochondrial biogenesis and respiratory function were impaired in the hippocampus of mice chronically exposed to $CuCl_2$ [232].

3.4. Cadmium (Cd)

Cd is a heavy metal that has no nutritional roles for humans. Cd-induced cellular damage is largely mediated by disruption of mitochondrial activity [236]. Elevation of ROS in the mitochondria and induction of mitochondria-derived apoptosis signaling are involved in Cd-induced neurotoxicity [237,238]. Mitochondrial protection afforded by antioxidants can attenuate Cd-induced neuronal damage [239].

An elevation in protein and lipid peroxidation, decrease in antioxidant capacity, and structural damage to the mitochondria were shown in the brains of rats chronically exposed to Cd [240]. The structural stability of mitochondria-associated ER membranes (MAMs) is critical for the proper function of the mitochondria. Recent studies show that MAMs are not only the physical bridge to facilitate communication between the ER and mitochondria, but they are also indispensable for cellular homeostasis processes such as autophagy, lipid metabolism, and Ca^{2+} transport [241]. Cd exposure induced increased production of ROS in the mitochondria, leading to impairment of MAMs [242]. The shapes of mitochondria are subjected to transformations in response to cellular stress, which is driven by two closely related processes: mitochondrial fusion and fission. Mitochondrial fusion and fission are required for proper intracellular distribution and quality control of the organelle [243]. Mitofusin 2 (Mfn2) is a mitochondrial outer membrane-localized GTPase that is essential for mitochondrial fusion. Cd-induced neuronal necroptosis was associated with ROS-induced S-glutathionylation of Mfn2 [242]. Increased ROS levels are detrimental to the activity of key enzymes involved in lipid metabolism. Cd exposure altered the lipid profile in a rat brain, resulting in an increased level of cholesterol (CHL) in the mitochondria [244]. Furthermore, Cd exposure promotes lipid peroxidation (LPO), which is mediated by the increased level of oxygen free radicals [245]. The mitochondria are both a storage site for cellular calcium ions and regulators for calcium ion homeostasis. Cd can competitively bind receptors and ion channels that regulate calcium ion influx, modulating calcium-dependent cellular activity [246]. The Ca^{2+}/calmodulin-dependent protein kinase II (CaMK-II) regulates cytoskeletal dynamics and apoptotic cell death. Recent advances show that CaMK-II mediates the effects of Cd exposure on actin depolymerization microtubules and cadherin junctions, which are the underlying mechanisms of Cd-induced cytoskeletal disruption and alterations in cellular morphology [246]. Nutritional trace metals, such as Zn and Se, can mitigate Cd-induced mitochondrial toxicity. For example, in a cellular toxicity model of PC12 cells, Cd exposure led to depletion of cellular GSH and oxidative damage to the mitochondria, which can be attenuated by Zn supplementation [247]. Additionally, Se supplementation suppressed Cd-induced oxidative stress and the mitochondrial apoptosis pathway [237].

3.5. Mercury (Hg)

Mercury is a naturally occurring element that is found in various inorganic and organic forms [248,249]. Both organic and inorganic mercury are neurotoxic. Methylmercury (MeHg) is of special concern as it is an ubiquitous environmental contaminant and its consumption in fish can lead to a devastating neurological disorder, referred to as Minamata disease [250]. Numerous studies have shown that mercury causes brain mitochondrial dysfunction, playing a key role in Hg-induced brain damage and neurological disorders.

As early as 1974, Chang and Hartmann found that mercury was present both in neurons and in glia after MeHg or mercuric bichloride ($HgCl_2$) administrated to rats orally or subcutaneously [251]. Notably, mitochondria accumulate mercury, mostly because of their abundance of thiol (–SH) groups. Although mercury initiates multiple additive or synergistic disruptive effects, a key mechanism of disruption of mitochondrial function is associated with the production of ROS. $HgCl_2$ and/or MeHg exposure enhance ROS formation in the CNS, evidenced by both in vivo [252] and in vitro models, including primary rat cortical neuron [253], rat cortical astrocyte [254,255], cerebellar granule neurons and astrocytes [256], and microglia [257], as well as in mixed primary neuron–astrocyte culture [258]. ROS overgeneration leads to consequent oxidative stress [259] and mitochondria-mediated apoptosis. For example, MeHg exposure results in cytochrome c release, caspase-3 and caspase-9 activation, and apoptosis-induced factors (AIF) increase in primary rat cortical neuron [253]. Mitochondria-mediated apoptosis in brain cells is secondary to alteration of mitochondrial membrane potential (MMP) and transition of mitochondrial permeability [260], which have been observed in neuron/astrocyte mixed-culture [258] and astrocyte mono-culture [261,262] after mercury exposure. In addition, the mitochondrial dysfunction evoked by mercury was correlated with damage in mitochondrial bioenergetics. Mercury has been found to act as an inhibitor of the enzymatic activities of mitochondrial respiratory complexes, impairing ATP synthesis in rat hippocampal mitochondria [263]. MeHg exposure reduced GSH levels in astrocytes, increasing the vulnerability to oxidative stress [264]. Apart from a series of biochemical impairments in mitochondria induced by mercury exposure, pathological changes in mitochondrial morphology have also been demonstrated. Li et al. [265] found that a low dose of mercury, lead, and cadmium caused dose-dependent mitochondrial depletion, as well as ridge and matrix dissolution in the hippocampal neurons of rats. Additionally, an in vivo study observed that MeHg induced mitochondrial swelling in the hippocampus of MeHg-exposed F1 generation rats, and enlarged and fused mitochondria in mice cerebral cortex [263].

Dreiem and Seegal [266] found that antioxidant Trolox significantly reduced MeHg-induced ROS, while failing to restore mitochondrial function in rat striatal synaptosomes. The authors revealed that MeHg increased mitochondrial calcium levels, which are fundamental to mitochondrial function. If mitochondria take up too much Ca^{2+}, it delays the rise in cytoplasmic Ca^{2+} [267] and the opening of the MPT pore, which may promote the release of cytochrome c and other pro-apoptotic factors, culminating in apoptosis [268]. The modulatory effect of cellular calcium homeostasis by MeHg in mouse spinal motor neurons was also found [269]. In addition, proteomic analysis revealed that many mitochondrial proteins were deregulated by mercury exposure in primary mouse cerebellar granule neuron and astrocytes [256,270], as well as in rat hippocampus [271], thus impairing mitochondrial function associated with cellular metabolism and energy production.

3.6. Lead (Pb)

Pb is an environmentally abundant metal pollutant with human exposure mainly through air inhalation and food and water intake. Pb is a strong toxicant for the developmental CNS [272,273]. Pb intoxication in children, even at low doses, is found to impair learning and memory and affect cognitive functions and intellectual development [274,275]. The brain is the primary target of Pb toxicity. Mitochondria play a key role in Pb-induced impairment of nervous system function.

An in vivo study found that the activity or levels of several mitochondrial enzymes were inhibited by Pb exposure. For example, lead acetate (PbAc) exposure in drinking water decreased aldehyde dehydrogenase (ALDH2) expression in brain nucleus accumbens [276], and PbAc exposure from postnatal day 1 (PND1) through PND21 in drinking water of the mother significantly decreased offspring activity of mitochondrial monoamine oxidase (MAO) in all brain regions, including cerebral cortex, hippocampus, and cerebellum, in a dose- and age-dependent manner [277], attributed to the high affinity of Pb for the -SH groups in enzymes, consequently damaging mitochondrial activity and function. In addition, pre- and neonatal exposure to a low dose of Pb (Pb concentration in whole blood < 10 μg/dL) induced synaptic ultrastructural abnormalities in mitochondria including elongated, swollen, and shrunken changes in mitochondria [278], indicating the mitochondrial morphological disruption induced by Pb. Mitochondria-mediated apoptosis has also been shown in Pb-induced neuronal death. PbAc intoxication caused cognitive dysfunction and anxiety-like behavior, along with altered Bax/Bcl-1 expression and increased cytochrome c release from mitochondria in rat brain [279]. In addition, $(CH_3COO)_2Pb$ exposure induced apoptosis via the mitochondrial pathway in embryonic neurocytes isolated from chicken [280]. Similarly, the combined treatment (As+Cd+Pb) in individual lethal concentration (LC)-5 induced a toxic effect on C6-glioma cells derived from rat glioma, via mitochondria-mediated apoptosis, including caspase-9 activation and Bax/Bcl-2 changes [281]. Notably, Zhu et al. found that MPT pore opening plays an important role in Pb-induced neurotoxicity. In SH-SY5Y cells, PbAc exposure significantly impaired mitochondrial function, evidenced by ATP decrease, MMP collapse, ROS production, mitochondrial apoptosis, and morphology changes (swelling and rupture). PbAc treatment significantly increased the protein level of Cyp D, a component of MPT, and induced MPT pore opening in both PC12 and SH-SY5Y cells. Inhibitor of Cyp D significantly reversed mitochondrial damages and cell death induced by Pb [282].

3.7. Zinc (Zn)

Zinc is an essential trace element that is required for the function of numerous enzymes and DNA-binding transcription factors. Excess zinc influx has been manifested to play a role in neuronal damage and death associated with traumatic brain injury, stroke, seizures, and neurodegenerative diseases [283,284]. Mitochondria have been identified as targets of the neurotoxic effects of zinc by reducing ATP production and increasing ROS.

Zinc exposure reduced the cellular nicotinamideademine dinucleotide (NAD+) in cultured mouse cortical neurons, followed with a progressive loss of ATP levels and subsequent cell death [285–287], indicating the potential inhibition of mitochondrial respiration enzyme. Indeed, several mitochondrial enzymes, including α-ketoglutarate dehydrogenase, NAD+-dependent isocitrate dehydrogenase, succinate dehydrogenase, and cytochrome c oxidase, have been demonstrated to be inhibited by zinc exposure in liver mitochondria [288,289]. Notably, by using bovine heart mitochondria, complex III, specifically the bc 1 complex, was identified as the site of Zn^{2+} binding and inhibition [290,291]. ROS generation has been found to be critical in zinc-induced neurotoxicity, demonstrated in diverse brain cell models [292,293]. As mitochondria are an important source of cellular ROS production, the influx of Zn^{2+} through Ca^{2+}-permeable AMPA/kainate channels also triggers rapid mitochondrial depolarization, leading to prolonged production of mitochondrial superoxide in cortical neurons [294].

In addition, several other mechanisms have been involved in the zinc-induced mitochondrial dysfunction. For example, extracellular zinc application stimulates the Ras/MEK/ERK pathway, which leads to zinc-induced mitochondrial dysfunction and consequent cell death in rat neurons [295]. An immediate early transcription factor, egr-1, was found to act downstream of ERK 1/2 to induce neuronal death after zinc exposure [296]. Furthermore, elevated intra-neuronal zinc impairs mitochondrial trafficking without altering morphology, which was restored by PI3k inhibitors, suggesting the role of PI3k activation in zinc-inhibited mitochondrial movement in neurons [295]. Apart from the adverse effects on neurons and glia, zinc overload also critically induced ROS formation in mitochondria and degradation of mitochondrial network in cerebral microvessels, which were mediated through Drp-1-dependent mitochondrial fission pathway, thus contributing to increased permeability of the BBB after cerebral ischemia.

Not only zinc overload, but also zinc defficiency, may impair neurological functions [297] and cause neuronal apoptosis via an intrinsic (mitochondrial) pathway in human neuroblastoma IMR-32 cells and primary rat cortical neurons [298]. Researchers have identified that the transposition of phosphorylated p53 into the mitochondria mediated zinc deficiency-induced mitochondrial alterations and apoptosis in neuronal precursor cell (NT-2 cell line) [299].

3.8. Iron (Fe)

Iron is a crucial trace metal for life and is the most abundant transition metal in the brain. It acts as a catalytic center for multiple enzymes and supports many elementary biological processes, including DNA synthesis and repair, oxygen transport, mitochondrial respiration, and neurotransmitter metabolism. Oxidative stress, iron deposition, and mitochondrial dysfunction have been considered as hallmarks of many neurodegenerative diseases, including PD, HD, and AD [300,301], and a positive feedback loop among these three factors seems to exist in neurological disorders.

Upregulation of cellular redox-active iron is directly related to increased ROS and with changes in intracellular reduction potential [302,303]. In the presence of H_2O_2, which is mainly produced by mitochondrial ETC, Fe^{2+} generates hydroxyl radicals (OH) via the Fenton reaction. The hydroxyl radical is considered to be one of the most reactive substances in biological systems because its reaction rate is limited only by its diffusion. This free radical can attack proteins, DNA, and lipid membranes, thus disrupting mitochondrial function and cellular integrity, and eventually leading to oxidative stress and cell apoptosis [304]. Iron overload promotes the production of mitochondrial ROS in SH-SY5Y cells, in an AMP-activated protein kinase (AMPK)-dependent manner [305], and caused ATP production defects, mitochondrial complex I inhibition, and mitochondrial apoptosis in primary cortical neurons [306]. In addition, mitochondria-targeted iron chelators showed protective effects against mitochondrial oxidative damage and neuronal death, both in rotenone-treated SH-SY5Y cells and the dopamine neurons from MPTP-intoxicated mice, which indirectly suggested that iron accumulation in mitochondria induced mitochondrial oxidative damages in neurons and consequent cell death [307]. Moreover, iron overload may induce Drp-1-dependent mitochondrial fragmentation by upregulating intracellular calcium. Lee et al. [308] found that in ferric ammonium citrate (FAC)-stimulated HT-22 hippocampal neuron cells, mitochondria were fragmentated by dephosphorylation of Drp1 (Ser637) and apoptotic neuronal death was increased. Notably, FAC-induced iron overload leads to intracellular calcium elevation and further activation of calcineurin, while inhibition of Ca^{2+} signals related to calcineurin prevents iron overload-induced mitochondrial fragmentation and neuronal cell death. Redox-sensitive ryanodine receptor (RyR)-mediated Ca^{2+} release also was shown to underlie the iron-induced mitochondrial fission in primary hippocampal neurons [309].

Recently, a new iron-dependent programmed cell death, namely ferroptosis, has been found to be a main driver of many neurodegenerative diseases. It is characterized by the accumulation of lipid peroxidation products and lethal ROS derived from iron metabolism and can be pharmacologically inhibited by iron chelators. Although the detailed mechanism by which iron overload promotes ferroptosis has yet to determined, it is reasonable to hypothesize that iron overload may drive the generation of hydroxyl radicals, which further react with liposomes to produce lipid peroxidation products and cause mitochondrial dysfunction, and eventually ferroptosis [310–312]. Although mitochondria have been shown to be vital regulators of iron homeostasis and ferroptosis in neurodegenerative diseases [313], more direct evidence targeting iron overload, mitochondrial dysfunction, and ferroptosis is still required. The mitochondria are also the site for the synthesis of iron–sulfur cluster biogenesis (ISCs) and heme prosthetic groups. There is evidence that mitochondrial ISC assembly defects may cause iron overload and consequent negative effects on cellular or mitochondrial function [314,315].

Therefore, iron accumulation induced by direct excessive iron exposure or secondary to iron overload has been demonstrated to play an important role in neurological diseases, via impairing mitochondrial function and inducing oxidative stress. Targeting chelatable iron and the consequent ROS, especially in mitochondria, appear as possible therapeutic options for age-related neurodegenerative conditions [316].

3.9. Manganese (Mn)

Mn is the 12th most abundant mineral element in the earth crust, and is both nutritionally essential and toxic in excess. Mn is an essential metal for normal growth, development, and cellular homeostasis, as well as a cofactor for multiple enzymes; for example, Mn-superoxide dismutase (Mn-SOD), pyruvate carboxylase, arginase, and glutamine synthase (GS). Manganese preferentially accumulates in tissues rich in mitochondria [317,318], and it is taken up by brain mitochondria via mitochondria Ca^{2+} uniporter [319].

Mn is known to induce mitochondrial dysfunction in the nervous system [320], including the inhibition of the enzymes of the tricarboxylic acid (TCA) cycle in human neuroblastoma (SK-N-SH) and astrocytoma (U87) cells [321] and a reduction in the activities of ETC in rat primary striatal neurons [322] and in PC12 cells [323], ultimately resulting in ATP depletion [324–326] and mitochondria-mediated apoptosis [327–329]. Notably, these mitochondrial impairments have been found to be rescued by some antioxidants [324,325,330], indicating that oxidative stress is primarily involved in the mechanism of Mn-induced mitochondrial dysfunction.

Another cause of mitochondria-mediated apoptosis induced by Mn exposure is the induction of the MPT [331]. This process causes unrestricted proton movement across the inner mitochondrial membrane, resulting in mitochondrial swelling, mitochondrial membrane potential destruction, further production of ROS, and cellular apoptosis [324,332].

4. Conclusions

Long-term exposure to various metals, both essential and non-essential, has become increasingly common among the public as accelerated human activities release excess metals into the environment. Studies over the past several decades have greatly improved understanding of the neurodegenerative diseases associated with metals exposure and refined the molecular mechanisms of metal-induced nerve cell damage. Excessive exposure to both essential and non-essential metals may cause neurotoxicity, but deficiency in some essential metals, such as Zn and Fe, can also aggravate neurodegenerative diseases. As reviewed herein, metals may play a role in some neurodegenerative diseases, including AD, PD, HD, autism, and ALS, all of which rely on similar intracellular mechanisms, including metal dyshomeostasis, mitochondrial dysfunction, oxidative stress, and protein aggregation. Of all organelles, mitochondria produce the most intracellular ROS [102]. Excessive production of ROS and/or inhibition of the antioxidant system in mitochondria may cause oxidative stress, damage the mitochondrial structure, and induce apoptosis, which plays an important role in neurodegeneration. Upstream mechanisms of ROS generation are nonetheless not well characterized, and additional studies are required. For example, Nrf2 is known to be activated by methylmercury, copper, and other metals, but other Nrf2-independent means for ROS mitigation have also been described.

To date, most studies have focused on the neurotoxic mechanisms of single metals; however, in real life, the exposure environment of human metal exposure is complex, and metals may interact with each other and influence their homeostasis. It is therefore imperative to further explore the effects of metal mixtures in the etiology of neurological disorders.

Author Contributions: Conceptualization, P.C. and H.C.; validation, P.C. and H.C.; investigation, H.C., B.Y., T.K. and S.L.; resources, P.C. and X.Y.; data curation, P.C.; writing—original draft preparation, H.C. wrote the Neurological disorders with mitochondrial dysfunction and oxidative stress part, B.Y. wrote the Mercury, Lead, Zinc, Iron, and Manganese parts, T.K. wrote the Aluminum, Copper and Cadmium parts, and S.L. wrote the Arsenic part; writing—review and editing, P.C. and M.A.; visualization, H.C.; supervision, X.Y. and P.C.; project administration, P.C.; funding acquisition, P.C. and X.Y. All authors have read and agreed to the published version of the manuscript.

Funding: This research was funded by the National Institute of Environmental Health Sciences, grant number NIEHS R21ES031315, NIEHS R01ES10563, and NIEHS R01ES07331; National Natural Science Foundation of China, grant number 81860573 and 82073504; and Guangxi Natural Science Foundation for Innovation Research Team, grant number 2017GXNSFGA198003.

Institutional Review Board Statement: Not applicable.

Informed Consent Statement: Not applicable.

Data Availability Statement: Not applicable.

Conflicts of Interest: The authors declare no conflict of interest.

References

1. Smith, E.F.; Shaw, P.J.; De Vos, K.J. The role of mitochondria in amyotrophic lateral sclerosis. *Neurosci. Lett.* **2019**, *710*, 132933. [CrossRef] [PubMed]
2. Cadonic, C.; Sabbir, M.G.; Albensi, B.C. Mechanisms of Mitochondrial Dysfunction in Alzheimer's Disease. *Mol. Neurobiol.* **2016**, *53*, 6078–6090. [CrossRef] [PubMed]
3. Correia-Melo, C.; Passos, J.F. Mitochondria: Are they causal players in cellular senescence? *Biochim. Biophys. Acta. Bioenerg.* **2015**, *1847*, 1373–1379. [CrossRef] [PubMed]
4. Richter, C. Oxidative damage to mitochondrial DNA and its relationship to ageing. *Int. J. Biochem. Cell. Biol.* **1995**, *27*, 647–653. [CrossRef]
5. Mecocci, P.; Fano, G.; Fulle, S.; MacGarvey, U.; Shinobu, L.; Polidori, M.C.; Cherubini, A.; Vecchiet, J.; Senin, U.; Beal, M.F. Age-dependent increases in oxidative damage to DNA, lipids, and proteins in human skeletal muscle. *Free Radic. Biol. Med.* **1999**, *26*, 303–308. [CrossRef]
6. Raha, S.; Robinson, B.H. Mitochondria, oxygen free radicals, disease and ageing. *Trends Biochem. Sci.* **2000**, *25*, 502–508. [CrossRef]
7. Engl, E.; Attwell, D. Non-signalling energy use in the brain. *J. Physiol.* **2015**, *593*, 3419–3429. [CrossRef]
8. Nicholls, D.G.; Budd, S.L. Mitochondria and neuronal survival. *Physiol. Rev.* **2000**, *80*, 315–360. [CrossRef]
9. Rizzuto, R.; De Stefani, D.; Raffaello, A.; Mammucari, C. Mitochondria as sensors and regulators of calcium signalling. *Nat. Rev. Mol. Cell. Biol.* **2012**, *13*, 566–578. [CrossRef]
10. Lezi, E.; Swerdlow, R.H. Mitochondria in neurodegeneration. *Adv. Exp. Med. Biol.* **2012**, *942*, 269–286. [CrossRef]
11. Bjorklund, G.; Skalny, A.V.; Rahman, M.M.; Dadar, M.; Yassa, H.A.; Aaseth, J.; Chirumbolo, S.; Skalnaya, M.G.; Tinkov, A.A. Toxic metal(loid)-based pollutants and their possible role in autism spectrum disorder. *Environ. Res.* **2018**, *166*, 234–250. [CrossRef]
12. Liddell, J.R. Targeting mitochondrial metal dyshomeostasis for the treatment of neurodegeneration. *Neurodegener. Dis. Manag.* **2015**, *5*, 345–364. [CrossRef]
13. UN Environment Programme. *UNEP Year Book 2011: Emerging Issues in Our Global Environment*; UN Environment Programme: Nairobi, Kenya, 2011.
14. Prüss-Ustün, A.; Wolf, J.; Corván, C.; Bos, R.; Neira, M. *Preventing Disease through Healthy Environments: A Global Assessment of the Burden of Disease from Environmental Risks*; World Health Organization: Geneva, Switzerland, 2016.
15. Bowman, A.B.; Kwakye, G.F.; Herrero Hernandez, E.; Aschner, M. Role of manganese in neurodegenerative diseases. *J. Trace Elem. Med. Biol.* **2011**, *25*, 191–203. [CrossRef] [PubMed]
16. Dusek, P.; Jankovic, J.; Le, W. Iron dysregulation in movement disorders. *Neurobiol. Dis.* **2012**, *46*, 1–18. [CrossRef]
17. Brini, M.; Cali, T.; Ottolini, D.; Carafoli, E. Neuronal calcium signaling: Function and dysfunction. *Cell. Mol. Life Sci.* **2014**, *71*, 2787–2814. [CrossRef]
18. Gatz, M.; Reynolds, C.A.; Fratiglioni, L.; Johansson, B.; Mortimer, J.A.; Berg, S.; Fiske, A.; Pedersen, N.L. Role of genes and environments for explaining Alzheimer disease. *Arch. Gen. Psychiatry* **2006**, *63*, 168–174. [CrossRef]
19. Francis, P.T.; Palmer, A.M.; Snape, M.; Wilcock, G.K. The cholinergic hypothesis of Alzheimer's disease: A review of progress. *J. Neurol. Neurosurg. Psychiatry* **1999**, *66*, 137–147. [CrossRef] [PubMed]
20. Hardy, J.; Allsop, D. Amyloid deposition as the central event in the aetiology of Alzheimer's disease. *Trends Pharmacol. Sci.* **1991**, *12*, 383–388. [CrossRef]
21. Mudher, A.; Lovestone, S. Alzheimer's disease-do tauists and baptists finally shake hands? *Trends Neurosci.* **2002**, *25*, 22–26. [CrossRef]
22. Shoshan-Barmatz, V.; Nahon-Crystal, E.; Shteinfer-Kuzmine, A.; Gupta, R. VDAC1, mitochondrial dysfunction, and Alzheimer's disease. *Pharmacol. Res.* **2018**, *131*, 87–101. [CrossRef]
23. De Strooper, B.; Scorrano, L. Close encounter: Mitochondria, endoplasmic reticulum and Alzheimer's disease. *EMBO J.* **2012**, *31*, 4095–4097. [CrossRef]
24. Vance, J.E. MAM (mitochondria-associated membranes) in mammalian cells: Lipids and beyond. *Biochim. Biophys. Acta* **2014**, *1841*, 595–609. [CrossRef]
25. Coskun, P.E.; Beal, M.F.; Wallace, D.C. Alzheimer's brains harbor somatic mtDNA control-region mutations that suppress mitochondrial transcription and replication. *Proc. Natl. Acad. Sci. USA* **2004**, *101*, 10726–10731. [CrossRef]
26. Hamblet, N.S.; Ragland, B.; Ali, M.; Conyers, B.; Castora, F.J. Mutations in mitochondrial-encoded cytochrome c oxidase subunits I, II, and III genes detected in Alzheimer's disease using single-strand conformation polymorphism. *Electrophoresis* **2006**, *27*, 398–408. [CrossRef]
27. Agnihotri, A.; Aruoma, O.I. Alzheimer's Disease and Parkinson's Disease: A Nutritional Toxicology Perspective of the Impact of Oxidative Stress, Mitochondrial Dysfunction, Nutrigenomics and Environmental Chemicals. *J. Am. Coll. Nutr.* **2020**, *39*, 16–27. [CrossRef] [PubMed]

28. Basha, M.R.; Murali, M.; Siddiqi, H.K.; Ghosal, K.; Siddiqi, O.K.; Lashuel, H.A.; Ge, Y.W.; Lahiri, D.K.; Zawia, N.H. Lead (Pb) exposure and its effect on APP proteolysis and Abeta aggregation. *FASEB J.* **2005**, *19*, 2083–2084. [CrossRef]
29. Meleleo, D.; Notarachille, G.; Mangini, V.; Arnesano, F. Concentration-dependent effects of mercury and lead on Abeta42: Possible implications for Alzheimer's disease. *Eur. Biophys. J.* **2019**, *48*, 173–187. [CrossRef]
30. Walton, J.R.; Wang, M.X. APP expression, distribution and accumulation are altered by aluminum in a rodent model for Alzheimer's disease. *J. Inorg. Biochem.* **2009**, *103*, 1548–1554. [CrossRef] [PubMed]
31. Notarachille, G.; Arnesano, F.; Calo, V.; Meleleo, D. Heavy metals toxicity: Effect of cadmium ions on amyloid beta protein 1-42. Possible implications for Alzheimer's disease. *Biometals* **2014**, *27*, 371–388. [CrossRef] [PubMed]
32. Ashok, A.; Rai, N.K.; Tripathi, S.; Bandyopadhyay, S. Exposure to As-, Cd-, and Pb-mixture induces Abeta, amyloidogenic APP processing and cognitive impairments via oxidative stress-dependent neuroinflammation in young rats. *Toxicol. Sci.* **2015**, *143*, 64–80. [CrossRef] [PubMed]
33. Nino, S.A.; Morales-Martinez, A.; Chi-Ahumada, E.; Carrizales, L.; Salgado-Delgado, R.; Perez-Severiano, F.; Diaz-Cintra, S.; Jimenez-Capdeville, M.E.; Zarazua, S. Arsenic Exposure Contributes to the Bioenergetic Damage in an Alzheimer's Disease Model. *ACS Chem. Neurosci.* **2019**, *10*, 323–336. [CrossRef]
34. Oteiza, P.I. A mechanism for the stimulatory effect of aluminum on iron-induced lipid peroxidation. *Arch. Biochem. Biophys.* **1994**, *308*, 374–379. [CrossRef]
35. Kim, Y.; Olivi, L.; Cheong, J.H.; Maertens, A.; Bressler, J.P. Aluminum stimulates uptake of non-transferrin bound iron and transferrin bound iron in human glial cells. *Toxicol. Appl. Pharmacol.* **2007**, *220*, 349–356. [CrossRef]
36. Bonda, D.J.; Lee, H.G.; Blair, J.A.; Zhu, X.; Perry, G.; Smith, M.A. Role of metal dyshomeostasis in Alzheimer's disease. *Metallomics* **2011**, *3*, 267–270. [CrossRef]
37. Barnham, K.J.; Bush, A.I. Biological metals and metal-targeting compounds in major neurodegenerative diseases. *Chem. Soc. Rev.* **2014**, *43*, 6727–6749. [CrossRef]
38. Miller, L.M.; Wang, Q.; Telivala, T.P.; Smith, R.J.; Lanzirotti, A.; Miklossy, J. Synchrotron-based infrared and X-ray imaging shows focalized accumulation of Cu and Zn co-localized with beta-amyloid deposits in Alzheimer's disease. *J. Struct. Biol.* **2006**, *155*, 30–37. [CrossRef] [PubMed]
39. Halliwell, B. Oxidative stress and neurodegeneration: Where are we now? *J. Neurochem.* **2006**, *97*, 1634–1658. [CrossRef]
40. Jeganathan, S.; Hascher, A.; Chinnathambi, S.; Biernat, J.; Mandelkow, E.M.; Mandelkow, E. Proline-directed pseudo-phosphorylation at AT8 and PHF1 epitopes induces a compaction of the paperclip folding of Tau and generates a pathological (MC-1) conformation. *J. Biol. Chem.* **2008**, *283*, 32066–32076. [CrossRef] [PubMed]
41. Su, X.Y.; Wu, W.H.; Huang, Z.P.; Hu, J.; Lei, P.; Yu, C.H.; Zhao, Y.F.; Li, Y.M. Hydrogen peroxide can be generated by tau in the presence of Cu(II). *Biochem. Biophys. Res. Commun.* **2007**, *358*, 661–665. [CrossRef] [PubMed]
42. Wang, P.; Wu, Q.; Wu, W.; Li, H.; Guo, Y.; Yu, P.; Gao, G.; Shi, Z.; Zhao, B.; Chang, Y.Z. Mitochondrial Ferritin Deletion Exacerbates beta-Amyloid-Induced Neurotoxicity in Mice. *Oxid. Med. Cell. Longev.* **2017**, *2017*, 1020357. [CrossRef] [PubMed]
43. Garzon-Rodriguez, W.; Yatsimirsky, A.K.; Glabe, C.G. Binding of Zn(II), Cu(II), and Fe(II) ions to Alzheimer's A beta peptide studied by fluorescence. *Bioorg. Med. Chem. Lett.* **1999**, *9*, 2243–2248. [CrossRef]
44. Huang, X.; Atwood, C.S.; Moir, R.D.; Hartshorn, M.A.; Tanzi, R.E.; Bush, A.I. Trace metal contamination initiates the apparent auto-aggregation, amyloidosis, and oligomerization of Alzheimer's Abeta peptides. *J. Biol. Inorg. Chem.* **2004**, *9*, 954–960. [CrossRef] [PubMed]
45. Boopathi, S.; Kolandaivel, P. Fe^{2+} binding on amyloid beta-peptide promotes aggregation. *Proteins* **2016**, *84*, 1257–1274. [CrossRef]
46. Huat, T.J.; Camats-Perna, J.; Newcombe, E.A.; Valmas, N.; Kitazawa, M.; Medeiros, R. Metal Toxicity Links to Alzheimer's Disease and Neuroinflammation. *J. Mol. Biol.* **2019**, *431*, 1843–1868. [CrossRef]
47. Aizenman, E.; Stout, A.K.; Hartnett, K.A.; Dineley, K.E.; McLaughlin, B.; Reynolds, I.J. Induction of neuronal apoptosis by thiol oxidation: Putative role of intracellular zinc release. *J. Neurochem.* **2000**, *75*, 1878–1888. [CrossRef] [PubMed]
48. Bossy-Wetzel, E.; Talantova, M.V.; Lee, W.D.; Scholzke, M.N.; Harrop, A.; Mathews, E.; Gotz, T.; Han, J.; Ellisman, M.H.; Perkins, G.A.; et al. Crosstalk between nitric oxide and zinc pathways to neuronal cell death involving mitochondrial dysfunction and p38-activated K_+ channels. *Neuron* **2004**, *41*, 351–365. [CrossRef]
49. Jomova, K.; Vondrakova, D.; Lawson, M.; Valko, M. Metals, oxidative stress and neurodegenerative disorders. *Mol. Cell. Biochem.* **2010**, *345*, 91–104. [CrossRef]
50. Cuajungco, M.P.; Lees, G.J. Zinc metabolism in the brain: Relevance to human neurodegenerative disorders. *Neurobiol. Dis.* **1997**, *4*, 137–169. [CrossRef]
51. Frederickson, C.J.; Koh, J.Y.; Bush, A.I. The neurobiology of zinc in health and disease. *Nat. Rev. Neurosci.* **2005**, *6*, 449–462. [CrossRef]
52. Lee, M.C.; Yu, W.C.; Shih, Y.H.; Chen, C.Y.; Guo, Z.H.; Huang, S.J.; Chan, J.C.C.; Chen, Y.R. Zinc ion rapidly induces toxic, off-pathway amyloid-beta oligomers distinct from amyloid-beta derived diffusible ligands in Alzheimer's disease. *Sci. Rep.* **2018**, *8*, 4772. [CrossRef]
53. Sensi, S.L.; Ton-That, D.; Sullivan, P.G.; Jonas, E.A.; Gee, K.R.; Kaczmarek, L.K.; Weiss, J.H. Modulation of mitochondrial function by endogenous Zn^{2+} pools. *Proc. Natl. Acad. Sci. USA* **2003**, *100*, 6157–6162. [CrossRef]
54. Sensi, S.L.; Rapposelli, I.G.; Frazzini, V.; Mascetra, N. Altered oxidant-mediated intraneuronal zinc mobilization in a triple transgenic mouse model of Alzheimer's disease. *Exp. Gerontol.* **2008**, *43*, 488–492. [CrossRef]

55. Du, K.; Liu, M.; Pan, Y.; Zhong, X.; Wei, M. Association of Serum Manganese Levels with Alzheimer's Disease and Mild Cognitive Impairment: A Systematic Review and Meta-Analysis. *Nutrients* **2017**, *9*, 231. [CrossRef] [PubMed]
56. Marcus, D.L.; Strafaci, J.A.; Freedman, M.L. Differential neuronal expression of manganese superoxide dismutase in Alzheimer's disease. *Med. Sci. Monit.* **2006**, *12*, BR8–BR14. [PubMed]
57. Cuenca, L.; Gil-Martinez, A.L.; Cano-Fernandez, L.; Sanchez-Rodrigo, C.; Estrada, C.; Fernandez-Villalba, E.; Herrero, M.T. Parkinson's disease: A short story of 200 years. *Histol. Histopathol.* **2019**, *34*, 573–591. [CrossRef]
58. Singh, A.; Zhi, L.; Zhang, H. LRRK2 and mitochondria: Recent advances and current views. *Brain Res.* **2019**, *1702*, 96–104. [CrossRef]
59. Grubman, A.; White, A.R.; Liddell, J.R. Mitochondrial metals as a potential therapeutic target in neurodegeneration. *Br. J. Pharmacol.* **2014**, *171*, 2159–2173. [CrossRef]
60. Kalia, L.V.; Lang, A.E. Parkinson's disease. *Lancet* **2015**, *386*, 896–912. [CrossRef]
61. Schapira, A.H.; Cooper, J.M.; Dexter, D.; Clark, J.B.; Jenner, P.; Marsden, C.D. Mitochondrial complex I deficiency in Parkinson's disease. *J. Neurochem.* **1990**, *54*, 823–827. [CrossRef] [PubMed]
62. Bose, A.; Beal, M.F. Mitochondrial dysfunction in Parkinson's disease. *J. Neurochem.* **2016**, *139*, 216–231. [CrossRef]
63. Ferrer, I.; Perez, E.; Dalfo, E.; Barrachina, M. Abnormal levels of prohibitin and ATP synthase in the substantia nigra and frontal cortex in Parkinson's disease. *Neurosci. Lett.* **2007**, *415*, 205–209. [CrossRef]
64. Vande Walle, L.; Lamkanfi, M.; Vandenabeele, P. The mitochondrial serine protease HtrA2/Omi: An overview. *Cell. Death Differ.* **2008**, *15*, 453–460. [CrossRef]
65. Freed, J.; Chakrabarti, L. Defining a role for hemoglobin in Parkinson's disease. *NPJ Parkinsons Dis.* **2016**, *2*, 16021. [CrossRef] [PubMed]
66. Chaudhuri, A.D.; Choi, D.C.; Kabaria, S.; Tran, A.; Junn, E. MicroRNA-7 Regulates the Function of Mitochondrial Permeability Transition Pore by Targeting VDAC1 Expression. *J. Biol. Chem.* **2016**, *291*, 6483–6493. [CrossRef]
67. Poewe, W.; Seppi, K.; Tanner, C.M.; Halliday, G.M.; Brundin, P.; Volkmann, J.; Schrag, A.E.; Lang, A.E. Parkinson disease. *Nat. Rev. Dis. Primers* **2017**, *3*, 17013. [CrossRef] [PubMed]
68. Zhang, J.; Culp, M.L.; Craver, J.G.; Darley-Usmar, V. Mitochondrial function and autophagy: Integrating proteotoxic, redox, and metabolic stress in Parkinson's disease. *J. Neurochem.* **2018**, *144*, 691–709. [CrossRef]
69. Surmeier, D.J. Determinants of dopaminergic neuron loss in Parkinson's disease. *FEBS J.* **2018**, *285*, 3657–3668. [CrossRef] [PubMed]
70. Dias, V.; Junn, E.; Mouradian, M.M. The role of oxidative stress in Parkinson's disease. *J. Parkinsons Dis.* **2013**, *3*, 461–491. [CrossRef] [PubMed]
71. Zhou, C.; Huang, Y.; Przedborski, S. Oxidative stress in Parkinson's disease: A mechanism of pathogenic and therapeutic significance. *Ann. N.Y. Acad. Sci.* **2008**, *1147*, 93–104. [CrossRef]
72. Blesa, J.; Trigo-Damas, I.; Quiroga-Varela, A.; Jackson-Lewis, V.R. Oxidative stress and Parkinson's disease. *Front. Neuroanat.* **2015**, *9*, 91. [CrossRef]
73. Cerri, S.; Milanese, C.; Mastroberardino, P.G. Endocytic iron trafficking and mitochondria in Parkinson's disease. *Int. J. Biochem. Cell. Biol.* **2019**, *110*, 70–74. [CrossRef]
74. Montgomery, E.B., Jr. Heavy metals and the etiology of Parkinson's disease and other movement disorders. *Toxicology* **1995**, *97*, 3–9. [CrossRef]
75. Bjorklund, G.; Stejskal, V.; Urbina, M.A.; Dadar, M.; Chirumbolo, S.; Mutter, J. Metals and Parkinson's Disease: Mechanisms and Biochemical Processes. *Curr. Med. Chem.* **2018**, *25*, 2198–2214. [CrossRef]
76. Zayed, J.; Ducic, S.; Campanella, G.; Panisset, J.C.; Andre, P.; Masson, H.; Roy, M. Environmental factors in the etiology of Parkinson's disease. *Can. J. Neurol. Sci.* **1990**, *17*, 286–291. [CrossRef] [PubMed]
77. Miller, K.; Ochudlo, S.; Opala, G.; Smolicha, W.; Siuda, J. Parkinsonism in chronic occupational metallic mercury intoxication. *Neurol. Neurochir. Pol.* **2003**, *37*, 31–38.
78. Chin-Chan, M.; Navarro-Yepes, J.; Quintanilla-Vega, B. Environmental pollutants as risk factors for neurodegenerative disorders: Alzheimer and Parkinson diseases. *Front. Cell. Neurosci.* **2015**, *9*, 124. [CrossRef] [PubMed]
79. Reddy, P.H.; Reddy, T.P. Mitochondria as a therapeutic target for aging and neurodegenerative diseases. *Curr. Alzheimer Res.* **2011**, *8*, 393–409. [CrossRef] [PubMed]
80. Salazar, J.G.; Ribes, D.; Cabre, M.; Domingo, J.L.; Sanchez-Santed, F.; Colomina, M.T. Amyloid beta peptide levels increase in brain of AbetaPP Swedish mice after exposure to chlorpyrifos. *Curr. Alzheimer. Res.* **2011**, *8*, 732–740. [CrossRef]
81. Yegambaram, M.; Manivannan, B.; Beach, T.G.; Halden, R.U. Role of environmental contaminants in the etiology of Alzheimer's disease: A review. *Curr. Alzheimer Res.* **2015**, *12*, 116–146. [CrossRef]
82. Sun, H. Association of soil selenium, strontium, and magnesium concentrations with Parkinson's disease mortality rates in the USA. *Environ. Geochem. Health* **2018**, *40*, 349–357. [CrossRef]
83. Xu, J.; Kao, S.Y.; Lee, F.J.; Song, W.; Jin, L.W.; Yankner, B.A. Dopamine-dependent neurotoxicity of alpha-synuclein: A mechanism for selective neurodegeneration in Parkinson disease. *Nat. Med.* **2002**, *8*, 600–606. [CrossRef]
84. Xu, F.; Farkas, S.; Kortbeek, S.; Zhang, F.X.; Chen, L.; Zamponi, G.W.; Syed, N.I. Mercury-induced toxicity of rat cortical neurons is mediated through N-Methyl-D-Aspartate receptors. *Mol. Brain* **2012**, *5*, 30. [CrossRef] [PubMed]

85. Cariccio, V.L.; Sama, A.; Bramanti, P.; Mazzon, E. Mercury Involvement in Neuronal Damage and in Neurodegenerative Diseases. *Biol. Trace Elem. Res.* **2019**, *187*, 341–356. [CrossRef]
86. Bharath, S.; Hsu, M.; Kaur, D.; Rajagopalan, S.; Andersen, J.K. Glutathione, iron and Parkinson's disease. *Biochem. Pharmacol.* **2002**, *64*, 1037–1048. [CrossRef]
87. Ben-Shachar, D.; Zuk, R.; Glinka, Y. Dopamine neurotoxicity: Inhibition of mitochondrial respiration. *J. Neurochem.* **1995**, *64*, 718–723. [CrossRef]
88. Paik, S.R.; Shin, H.J.; Lee, J.H. Metal-catalyzed oxidation of alpha-synuclein in the presence of Copper(II) and hydrogen peroxide. *Arch. Biochem. Biophys.* **2000**, *378*, 269–277. [CrossRef] [PubMed]
89. Wang, C.; Liu, L.; Zhang, L.; Peng, Y.; Zhou, F. Redox reactions of the alpha-synuclein-Cu^{2+} complex and their effects on neuronal cell viability. *Biochemistry* **2010**, *49*, 8134–8142. [CrossRef]
90. Schildknecht, S.; Gerding, H.R.; Karreman, C.; Drescher, M.; Lashuel, H.A.; Outeiro, T.F.; Di Monte, D.A.; Leist, M. Oxidative and nitrative alpha-synuclein modifications and proteostatic stress: Implications for disease mechanisms and interventions in synucleinopathies. *J. Neurochem.* **2013**, *125*, 491–511. [CrossRef] [PubMed]
91. Miotto, M.C.; Rodriguez, E.E.; Valiente-Gabioud, A.A.; Torres-Monserrat, V.; Binolfi, A.; Quintanar, L.; Zweckstetter, M.; Griesinger, C.; Fernandez, C.O. Site-specific copper-catalyzed oxidation of alpha-synuclein: Tightening the link between metal binding and protein oxidative damage in Parkinson's disease. *Inorg. Chem.* **2014**, *53*, 4350–4358. [CrossRef]
92. Gardner, B.; Dieriks, B.V.; Cameron, S.; Mendis, L.H.S.; Turner, C.; Faull, R.L.M.; Curtis, M.A. Metal concentrations and distributions in the human olfactory bulb in Parkinson's disease. *Sci. Rep.* **2017**, *7*, 10454. [CrossRef]
93. Langley, M.R.; Ghaisas, S.; Ay, M.; Luo, J.; Palanisamy, B.N.; Jin, H.; Anantharam, V.; Kanthasamy, A.; Kanthasamy, A.G. Manganese exposure exacerbates progressive motor deficits and neurodegeneration in the MitoPark mouse model of Parkinson's disease: Relevance to gene and environment interactions in metal neurotoxicity. *Neurotoxicology* **2018**, *64*, 240–255. [CrossRef]
94. Veeresh, P.; Kaur, H.; Sarmah, D.; Mounica, L.; Verma, G.; Kotian, V.; Kesharwani, R.; Kalia, K.; Borah, A.; Wang, X.; et al. Endoplasmic reticulum-mitochondria crosstalk: From junction to function across neurological disorders. *Ann. N. Y. Acad. Sci.* **2019**, *1457*, 41–60. [CrossRef] [PubMed]
95. The Huntington's Disease Collaborative Research Group. A novel gene containing a trinucleotide repeat that is expanded and unstable on Huntington's disease chromosomes. *Cell* **1993**, *72*, 971–983. [CrossRef]
96. Vonsattel, J.P.; Di Figlia, M. Huntington disease. *J. Neuropathol. Exp. Neurol.* **1998**, *57*, 369–384. [CrossRef] [PubMed]
97. Gu, M.; Gash, M.T.; Mann, V.M.; Javoy-Agid, F.; Cooper, J.M.; Schapira, A.H. Mitochondrial defect in Huntington's disease caudate nucleus. *Ann. Neurol.* **1996**, *39*, 385–389. [CrossRef]
98. Brennan, W.A., Jr.; Bird, E.D.; Aprille, J.R. Regional mitochondrial respiratory activity in Huntington's disease brain. *J. Neurochem.* **1985**, *44*, 1948–1950. [CrossRef] [PubMed]
99. Guidetti, P.; Charles, V.; Chen, E.Y.; Reddy, P.H.; Kordower, J.H.; Whetsell, W.O., Jr.; Schwarcz, R.; Tagle, D.A. Early degenerative changes in transgenic mice expressing mutant huntingtin involve dendritic abnormalities but no impairment of mitochondrial energy production. *Exp. Neurol.* **2001**, *169*, 340–350. [CrossRef] [PubMed]
100. Oliveira, J.M. Nature and cause of mitochondrial dysfunction in Huntington's disease: Focusing on huntingtin and the striatum. *J. Neurochem.* **2010**, *114*, 1–12. [CrossRef]
101. Djousse, L.; Knowlton, B.; Cupples, L.A.; Marder, K.; Shoulson, I.; Myers, R.H. Weight loss in early stage of Huntington's disease. *Neurology* **2002**, *59*, 1325–1330. [CrossRef]
102. Bryan, M.R.; Bowman, A.B. Manganese and the Insulin-IGF Signaling Network in Huntington's Disease and Other Neurodegenerative Disorders. *Adv. Neurobiol.* **2017**, *18*, 113–142. [CrossRef]
103. Duan, W.; Jiang, M.; Jin, J. Metabolism in HD: Still a relevant mechanism? *Mov. Disord.* **2014**, *29*, 1366–1374. [CrossRef] [PubMed]
104. Shirendeb, U.P.; Calkins, M.J.; Manczak, M.; Anekonda, V.; Dufour, B.; McBride, J.L.; Mao, P.; Reddy, P.H. Mutant huntingtin's interaction with mitochondrial protein Drp1 impairs mitochondrial biogenesis and causes defective axonal transport and synaptic degeneration in Huntington's disease. *Hum. Mol. Genet.* **2012**, *21*, 406–420. [CrossRef] [PubMed]
105. Johri, A.; Chandra, A.; Flint Beal, M. PGC-1alpha, mitochondrial dysfunction, and Huntington's disease. *Free Radic. Biol. Med.* **2013**, *62*, 37–46. [CrossRef] [PubMed]
106. Mena, N.P.; Urrutia, P.J.; Lourido, F.; Carrasco, C.M.; Nunez, M.T. Mitochondrial iron homeostasis and its dysfunctions in neurodegenerative disorders. *Mitochondrion* **2015**, *21*, 92–105. [CrossRef] [PubMed]
107. Muller, M.; Leavitt, B.R. Iron dysregulation in Huntington's disease. *J. Neurochem.* **2014**, *130*, 328–350. [CrossRef]
108. Agrawal, S.; Fox, J.; Thyagarajan, B.; Fox, J.H. Brain mitochondrial iron accumulates in Huntington's disease, mediates mitochondrial dysfunction, and can be removed pharmacologically. *Free Radic. Biol. Med.* **2018**, *120*, 317–329. [CrossRef]
109. Fox, J.H.; Kama, J.A.; Lieberman, G.; Chopra, R.; Dorsey, K.; Chopra, V.; Volitakis, I.; Cherny, R.A.; Bush, A.I.; Hersch, S. Mechanisms of copper ion mediated Huntington's disease progression. *PLoS ONE* **2007**, *2*, e334. [CrossRef]
110. Sheline, C.T.; Zhu, J.; Zhang, W.; Shi, C.; Cai, A.L. Mitochondrial inhibitor models of Huntington's disease and Parkinson's disease induce zinc accumulation and are attenuated by inhibition of zinc neurotoxicity in vitro or in vivo. *Neurodegener. Dis.* **2013**, *11*, 49–58. [CrossRef]
111. Williams, B.B.; Kwakye, G.F.; Wegrzynowicz, M.; Li, D.; Aschner, M.; Erikson, K.M.; Bowman, A.B. Altered manganese homeostasis and manganese toxicity in a Huntington's disease striatal cell model are not explained by defects in the iron transport system. *Toxicol. Sci.* **2010**, *117*, 169–179. [CrossRef]

112. Williams, B.B.; Li, D.; Wegrzynowicz, M.; Vadodaria, B.K.; Anderson, J.G.; Kwakye, G.F.; Aschner, M.; Erikson, K.M.; Bowman, A.B. Disease-toxicant screen reveals a neuroprotective interaction between Huntington's disease and manganese exposure. *J. Neurochem.* **2010**, *112*, 227–237. [CrossRef]
113. Kwakye, G.F.; Li, D.; Bowman, A.B. Novel high-throughput assay to assess cellular manganese levels in a striatal cell line model of Huntington's disease confirms a deficit in manganese accumulation. *Neurotoxicology* **2011**, *32*, 630–639. [CrossRef] [PubMed]
114. Stansfield, K.H.; Bichell, T.J.; Bowman, A.B.; Guilarte, T.R. BDNF and Huntingtin protein modifications by manganese: Implications for striatal medium spiny neuron pathology in manganese neurotoxicity. *J. Neurochem.* **2014**, *131*, 655–666. [CrossRef] [PubMed]
115. Tidball, A.M.; Bryan, M.R.; Uhouse, M.A.; Kumar, K.K.; Aboud, A.A.; Feist, J.E.; Ess, K.C.; Neely, M.D.; Aschner, M.; Bowman, A.B. A novel manganese-dependent ATM-p53 signaling pathway is selectively impaired in patient-based neuroprogenitor and murine striatal models of Huntington's disease. *Hum. Mol. Genet.* **2015**, *24*, 1929–1944. [CrossRef] [PubMed]
116. Pfalzer, A.C.; Wilcox, J.M.; Codreanu, S.G.; Totten, M.; Bichell, T.J.V.; Halbesma, T.; Umashanker, P.; Yang, K.L.; Parmalee, N.L.; Sherrod, S.D.; et al. Huntington's disease genotype suppresses global manganese-responsive processes in pre-manifest and manifest YAC128 mice. *Metallomics* **2020**, *12*, 1118–1130. [CrossRef] [PubMed]
117. Zhang, L.; Wei, P.F.; Song, Y.H.; Dong, L.; Wu, Y.D.; Hao, Z.Y.; Fan, S.; Tai, S.; Meng, J.L.; Lu, Y.; et al. MnFe2O4 nanoparticles accelerate the clearance of mutant huntingtin selectively through ubiquitin-proteasome system. *Biomaterials* **2019**, *216*, 119248. [CrossRef] [PubMed]
118. Bryan, M.R.; O'Brien, M.T.; Nordham, K.D.; Rose, D.I.R.; Foshage, A.M.; Joshi, P.; Nitin, R.; Uhouse, M.A.; Di Pardo, A.; Zhang, Z.; et al. Acute manganese treatment restores defective autophagic cargo loading in Huntington's disease cell lines. *Hum. Mol. Genet.* **2019**, *28*, 3825–3841. [CrossRef]
119. Warren, E.B.; Bryan, M.R.; Morcillo, P.; Hardeman, K.N.; Aschner, M.; Bowman, A.B. Manganese-induced Mitochondrial Dysfunction Is Not Detectable at Exposures Below the Acute Cytotoxic Threshold in Neuronal Cell Types. *Toxicol. Sci.* **2020**, *176*, 446–459. [CrossRef]
120. Kwakye, G.F.; Jimenez, J.A.; Thomas, M.G.; Kingsley, B.A.; Mc, I.M.; Saito, M.A.; Korley, E.M. Heterozygous huntingtin promotes cadmium neurotoxicity and neurodegeneration in striatal cells via altered metal transport and protein kinase C delta dependent oxidative stress and apoptosis signaling mechanisms. *Neurotoxicology* **2019**, *70*, 48–61. [CrossRef]
121. Napolioni, V.; Persico, A.M.; Porcelli, V.; Palmieri, L. The mitochondrial aspartate/glutamate carrier AGC1 and calcium homeostasis: Physiological links and abnormalities in autism. *Mol. Neurobiol.* **2011**, *44*, 83–92. [CrossRef]
122. Castora, F.J. Mitochondrial function and abnormalities implicated in the pathogenesis of ASD. *Prog. Neuropsychopharmacol. Biol. Psychiatry* **2019**, *92*, 83–108. [CrossRef]
123. Kolevzon, A.; Gross, R.; Reichenberg, A. Prenatal and perinatal risk factors for autism: A review and integration of findings. *Arch. Pediatr. Adolesc. Med.* **2007**, *161*, 326–333. [CrossRef] [PubMed]
124. Meguid, N.A.; Anwar, M.; Bjorklund, G.; Hashish, A.; Chirumbolo, S.; Hemimi, M.; Sultan, E. Dietary adequacy of Egyptian children with autism spectrum disorder compared to healthy developing children. *Metab. Brain. Dis.* **2017**, *32*, 607–615. [CrossRef]
125. Gu, F.; Chauhan, V.; Chauhan, A. Impaired synthesis and antioxidant defense of glutathione in the cerebellum of autistic subjects: Alterations in the activities and protein expression of glutathione-related enzymes. *Free Radic. Biol. Med.* **2013**, *65*, 488–496. [CrossRef]
126. Kern, J.K.; Haley, B.E.; Geier, D.A.; Sykes, L.K.; King, P.G.; Geier, M.R. Thimerosal exposure and the role of sulfation chemistry and thiol availability in autism. *Int. J. Environ. Res. Public Health* **2013**, *10*, 3771–3800. [CrossRef]
127. El-Ansary, A. Data of multiple regressions analysis between selected biomarkers related to glutamate excitotoxicity and oxidative stress in Saudi autistic patients. *Data Brief* **2016**, *7*, 111–116. [CrossRef] [PubMed]
128. Rose, S.; Melnyk, S.; Pavliv, O.; Bai, S.; Nick, T.G.; Frye, R.E.; James, S.J. Evidence of oxidative damage and inflammation associated with low glutathione redox status in the autism brain. *Transl. Psychiatry* **2012**, *2*, e134. [CrossRef] [PubMed]
129. Arora, M.; Reichenberg, A.; Willfors, C.; Austin, C.; Gennings, C.; Berggren, S.; Lichtenstein, P.; Anckarsater, H.; Tammimies, K.; Bolte, S. Fetal and postnatal metal dysregulation in autism. *Nat. Commun.* **2017**, *8*, 15493. [CrossRef]
130. Wang, M.; Hossain, F.; Sulaiman, R.; Ren, X. Exposure to Inorganic Arsenic and Lead and Autism Spectrum Disorder in Children: A Systematic Review and Meta-Analysis. *Chem. Res. Toxicol.* **2019**, *32*, 1904–1919. [CrossRef] [PubMed]
131. Sulaiman, R.; Wang, M.; Ren, X. Exposure to Aluminum, Cadmium, and Mercury and Autism Spectrum Disorder in Children: A Systematic Review and Meta-Analysis. *Chem. Res. Toxicol.* **2020**, *33*, 2699–2718. [CrossRef]
132. Wang, J.; Wu, J.; Zhang, Z. Oxidative stress in mouse brain exposed to lead. *Ann. Occup. Hyg.* **2006**, *50*, 405–409. [CrossRef] [PubMed]
133. Fiore, M.; Barone, R.; Copat, C.; Grasso, A.; Cristaldi, A.; Rizzo, R.; Ferrante, M. Metal and essential element levels in hair and association with autism severity. *J. Trace Elem. Med. Biol.* **2020**, *57*, 126409. [CrossRef]
134. Bener, A.; Khattab, A.O.; Bhugra, D.; Hoffmann, G.F. Iron and vitamin D levels among autism spectrum disorders children. *Ann. Afr. Med.* **2017**, *16*, 186–191. [CrossRef]
135. Bener, A.; Kamal, M.; Bener, H.; Bhugra, D. Higher prevalence of iron deficiency as strong predictor of attention deficit hyperactivity disorder in children. *Ann. Med. Health Sci. Res.* **2014**, *4*, S291–S297. [CrossRef]
136. Yoshida, M.; Takahashi, Y.; Koike, A.; Fukuda, Y.; Goto, J.; Tsuji, S. A mutation database for amyotrophic lateral sclerosis. *Hum. Mutat.* **2010**, *31*, 1003–1010. [CrossRef] [PubMed]

137. Hayashi, Y.; Homma, K.; Ichijo, H. SOD1 in neurotoxicity and its controversial roles in SOD1 mutation-negative ALS. *Adv. Biol. Regul.* **2016**, *60*, 95–104. [CrossRef]
138. Sheykhansari, S.; Kozielski, K.; Bill, J.; Sitti, M.; Gemmati, D.; Zamboni, P.; Singh, A.V. Redox metals homeostasis in multiple sclerosis and amyotrophic lateral sclerosis: A review. *Cell Death Dis.* **2018**, *9*, 348. [CrossRef] [PubMed]
139. Onesto, E.; Colombrita, C.; Gumina, V.; Borghi, M.O.; Dusi, S.; Doretti, A.; Fagiolari, G.; Invernizzi, F.; Moggio, M.; Tiranti, V.; et al. Gene-specific mitochondria dysfunctions in human TARDBP and C9ORF72 fibroblasts. *Acta Neuropathol. Commun.* **2016**, *4*, 47. [CrossRef]
140. Li, S.; Wu, Z.; Tantray, I.; Li, Y.; Chen, S.; Dong, J.; Glynn, S.; Vogel, H.; Snyder, M.; Lu, B. Quality-control mechanisms targeting translationally stalled and C-terminally extended poly(GR) associated with ALS/FTD. *Proc. Natl. Acad. Sci. USA* **2020**, *117*, 25104–25115. [CrossRef] [PubMed]
141. Bozzoni, V.; Pansarasa, O.; Diamanti, L.; Nosari, G.; Cereda, C.; Ceroni, M. Amyotrophic lateral sclerosis and environmental factors. *Funct. Neurol.* **2016**, *31*, 7–19. [CrossRef]
142. Meng, E.; Mao, Y.; Yao, Q.; Han, X.; Li, X.; Zhang, K.; Jin, W. Population-based study of environmental/occupational lead exposure and amyotrophic lateral sclerosis: A systematic review and meta-analysis. *Neurol. Sci.* **2020**, *41*, 35–40. [CrossRef]
143. Pamphlett, R.; Png, F.Y. Shrinkage of motor axons following systemic exposure to inorganic mercury. *J. Neuropathol. Exp. Neurol.* **1998**, *57*, 360–366. [CrossRef]
144. Ash, P.E.A.; Dhawan, U.; Boudeau, S.; Lei, S.; Carlomagno, Y.; Knobel, M.; Al Mohanna, L.F.A.; Boomhower, S.R.; Newland, M.C.; Sherr, D.H.; et al. Heavy Metal Neurotoxicants Induce ALS-Linked TDP-43 Pathology. *Toxicol. Sci.* **2019**, *167*, 105–115. [CrossRef] [PubMed]
145. Johnson, F.O.; Yuan, Y.; Hajela, R.K.; Chitrakar, A.; Parsell, D.M.; Atchison, W.D. Exposure to an environmental neurotoxicant hastens the onset of amyotrophic lateral sclerosis-like phenotype in human Cu^{2+}/Zn^{2+} superoxide dismutase 1 G93A mice: Glutamate-mediated excitotoxicity. *J. Pharmacol. Exp. Ther.* **2011**, *338*, 518–527. [CrossRef] [PubMed]
146. Vinceti, M.; Guidetti, D.; Bergomi, M.; Caselgrandi, E.; Vivoli, R.; Olmi, M.; Rinaldi, L.; Rovesti, S.; Solime, F. Lead, cadmium, and selenium in the blood of patients with sporadic amyotrophic lateral sclerosis. *Ital. J. Neurol. Sci.* **1997**, *18*, 87–92. [CrossRef] [PubMed]
147. Gellein, K.; Garruto, R.M.; Syversen, T.; Sjobakk, T.E.; Flaten, T.P. Concentrations of Cd, Co, Cu, Fe, Mn, Rb, V, and Zn in formalin-fixed brain tissue in amyotrophic lateral sclerosis and Parkinsonism-dementia complex of Guam determined by High-resolution ICP-MS. *Biol. Trace Elem. Res.* **2003**, *96*, 39–60. [CrossRef]
148. Roos, P.M.; Vesterberg, O.; Syversen, T.; Flaten, T.P.; Nordberg, M. Metal concentrations in cerebrospinal fluid and blood plasma from patients with amyotrophic lateral sclerosis. *Biol. Trace Elem. Res.* **2013**, *151*, 159–170. [CrossRef]
149. Peters, T.L.; Beard, J.D.; Umbach, D.M.; Allen, K.; Keller, J.; Mariosa, D.; Sandler, D.P.; Schmidt, S.; Fang, F.; Ye, W.; et al. Blood levels of trace metals and amyotrophic lateral sclerosis. *Neurotoxicology* **2016**, *54*, 119–126. [CrossRef]
150. Oggiano, R.; Solinas, G.; Forte, G.; Bocca, B.; Farace, C.; Pisano, A.; Sotgiu, M.A.; Clemente, S.; Malaguarnera, M.; Fois, A.G.; et al. Trace elements in ALS patients and their relationships with clinical severity. *Chemosphere* **2018**, *197*, 457–466. [CrossRef]
151. Trumbull, K.A.; Beckman, J.S. A role for copper in the toxicity of zinc-deficient superoxide dismutase to motor neurons in amyotrophic lateral sclerosis. *Antioxid. Redox Signal.* **2009**, *11*, 1627–1639. [CrossRef] [PubMed]
152. Rivera-Mancia, S.; Perez-Neri, I.; Rios, C.; Tristan-Lopez, L.; Rivera-Espinosa, L.; Montes, S. The transition metals copper and iron in neurodegenerative diseases. *Chem. Biol. Interact.* **2010**, *186*, 184–199. [CrossRef] [PubMed]
153. Kokic, A.N.; Stevic, Z.; Stojanovic, S.; Blagojevic, D.P.; Jones, D.R.; Pavlovic, S.; Niketic, V.; Apostolski, S.; Spasic, M.B. Biotransformation of nitric oxide in the cerebrospinal fluid of amyotrophic lateral sclerosis patients. *Redox Rep.* **2005**, *10*, 265–270. [CrossRef]
154. Hozumi, I.; Hasegawa, T.; Honda, A.; Ozawa, K.; Hayashi, Y.; Hashimoto, K.; Yamada, M.; Koumura, A.; Sakurai, T.; Kimura, A.; et al. Patterns of levels of biological metals in CSF differ among neurodegenerative diseases. *J. Neurol. Sci.* **2011**, *303*, 95–99. [CrossRef] [PubMed]
155. Ignjatovic, A.; Stevic, Z.; Lavrnic, D.; Nikolic-Kokic, A.; Blagojevic, D.; Spasic, M.; Spasojevic, I. Inappropriately chelated iron in the cerebrospinal fluid of amyotrophic lateral sclerosis patients. *Amyotroph. Lateral. Scler.* **2012**, *13*, 357–362. [CrossRef]
156. States, J.C.; Barchowsky, A.; Cartwright, I.L.; Reichard, J.F.; Futscher, B.W.; Lantz, R.C. Arsenic toxicology: Translating between experimental models and human pathology. *Environ. Health Perspect.* **2011**, *119*, 1356–1363. [CrossRef] [PubMed]
157. Prakash, C.; Soni, M.; Kumar, V. Mitochondrial oxidative stress and dysfunction in arsenic neurotoxicity: A review. *J. Appl. Toxicol. JAT* **2016**, *36*, 179–188. [CrossRef] [PubMed]
158. Kuivenhoven, M.; Mason, K. Arsenic Toxicity. In *StatPearls*; StatPearls Publishing: Treasure Island, FL, USA, 2020.
159. Osuna-Martinez, C.C.; Armienta, M.A.; Berges-Tiznado, M.E.; Paez-Osuna, F. Arsenic in waters, soils, sediments, and biota from Mexico: An environmental review. *Sci. Total Environ.* **2021**, *752*, 142062. [CrossRef]
160. Brinkel, J.; Khan, M.H.; Kraemer, A. A systematic review of arsenic exposure and its social and mental health effects with special reference to Bangladesh. *Int. J. Environ. Res. Public Health* **2009**, *6*, 1609–1619. [CrossRef]
161. Tsuji, J.S.; Garry, M.R.; Perez, V.; Chang, E.T. Low-level arsenic exposure and developmental neurotoxicity in children: A systematic review and risk assessment. *Toxicology* **2015**, *337*, 91–107. [CrossRef]
162. Mochizuki, H. Arsenic Neurotoxicity in Humans. *Int. J. Mol. Sci.* **2019**, *20*, 3418. [CrossRef]

163. Xi, S.; Guo, L.; Qi, R.; Sun, W.; Jin, Y.; Sun, G. Prenatal and early life arsenic exposure induced oxidative damage and altered activities and mRNA expressions of neurotransmitter metabolic enzymes in offspring rat brain. *J. Biochem. Mol. Toxicol.* **2010**, *24*, 368–378. [CrossRef] [PubMed]
164. Pandey, R.; Rai, V.; Mishra, J.; Mandrah, K.; Kumar Roy, S.; Bandyopadhyay, S. From the Cover: Arsenic Induces Hippocampal Neuronal Apoptosis and Cognitive Impairments via an Up-Regulated BMP2/Smad-Dependent Reduced BDNF/TrkB Signaling in Rats. *Toxicol. Sci. Off. J. Soc. Toxicol.* **2017**, *159*, 137–158. [CrossRef]
165. Chandravanshi, L.P.; Gupta, R.; Shukla, R.K. Arsenic-Induced Neurotoxicity by Dysfunctioning Cholinergic and Dopaminergic System in Brain of Developing Rats. *Biol. Trace Elem. Res.* **2019**, *189*, 118–133. [CrossRef]
166. O'Bryant, S.E.; Edwards, M.; Menon, C.V.; Gong, G.; Barber, R. Long-term low-level arsenic exposure is associated with poorer neuropsychological functioning: A Project FRONTIER study. *Int. J. Environ. Res. Public Health* **2011**, *8*, 861–874. [CrossRef] [PubMed]
167. Tyler, C.R.; Allan, A.M. The Effects of Arsenic Exposure on Neurological and Cognitive Dysfunction in Human and Rodent Studies: A Review. *Curr. Environ. Health Rep.* **2014**, *1*, 132–147. [CrossRef]
168. Koseoglu, E.; Kutuk, B.; Nalbantoglu, O.U.; Koseoglu, R.; Kendirci, M. Arsenic and selenium measurements in nail and hair show important relationships to Alzheimer's disease in the elderly. *J. Trace Elem. Med. Biol. Organ Soc. Miner. Trace Elem.* **2020**, *64*, 126684. [CrossRef]
169. Li, X.L.; Zhan, R.Q.; Zheng, W.; Jiang, H.; Zhang, D.F.; Shen, X.L. Positive association between soil arsenic concentration and mortality from alzheimer's disease in mainland China. *J. Trace Elem. Med. Biol. Organ Soc. Miner. Trace Elem.* **2020**, *59*, 126452. [CrossRef]
170. Zarazua, S.; Burger, S.; Delgado, J.M.; Jimenez-Capdeville, M.E.; Schliebs, R. Arsenic affects expression and processing of amyloid precursor protein (APP) in primary neuronal cells overexpressing the Swedish mutation of human APP. *Off. J. Int. Soc. Dev. Neurosci.* **2011**, *29*, 389–396. [CrossRef]
171. Escudero-Lourdes, C. Toxicity mechanisms of arsenic that are shared with neurodegenerative diseases and cognitive impairment: Role of oxidative stress and inflammatory responses. *Neurotoxicology* **2016**, *53*, 223–235. [CrossRef]
172. Sun, X.; He, Y.; Guo, Y.; Li, S.; Zhao, H.; Wang, Y.; Zhang, J.; Xing, M. Arsenic affects inflammatory cytokine expression in Gallus gallus brain tissues. *BMC Vet. Res.* **2017**, *13*, 157. [CrossRef] [PubMed]
173. Yip, S.F.; Yeung, Y.M.; Tsui, E.Y. Severe neurotoxicity following arsenic therapy for acute promyelocytic leukemia: Potentiation by thiamine deficiency. *Blood* **2002**, *99*, 3481–3482. [CrossRef] [PubMed]
174. Singh, A.P.; Goel, R.K.; Kaur, T. Mechanisms pertaining to arsenic toxicity. *Toxicol. Int.* **2011**, *18*, 87–93. [CrossRef]
175. Haga, N.; Fujita, N.; Tsuruo, T. Involvement of mitochondrial aggregation in arsenic trioxide (As2O3)-induced apoptosis in human glioblastoma cells. *Cancer Sci.* **2005**, *96*, 825–833. [CrossRef] [PubMed]
176. Lu, T.H.; Tseng, T.J.; Su, C.C.; Tang, F.C.; Yen, C.C.; Liu, Y.Y.; Yang, C.Y.; Wu, C.C.; Chen, K.L.; Hung, D.Z.; et al. Arsenic induces reactive oxygen species-caused neuronal cell apoptosis through JNK/ERK-mediated mitochondria-dependent and GRP 78/CHOP-regulated pathways. *Toxicol. Lett.* **2014**, *224*, 130–140. [CrossRef] [PubMed]
177. Florea, A.M.; Splettstoesser, F.; Busselberg, D. Arsenic trioxide (As2O3) induced calcium signals and cytotoxicity in two human cell lines: SY-5Y neuroblastoma and 293 embryonic kidney (HEK). *Toxicol. Appl. Pharmacol.* **2007**, *220*, 292–301. [CrossRef]
178. Zhao, F.; Liao, Y.; Jin, Y.; Li, G.; Lv, X.; Sun, G. Effects of arsenite on glutamate metabolism in primary cultured astrocytes. *Toxicol. In Vitro Int. J. Publ. Assoc. BIBRA* **2012**, *26*, 24–31. [CrossRef]
179. Li, X.; Chan, L.; Zhang, H.; Zhang, H.; Niu, Q. Effects of arsenic poisoning on neuronal cell apoptosis and mRNA and protein expression of calpain 1, calpain 2, and cdk5/p25. *Chin. J. Ind. Hyg. Occup. Dis.* **2014**, *32*, 202–206.
180. Dwivedi, N.; Mehta, A.; Yadav, A.; Binukumar, B.K.; Gill, K.D.; Flora, S.J. MiADMSA reverses impaired mitochondrial energy metabolism and neuronal apoptotic cell death after arsenic exposure in rats. *Toxicol. Appl. Pharmacol.* **2011**, *256*, 241–248. [CrossRef] [PubMed]
181. Srivastava, P.; Yadav, R.S.; Chandravanshi, L.P.; Shukla, R.K.; Dhuriya, Y.K.; Chauhan, L.K.S.; Dwivedi, H.N.; Pant, A.B.; Khanna, V.K. Unraveling the mechanism of neuroprotection of curcumin in arsenic induced cholinergic dysfunctions in rats. *Toxicol. Appl. Pharmacol.* **2014**, *279*, 428–440. [CrossRef]
182. Wang, Y.; Tang, B.; Long, L.; Luo, P.; Xiang, W.; Li, X.; Wang, H.; Jiang, Q.; Tan, X.; Luo, S.; et al. Improvement of obesity-associated disorders by a small-molecule drug targeting mitochondria of adipose tissue macrophages. *Nat. Commun.* **2021**, *12*, 102. [CrossRef] [PubMed]
183. Yadav, R.S.; Sankhwar, M.L.; Shukla, R.K.; Chandra, R.; Pant, A.B.; Islam, F.; Khanna, V.K. Attenuation of arsenic neurotoxicity by curcumin in rats. *Toxicol. Appl. Pharmacol.* **2009**, *240*, 367–376. [CrossRef] [PubMed]
184. Ram Kumar, M.; Flora, S.J.; Reddy, G.R. Monoisoamyl 2,3-dimercaptosuccinic acid attenuates arsenic induced toxicity: Behavioral and neurochemical approach. *Environ. Toxicol. Pharmacol.* **2013**, *36*, 231–242. [CrossRef] [PubMed]
185. Prakash, C.; Soni, M.; Kumar, V. Biochemical and Molecular Alterations Following Arsenic-Induced Oxidative Stress and Mitochondrial Dysfunction in Rat Brain. *Biol. Trace Elem. Res.* **2015**, *167*, 121–129. [CrossRef] [PubMed]
186. Hughes, M.F. Arsenic toxicity and potential mechanisms of action. *Toxicol. Lett.* **2002**, *133*, 1–16. [CrossRef]
187. Falkenberg, M.; Larsson, N.G.; Gustafsson, C.M. DNA replication and transcription in mammalian mitochondria. *Ann. Rev. Biochem.* **2007**, *76*, 679–699. [CrossRef]

188. Hong, Y.; Piao, F.; Zhao, Y.; Li, S.; Wang, Y.; Liu, P. Subchronic exposure to arsenic decreased Sdha expression in the brain of mice. *Neurotoxicology* **2009**, *30*, 538–543. [CrossRef]
189. Gibson, G.E.; Chen, H.L.; Xu, H.; Qiu, L.; Xu, Z.; Denton, T.T.; Shi, Q. Deficits in the mitochondrial enzyme alpha-ketoglutarate dehydrogenase lead to Alzheimer's disease-like calcium dysregulation. *Neurobiol. Aging* **2012**, *33*, 1121.e13–1121.e24. [CrossRef]
190. Liu, J.; Zhao, H.; Wang, Y.; Shao, Y.; Zong, H.; Zeng, X.; Xing, M. Arsenic trioxide and/or copper sulfate induced apoptosis and autophagy associated with oxidative stress and perturbation of mitochondrial dynamics in the thymus of Gallus gallus. *Chemosphere* **2019**, *219*, 227–235. [CrossRef]
191. Guo, M.; Wang, Y.; Zhao, H.; Mu, M.; Yang, X.; Fei, D.; Liu, Y.; Zong, H.; Xing, M. Oxidative damage under As^{3+} and/or Cu^{2+} stress leads to apoptosis and autophagy and may be cross-talking with mitochondrial disorders in bursa of Fabricius. *J. Inorg. Biochem.* **2020**, *205*, 110989. [CrossRef]
192. Peraza, M.A.; Cromey, D.W.; Carolus, B.; Carter, D.E.; Gandolfi, A.J. Morphological and functional alterations in human proximal tubular cell line induced by low level inorganic arsenic: Evidence for targeting of mitochondria and initiated apoptosis. *J. Appl. Toxicol. JAT* **2006**, *26*, 356–367. [CrossRef]
193. Zhang, K.; Zhou, Q. Toxic effects of Al-based coagulants on Brassica chinensis and Raphanus sativus growing in acid and neutral conditions. *Environ. Toxicol.* **2005**, *20*, 179–187. [CrossRef]
194. Bertholf, R.L.; Herman, M.M.; Savory, J.; Carpenter, R.M.; Sturgill, B.C.; Katsetos, C.D.; Vandenberg, S.R.; Wills, M.R. A long-term intravenous model of aluminum maltol toxicity in rabbits: Tissue distribution, hepatic, renal, and neuronal cytoskeletal changes associated with systemic exposure. *Toxicol. Appl. Pharmacol.* **1989**, *98*, 58–74. [CrossRef]
195. Sahin, G.; Varol, I.; Temizer, A.; Benli, K.; Demirdamar, R.; Duru, S. Determination of aluminum levels in the kidney, liver, and brain of mice treated with aluminum hydroxide. *Biol. Trace Elem. Res.* **1994**, *41*, 129–135. [CrossRef]
196. Promyo, K.; Iqbal, F.; Chaidee, N.; Chetsawang, B. Aluminum chloride-induced amyloid beta accumulation and endoplasmic reticulum stress in rat brain are averted by melatonin. *Food Chem. Toxicol. Int. J. Pub. Br. Ind. Biol. Res. Assoc.* **2020**, *146*, 111829. [CrossRef] [PubMed]
197. Kumar, V.; Gill, K.D. Oxidative stress and mitochondrial dysfunction in aluminium neurotoxicity and its amelioration: A review. *Neurotoxicology* **2014**, *41*, 154–166. [CrossRef] [PubMed]
198. Al-Otaibi, S.S.; Arafah, M.M.; Sharma, B.; Alhomida, A.S.; Siddiqi, N.J. Synergistic Effect of Quercetin and alpha-Lipoic Acid on Aluminium Chloride Induced Neurotoxicity in Rats. *J. Toxicol.* **2018**, *2018*, 2817036. [CrossRef] [PubMed]
199. Nie, J.; Lv, S.; Fu, X.; Niu, Q. Effects of Al Exposure on Mitochondrial Dynamics in Rat Hippocampus. *Neurotox. Res.* **2019**, *36*, 334–346. [CrossRef]
200. Liu, H.; Zhang, W.; Fang, Y.; Yang, H.; Tian, L.; Li, K.; Lai, W.; Bian, L.; Lin, B.; Liu, X.; et al. Neurotoxicity of aluminum oxide nanoparticles and their mechanistic role in dopaminergic neuron injury involving p53-related pathways. *J. Hazard. Mater.* **2020**, *392*, 122312. [CrossRef]
201. Altmann, P.; Cunningham, J.; Dhanesha, U.; Ballard, M.; Thompson, J.; Marsh, F. Disturbance of cerebral function in people exposed to drinking water contaminated with aluminium sulphate: Retrospective study of the Camelford water incident. *BMJ* **1999**, *319*, 807–811. [CrossRef]
202. Kandimalla, R.; Vallamkondu, J.; Corgiat, E.B.; Gill, K.D. Understanding Aspects of Aluminum Exposure in Alzheimer's Disease Development. *Brain Pathol.* **2016**, *26*, 139–154. [CrossRef]
203. Yasui, M.; Kihira, T.; Ota, K. Calcium, magnesium and aluminum concentrations in Parkinson's disease. *Neurotoxicology* **1992**, *13*, 593–600.
204. Sanchez-Iglesias, S.; Mendez-Alvarez, E.; Iglesias-Gonzalez, J.; Munoz-Patino, A.; Sanchez-Sellero, I.; Labandeira-Garcia, J.L.; Soto-Otero, R. Brain oxidative stress and selective behaviour of aluminium in specific areas of rat brain: Potential effects in a 6-OHDA-induced model of Parkinson's disease. *J. Neurochem.* **2009**, *109*, 879–888. [CrossRef] [PubMed]
205. Maya, S.; Prakash, T.; Madhu, K.D.; Goli, D. Multifaceted effects of aluminium in neurodegenerative diseases: A review. *Biomed. Pharmacother.* **2016**, *83*, 746–754. [CrossRef] [PubMed]
206. Maya, S.; Prakash, T.; Goli, D. Evaluation of neuroprotective effects of wedelolactone and gallic acid on aluminium-induced neurodegeneration: Relevance to sporadic amyotrophic lateral sclerosis. *Eur. J. Pharmacol.* **2018**, *835*, 41–51. [CrossRef] [PubMed]
207. McLachlan, D.R.C.; Bergeron, C.; Alexandrov, P.N.; Walsh, W.J.; Pogue, A.I.; Percy, M.E.; Kruck, T.P.A.; Fang, Z.; Sharfman, N.M.; Jaber, V.; et al. Aluminum in Neurological and Neurodegenerative Disease. *Mol. Neurobiol.* **2019**, *56*, 1531–1538. [CrossRef]
208. Huang, T.; Guo, W.; Wang, Y.; Chang, L.; Shang, N.; Chen, J.; Fan, R.; Zhang, L.; Gao, X.; Niu, Q.; et al. Involvement of Mitophagy in Aluminum Oxide Nanoparticle-Induced Impairment of Learning and Memory in Mice. *Neurotox. Res.* **2020**, *39*, 378–391. [CrossRef]
209. Rao, K.S.; Rao, G.V. Effect of aluminium (Al) on brain mitochondrial monoamine oxidase-A (MAO-A) activity—An in vitro kinetic study. *Mol. Cell. Biochem.* **1994**, *137*, 57–60. [CrossRef]
210. Bosetti, F.; Solaini, G.; Tendi, E.A.; Chikhale, E.G.; Chandrasekaran, K.; Rapoport, S.I. Mitochondrial cytochrome c oxidase subunit III is selectively down-regulated by aluminum exposure in PC12S cells. *Neuroreport* **2001**, *12*, 721–724. [CrossRef]
211. Iranpak, F.; Saberzadeh, J.; Vessal, M.; Takhshid, M.A. Sodium valproate ameliorates aluminum-induced oxidative stress and apoptosis of PC12 cells. *Iran. J. Basic Med. Sci.* **2019**, *22*, 1353–1358. [CrossRef]
212. Rahmani, S.; Saberzadeh, J.; Takhshid, M.A. The Hydroalcoholic Extract of Saffron Protects PC12 Cells against Aluminum-Induced Cell Death and Oxidative Stress in Vitro. *Iran. J. Med. Sci.* **2020**, *45*, 59–66. [CrossRef]

213. Wang, H.; Shao, B.; Yu, H.; Xu, F.; Wang, P.; Yu, K.; Han, Y.; Song, M.; Li, Y.; Cao, Z. Neuroprotective role of hyperforin on aluminum maltolate-induced oxidative damage and apoptosis in PC12 cells and SH-SY5Y cells. *Chem. Biol. Interact.* **2019**, *299*, 15–26. [CrossRef]
214. Tsialtas, I.; Gorgogietas, V.A.; Michalopoulou, M.; Komninou, A.; Liakou, E.; Georgantopoulos, A.; Kalousi, F.D.; Karra, A.G.; Protopapa, E.; Psarra, A.G. Neurotoxic effects of aluminum are associated with its interference with estrogen receptors signaling. *Neurotoxicology* **2020**, *77*, 114–126. [CrossRef]
215. Tuneva, J.; Chittur, S.; Boldyrev, A.A.; Birman, I.; Carpenter, D.O. Cerebellar granule cell death induced by aluminum. *Neurotox. Res.* **2006**, *9*, 297–304. [CrossRef] [PubMed]
216. Rui, D.; Yongjian, Y. Aluminum chloride induced oxidative damage on cells derived from hippocampus and cortex of ICR mice. *Brain Res.* **2010**, *1324*, 96–102. [CrossRef]
217. Stevanovic, I.D.; Jovanovic, M.D.; Colic, M.; Ninkovic, M.; Jelenkovic, A.; Mihajlovic, R. Cytochrome c oxidase activity and nitric oxide synthase in the rat brain following aluminium intracerebral application. *Folia Neuropathol.* **2013**, *51*, 140–146. [CrossRef]
218. Ghribi, O.; Herman, M.M.; Forbes, M.S.; DeWitt, D.A.; Savory, J. GDNF protects against aluminum-induced apoptosis in rabbits by upregulating Bcl-2 and Bcl-XL and inhibiting mitochondrial Bax translocation. *Neurobiol. Dis.* **2001**, *8*, 764–773. [CrossRef]
219. Kumar, V.; Bal, A.; Gill, K.D. Impairment of mitochondrial energy metabolism in different regions of rat brain following chronic exposure to aluminium. *Brain Res.* **2008**, *1232*, 94–103. [CrossRef]
220. Kumar, V.; Bal, A.; Gill, K.D. Susceptibility of mitochondrial superoxide dismutase to aluminium induced oxidative damage. *Toxicology* **2009**, *255*, 117–123. [CrossRef] [PubMed]
221. Sharma, D.R.; Wani, W.Y.; Sunkaria, A.; Kandimalla, R.J.; Sharma, R.K.; Verma, D.; Bal, A.; Gill, K.D. Quercetin attenuates neuronal death against aluminum-induced neurodegeneration in the rat hippocampus. *Neuroscience* **2016**, *324*, 163–176. [CrossRef] [PubMed]
222. Prakash, A.; Shur, B.; Kumar, A. Naringin protects memory impairment and mitochondrial oxidative damage against aluminum-induced neurotoxicity in rats. *Int. J. Neurosci.* **2013**, *123*, 636–645. [CrossRef] [PubMed]
223. Prakash, A.; Kumar, A. Mitoprotective effect of Centella asiatica against aluminum-induced neurotoxicity in rats: Possible relevance to its anti-oxidant and anti-apoptosis mechanism. *Neurol. Sci. Off. J. Ital. Neurol. Soc. Ital. Soc. Clin. Neurophysiol.* **2013**, *34*, 1403–1409. [CrossRef]
224. Kumar, A.; Dogra, S.; Prakash, A. Protective effect of curcumin (*Curcuma longa*), against aluminium toxicity: Possible behavioral and biochemical alterations in rats. *Behav. Brain Res.* **2009**, *205*, 384–390. [CrossRef] [PubMed]
225. Wang, C.; Cai, X.; Hu, W.; Li, Z.; Kong, F.; Chen, X.; Wang, D. Investigation of the neuroprotective effects of crocin via antioxidant activities in HT22 cells and in mice with Alzheimer's disease. *Int. J. Mol. Med.* **2019**, *43*, 956–966. [CrossRef]
226. Cobine, P.A.; Pierrel, F.; Winge, D.R. Copper trafficking to the mitochondrion and assembly of copper metalloenzymes. *Biochim. Biophys. Acta* **2006**, *1763*, 759–772. [CrossRef] [PubMed]
227. Zischka, H.; Einer, C. Mitochondrial copper homeostasis and its derailment in Wilson disease. *Int. J. Biochem. Cell. Biol.* **2018**, *102*, 71–75. [CrossRef] [PubMed]
228. Baker, Z.N.; Cobine, P.A.; Leary, S.C. The mitochondrion: A central architect of copper homeostasis. *Metallomics* **2017**, *9*, 1501–1512. [CrossRef]
229. Horn, D.; Barrientos, A. Mitochondrial copper metabolism and delivery to cytochrome c oxidase. *IUBMB Life* **2008**, *60*, 421–429. [CrossRef]
230. Borchard, S.; Bork, F.; Rieder, T.; Eberhagen, C.; Popper, B.; Lichtmannegger, J.; Schmitt, S.; Adamski, J.; Klingenspor, M.; Weiss, K.H.; et al. The exceptional sensitivity of brain mitochondria to copper. *Toxicol. In Vitro* **2018**, *51*, 11–22. [CrossRef]
231. Behzadfar, L.; Abdollahi, M.; Sabzevari, O.; Hosseini, R.; Salimi, A.; Naserzadeh, P.; Sharifzadeh, M.; Pourahmad, J. Potentiating role of copper on spatial memory deficit induced by beta amyloid and evaluation of mitochondrial function markers in the hippocampus of rats. *Metallomics* **2017**, *9*, 969–980. [CrossRef]
232. Chen, C.; Jiang, X.; Li, Y.; Yu, H.; Li, S.; Zhang, Z.; Xu, H.; Yang, Y.; Liu, G.; Zhu, F.; et al. Low-dose oral copper treatment changes the hippocampal phosphoproteomic profile and perturbs mitochondrial function in a mouse model of Alzheimer's disease. *Free Radic Biol. Med.* **2019**, *135*, 144–156. [CrossRef]
233. Liddell, J.R.; White, A.R. Nexus between mitochondrial function, iron, copper and glutathione in Parkinson's disease. *Neurochem. Int.* **2018**, *117*, 126–138. [CrossRef]
234. Cruces-Sande, A.; Méndez-Álvarez, E.; Soto-Otero, R. Copper increases the ability of 6-hydroxydopamine to generate oxidative stress and the ability of ascorbate and glutathione to potentiate this effect: Potential implications in Parkinson's disease. *J. Neurochem.* **2017**, *141*, 738–749. [CrossRef] [PubMed]
235. Reddy, P.V.; Rao, K.V.; Norenberg, M.D. The mitochondrial permeability transition, and oxidative and nitrosative stress in the mechanism of copper toxicity in cultured neurons and astrocytes. *Lab. Investig.* **2008**, *88*, 816–830. [CrossRef] [PubMed]
236. Cannino, G.; Ferruggia, E.; Luparello, C.; Rinaldi, A.M. Cadmium and mitochondria. *Mitochondrion* **2009**, *9*, 377–384. [CrossRef]
237. Binte Hossain, K.F.; Rahman, M.M.; Sikder, M.T.; Saito, T.; Hosokawa, T.; Kurasaki, M. Inhibitory effects of selenium on cadmium-induced cytotoxicity in PC12 cells via regulating oxidative stress and apoptosis. *Food Chem. Toxicol.* **2018**, *114*, 180–189. [CrossRef]

238. Xu, C.; Wang, X.; Zhu, Y.; Dong, X.; Liu, C.; Zhang, H.; Liu, L.; Huang, S.; Chen, L. Rapamycin ameliorates cadmium-induced activation of MAPK pathway and neuronal apoptosis by preventing mitochondrial ROS inactivation of PP2A. *Neuropharmacology* **2016**, *105*, 270–284. [CrossRef]
239. Gupta, R.; Shukla, R.K.; Chandravanshi, L.P.; Srivastava, P.; Dhuriya, Y.K.; Shanker, J.; Singh, M.P.; Pant, A.B.; Khanna, V.K. Protective Role of Quercetin in Cadmium-Induced Cholinergic Dysfunctions in Rat Brain by Modulating Mitochondrial Integrity and MAP Kinase Signaling. *Mol. Neurobiol.* **2017**, *54*, 4560–4583. [CrossRef]
240. Xu, M.Y.; Wang, P.; Sun, Y.J.; Yang, L.; Wu, Y.J. Joint toxicity of chlorpyrifos and cadmium on the oxidative stress and mitochondrial damage in neuronal cells. *Food Chem. Toxicol.* **2017**, *103*, 246–252. [CrossRef] [PubMed]
241. Yang, M.; Li, C.; Yang, S.; Xiao, Y.; Xiong, X.; Chen, W.; Zhao, H.; Zhang, Q.; Han, Y.; Sun, L. Mitochondria-Associated ER Membranes—The Origin Site of Autophagy. *Front. Cell. Dev. Biol.* **2020**, *8*, 595. [CrossRef]
242. Che, L.; Yang, C.L.; Chen, Y.; Wu, Z.L.; Du, Z.B.; Wu, J.S.; Gan, C.L.; Yan, S.P.; Huang, J.; Guo, N.J.; et al. Mitochondrial redox-driven mitofusin 2 S-glutathionylation promotes neuronal necroptosis via disrupting ER-mitochondria crosstalk in cadmium-induced neurotoxicity. *Chemosphere* **2021**, *262*, 127878. [CrossRef]
243. Xie, L.L.; Shi, F.; Tan, Z.; Li, Y.; Bode, A.M.; Cao, Y. Mitochondrial network structure homeostasis and cell death. *Cancer Sci.* **2018**, *109*, 3686–3694. [CrossRef]
244. Modi, H.R.; Katyare, S.S. Cadmium exposure-induced alterations in the lipid/phospholipids composition of rat brain microsomes and mitochondria. *Neurosci. Lett.* **2009**, *464*, 108–112. [CrossRef]
245. Kumar, R.; Agarwal, A.K.; Seth, P.K. Oxidative stress-mediated neurotoxicity of cadmium. *Toxicol. Lett.* **1996**, *89*, 65–69. [CrossRef]
246. Choong, G.; Liu, Y.; Templeton, D.M. Interplay of calcium and cadmium in mediating cadmium toxicity. *Chem. Biol. Interact.* **2014**, *211*, 54–65. [CrossRef] [PubMed]
247. Rahman, M.M.; Ukiana, J.; Uson-Lopez, R.; Sikder, M.T.; Saito, T.; Kurasaki, M. Cytotoxic effects of cadmium and zinc co-exposure in PC12 cells and the underlying mechanism. *Chem. Biol. Interact.* **2017**, *269*, 41–49. [CrossRef]
248. Clarkson, T.W. The three modern faces of mercury. *Environ. Health Perspect.* **2002**, *110*, 11–23. [CrossRef]
249. Clarkson, T.W.; Magos, L. The toxicology of mercury and its chemical compounds. *Crit. Rev. Toxicol.* **2006**, *36*, 609–662. [CrossRef] [PubMed]
250. Dórea, J.G. Persistent, bioaccumulative and toxic substances in fish: Human health considerations. *Sci. Total Environ.* **2008**, *400*, 93–114. [CrossRef] [PubMed]
251. Chang, L.W.; Hartmann, H.A. Electron microscopic histochemical study on the localization and distribution of mercury in the nervous system after mercury intoxication. *Exp. Neurol.* **1972**, *35*, 122–137. [CrossRef]
252. Oliveira, L.F.; Rodrigues, L.D.; Cardillo, G.M.; Nejm, M.B.; Guimarães-Marques, M.; Reyes-Garcia, S.Z.; Zuqui, K.; Vassallo, D.V.; Fiorini, A.C.; Scorza, C.A.; et al. Deleterious effects of chronic mercury exposure on in vitro LTP, memory process, and oxidative stress. *Environ. Sci. Pollut. Res. Int.* **2020**, *27*, 7559–7569. [CrossRef]
253. Liu, W.; Yang, T.; Xu, Z.; Xu, B.; Deng, Y. Methyl-mercury induces apoptosis through ROS-mediated endoplasmic reticulum stress and mitochondrial apoptosis pathways activation in rat cortical neurons. *Free Radic. Res.* **2019**, *53*, 26–44. [CrossRef]
254. Chang, J.; Yang, B.; Zhou, Y.; Yin, C.; Liu, T.; Qian, H.; Xing, G.; Wang, S.; Li, F.; Zhang, Y.; et al. Acute Methylmercury Exposure and the Hypoxia-Inducible Factor-1α Signaling Pathway under Normoxic Conditions in the Rat Brain and Astrocytes in Vitro. *Environ. Health Perspect.* **2019**, *127*, 127006. [CrossRef]
255. Yang, B.; Yin, C.; Zhou, Y.; Wang, Q.; Jiang, Y.; Bai, Y.; Qian, H.; Xing, G.; Wang, S.; Li, F.; et al. Curcumin protects against methylmercury-induced cytotoxicity in primary rat astrocytes by activating the Nrf2/ARE pathway independently of PKCδ. *Toxicology* **2019**, *425*, 152248. [CrossRef] [PubMed]
256. Shao, Y.; Wang, L.; Langlois, P.; Mironov, G.; Chan, H.M. Proteome changes in methylmercury-exposed mouse primary cerebellar granule neurons and astrocytes. *Toxicol. In Vitro Int. J. Pub. Assoc. BIBRA* **2019**, *57*, 96–104. [CrossRef] [PubMed]
257. Ni, M.; Li, X.; Yin, Z.; Sidoryk-Węgrzynowicz, M.; Jiang, H.; Farina, M.; Rocha, J.B.; Syversen, T.; Aschner, M. Comparative study on the response of rat primary astrocytes and microglia to methylmercury toxicity. *Glia* **2011**, *59*, 810–820. [CrossRef] [PubMed]
258. Zhang, J.; Zhang, X.; Wen, C.; Duan, Y.; Zhang, H. Lotus seedpod proanthocyanidins protect against neurotoxicity after methyl-mercuric chloride injury. *Ecotoxicol. Environ. Saf.* **2019**, *183*, 109560. [CrossRef]
259. Shanker, G.; Syversen, T.; Aschner, J.L.; Aschner, M. Modulatory effect of glutathione status and antioxidants on methylmercury-induced free radical formation in primary cultures of cerebral astrocytes. *Mol. Brain Res.* **2005**, *137*, 11–22. [CrossRef]
260. Ly, J.D.; Grubb, D.R.; Lawen, A. The mitochondrial membrane potential (deltapsi(m)) in apoptosis; an update. *Apoptosis* **2003**, *8*, 115–128. [CrossRef]
261. Yin, Z.; Milatovic, D.; Aschner, J.L.; Syversen, T.; Rocha, J.B.; Souza, D.O.; Sidoryk, M.; Albrecht, J.; Aschner, M. Methylmercury induces oxidative injury, alterations in permeability and glutamine transport in cultured astrocytes. *Brain Res.* **2007**, *1131*, 1–10. [CrossRef] [PubMed]
262. Yin, Z.; Lee, E.; Ni, M.; Jiang, H.; Milatovic, D.; Rongzhu, L.; Farina, M.; Rocha, J.B.; Aschner, M. Methylmercury-induced alterations in astrocyte functions are attenuated by ebselen. *Neurotoxicology* **2011**, *32*, 291–299. [CrossRef]
263. Jacob, S.; Thangarajan, S. Fisetin impedes developmental methylmercury neurotoxicity via downregulating apoptotic signalling pathway and upregulating Rho GTPase signalling pathway in hippocampus of F(1) generation rats. *Int. J. Dev. Neurosci. Off. J. Int. Soc. Dev. Neurosci.* **2018**, *69*, 88–96. [CrossRef] [PubMed]

264. Allen, J.W.; Shanker, G.; Tan, K.H.; Aschner, M. The Consequences of Methylmercury Exposure on Interactive Functions between Astrocytes and Neurons. *Neurotoxicology* **2002**, *23*, 755–759. [CrossRef]
265. Li, Z.G.; Zhou, F.K.; Yin, A.M.; Gao, Y.Y.; Jiang, X.; Liu, S.S.; Zhang, Y.Y.; Bo, D.D.; Xie, J.; Jia, Q.Y.; et al. Cellular damage of low-dose combined exposure to mercury, lead and cadmium on hippocampal neurons in rats. *Chin. J. Prev. Med.* **2018**, *52*, 976–982. [CrossRef]
266. Dreiem, A.; Seegal, R.F. Methylmercury-induced changes in mitochondrial function in striatal synaptosomes are calcium-dependent and ROS-independent. *Neurotoxicol.* **2007**, *28*, 720–726. [CrossRef]
267. Pivovarova, N.B.; Nguyen, H.V.; Winters, C.A.; Brantner, C.A.; Smith, C.L.; Andrews, S.B. Excitotoxic calcium overload in a subpopulation of mitochondria triggers delayed death in hippocampal neurons. *J. Neurosci. Off. J. Soc. Neurosci.* **2004**, *24*, 5611–5622. [CrossRef]
268. Calvo, M.; Villalobos, C.; Núñez, L. Calcium imaging in neuron cell death. *Methods Mol. Biol.* **2015**, *1254*, 73–85. [CrossRef]
269. Ramanathan, G.; Atchison, W.D. Ca^{2+} entry pathways in mouse spinal motor neurons in culture following in vitro exposure to methylmercury. *Neurotoxicology* **2011**, *32*, 742–750. [CrossRef] [PubMed]
270. Vendrell, I.; Carrascal, M.; Vilaró, M.T.; Abián, J.; Rodríguez-Farré, E.; Suñol, C. Cell viability and proteomic analysis in cultured neurons exposed to methylmercury. *Hum. Exp. Toxicol.* **2007**, *26*, 263–272. [CrossRef]
271. Bittencourt, L.O.; Dionizio, A.; Nascimento, P.C.; Puty, B.; Leão, L.K.R.; Luz, D.A.; Silva, M.C.F.; Amado, L.L.; Leite, A.; Buzalaf, M.R.; et al. Proteomic approach underlying the hippocampal neurodegeneration caused by low doses of methylmercury after long-term exposure in adult rats. *Metallomics* **2019**, *11*, 390–403. [CrossRef] [PubMed]
272. Sanders, T.; Liu, Y.; Buchner, V.; Tchounwou, P.B. Neurotoxic effects and biomarkers of lead exposure: A review. *Rev. Environ. Health* **2009**, *24*, 15–45. [CrossRef]
273. Bakulski, K.M.; Rozek, L.S.; Dolinoy, D.C.; Paulson, H.L.; Hu, H. Alzheimer's disease and environmental exposure to lead: The epidemiologic evidence and potential role of epigenetics. *Curr. Alzheimer Res.* **2012**, *9*, 563–573. [CrossRef]
274. Senut, M.C.; Cingolani, P.; Sen, A.; Kruger, A.; Shaik, A.; Hirsch, H.; Suhr, S.T.; Ruden, D. Epigenetics of early-life lead exposure and effects on brain development. *Epigenomics* **2012**, *4*, 665–674. [CrossRef]
275. Dórea, J.G. Environmental exposure to low-level lead (Pb) co-occurring with other neurotoxicants in early life and neurodevelopment of children. *Environ. Res.* **2019**, *177*, 108641. [CrossRef] [PubMed]
276. Mattalloni, M.S.; Deza-Ponzio, R.; Albrecht, P.A.; Fernandez-Hubeid, L.E.; Cancela, L.M.; Virgolini, M.B. Brain ethanol-metabolizing enzymes are differentially expressed in lead-exposed animals after voluntary ethanol consumption: Pharmacological approaches. *Neurotoxicology* **2019**, *75*, 174–185. [CrossRef]
277. Devi, C.B.; Reddy, G.H.; Prasanthi, R.P.; Chetty, C.S.; Reddy, G.R. Developmental lead exposure alters mitochondrial monoamine oxidase and synaptosomal catecholamine levels in rat brain. *Int. J. Dev. Neurosci. Off. J. Int. Soc. Dev. Neurosci.* **2005**, *23*, 375–381. [CrossRef]
278. Gąssowska, M.; Baranowska-Bosiacka, I.; Moczydłowska, J.; Frontczak-Baniewicz, M.; Gewartowska, M.; Strużyńska, L.; Gutowska, I.; Chlubek, D.; Adamczyk, A. Perinatal exposure to lead (Pb) induces ultrastructural and molecular alterations in synapses of rat offspring. *Toxicology* **2016**, *373*, 13–29. [CrossRef] [PubMed]
279. Thangarajan, S.; Vedagiri, A.; Somasundaram, S.; Sakthimanogaran, R.; Murugesan, M. Neuroprotective effect of morin on lead acetate-induced apoptosis by preventing cytochrome c translocation via regulation of Bax/Bcl-2 ratio. *Neurotoxicol. Teratol.* **2018**, *66*, 35–45. [CrossRef]
280. Zhu, Y.; Jiao, X.; An, Y.; Li, S.; Teng, X. Selenium against lead-induced apoptosis in chicken nervous tissues via mitochondrial pathway. *Oncotarget* **2017**, *8*, 108130–108145. [CrossRef]
281. He, W.; Li, Y.; Tian, J.; Jiang, N.; Du, B.; Peng, Y. Optimized mixture of As, Cd and Pb induce mitochondria-mediated apoptosis in C6-glioma via astroglial activation, inflammation and P38-MAPK. *Am. J. Cancer Res.* **2015**, *5*, 2396–2408.
282. Ye, F.; Li, X.; Li, F.; Li, J.; Chang, W.; Yuan, J.; Chen, J. Cyclosporin A protects against Lead neurotoxicity through inhibiting mitochondrial permeability transition pore opening in nerve cells. *Neurotoxicology* **2016**, *57*, 203–213. [CrossRef]
283. Szewczyk, B. Zinc homeostasis and neurodegenerative disorders. *Front. Aging Neurosci.* **2013**, *5*, 33. [CrossRef] [PubMed]
284. Kawahara, M.; Tanaka, K.I.; Kato-Negishi, M. Zinc, Carnosine, and Neurodegenerative Diseases. *Nutrients* **2018**, *10*, 147. [CrossRef] [PubMed]
285. Sheline, C.T.; Behrens, M.M.; Choi, D.W. Zinc-induced cortical neuronal death: Contribution of energy failure attributable to loss of NAD(+) and inhibition of glycolysis. *J. Neurosci. Off. J. Soc. Neurosci.* **2000**, *20*, 3139–3146. [CrossRef]
286. Cai, A.L.; Zipfel, G.J.; Sheline, C.T. Zinc neurotoxicity is dependent on intracellular NAD levels and the sirtuin pathway. *Eur. J. Neurosci.* **2006**, *24*, 2169–2176. [CrossRef] [PubMed]
287. Sheline, C.T.; Cai, A.L.; Zhu, J.; Shi, C. Serum or target deprivation-induced neuronal death causes oxidative neuronal accumulation of Zn^{2+} and loss of NAD^{+}. *Eur. J. Neurosci.* **2010**, *32*, 894–904. [CrossRef]
288. Brown, A.M.; Kristal, B.S.; Effron, M.S.; Shestopalov, A.I.; Ullucci, P.A.; Sheu, K.F.; Blass, J.P.; Cooper, A.J. Zn^{2+} inhibits alpha-ketoglutarate-stimulated mitochondrial respiration and the isolated alpha-ketoglutarate dehydrogenase complex. *J. Biol. Chem.* **2000**, *275*, 13441–13447. [CrossRef]
289. Lemire, J.; Mailloux, R.; Appanna, V.D. Zinc toxicity alters mitochondrial metabolism and leads to decreased ATP production in hepatocytes. *J. Appl. Toxicol. JAT* **2008**, *28*, 175–182. [CrossRef]

290. Lorusso, M.; Cocco, T.; Sardanelli, A.M.; Minuto, M.; Bonomi, F.; Papa, S. Interaction of Zn^{2+} with the bovine-heart mitochondrial bc1 complex. *Eur. J. Biochem.* **1991**, *197*, 555–561. [CrossRef] [PubMed]
291. Link, T.A.; von Jagow, G. Zinc ions inhibit the QP center of bovine heart mitochondrial bc1 complex by blocking a protonatable group. *J. Biol. Chem.* **1995**, *270*, 25001–25006. [CrossRef]
292. Manev, H.; Kharlamov, E.; Uz, T.; Mason, R.P.; Cagnoli, C.M. Characterization of zinc-induced neuronal death in primary cultures of rat cerebellar granule cells. *Exp. Neurol.* **1997**, *146*, 171–178. [CrossRef]
293. Kim, E.Y.; Koh, J.Y.; Kim, Y.H.; Sohn, S.; Joe, E.; Gwag, B.J. Zn^{2+} entry produces oxidative neuronal necrosis in cortical cell cultures. *Eur. J. Neurosci.* **1999**, *11*, 327–334. [CrossRef]
294. Sensi, S.L.; Yin, H.Z.; Carriedo, S.G.; Rao, S.S.; Weiss, J.H. Preferential Zn^{2+} influx through Ca2+-permeable AMPA/kainate channels triggers prolonged mitochondrial superoxide production. *Proc. Natl. Acad. Sci. USA* **1999**, *96*, 2414–2419. [CrossRef]
295. He, K.; Aizenman, E. ERK signaling leads to mitochondrial dysfunction in extracellular zinc-induced neurotoxicity. *J. Neurochem.* **2010**, *114*, 452–461. [CrossRef]
296. Park, J.A.; Koh, J.Y. Induction of an immediate early gene egr-1 by zinc through extracellular signal-regulated kinase activation in cortical culture: Its role in zinc-induced neuronal death. *J. Neurochem.* **1999**, *73*, 450–456. [CrossRef]
297. Prasad, A.S. Discovery of human zinc deficiency: Its impact on human health and disease. *Adv. Nutr.* **2013**, *4*, 176–190. [CrossRef] [PubMed]
298. Adamo, A.M.; Zago, M.P.; Mackenzie, G.G.; Aimo, L.; Keen, C.L.; Keenan, A.; Oteiza, P.I. The role of zinc in the modulation of neuronal proliferation and apoptosis. *Neurotox. Res.* **2010**, *17*, 1–14. [CrossRef]
299. Seth, R.; Corniola, R.S.; Gower-Winter, S.D.; Morgan, T.J., Jr.; Bishop, B.; Levenson, C.W. Zinc deficiency induces apoptosis via mitochondrial p53- and caspase-dependent pathways in human neuronal precursor cells. *J. Trace Elem. Med. Biol. Organ Soc. Miner. Trace Elem.* **2015**, *30*, 59–65. [CrossRef]
300. Li, L.B.; Chai, R.; Zhang, S.; Xu, S.F.; Zhang, Y.H.; Li, H.L.; Fan, Y.G.; Guo, C. Iron Exposure and the Cellular Mechanisms Linked to Neuron Degeneration in Adult Mice. *Cells* **2019**, *8*, 198. [CrossRef] [PubMed]
301. Jiang, H.; Wang, J.; Rogers, J.; Xie, J. Brain Iron Metabolism Dysfunction in Parkinson's Disease. *Mol. Neurobiol.* **2017**, *54*, 3078–3101. [CrossRef] [PubMed]
302. Kruszewski, M. Labile iron pool: The main determinant of cellular response to oxidative stress. *Mutat. Res.* **2003**, *531*, 81–92. [CrossRef]
303. Núñez, M.T.; Gallardo, V.; Muñoz, P.; Tapia, V.; Esparza, A.; Salazar, J.; Speisky, H. Progressive iron accumulation induces a biphasic change in the glutathione content of neuroblastoma cells. *Free Radic. Biol. Med.* **2004**, *37*, 953–960. [CrossRef]
304. Lipinski, B. Hydroxyl radical and its scavengers in health and disease. *Oxid. Med. Cell. Longev.* **2011**, *2011*, 809696. [CrossRef] [PubMed]
305. Huang, H.; Chen, J.; Lu, H.; Zhou, M.; Chai, Z.; Hu, Y. Iron-induced generation of mitochondrial ROS depends on AMPK activity. *Biomet. Int. J. Met. Ions Biol. Biochem. Med.* **2017**, *30*, 623–628. [CrossRef] [PubMed]
306. Huang, X.T.; Liu, X.; Ye, C.Y.; Tao, L.X.; Zhou, H.; Zhang, H.Y. Iron-induced energy supply deficiency and mitochondrial fragmentation in neurons. *J. Neurochem.* **2018**, *147*, 816–830. [CrossRef]
307. Mena, N.P.; García-Beltrán, O.; Lourido, F.; Urrutia, P.J.; Mena, R.; Castro-Castillo, V.; Cassels, B.K.; Núñez, M.T. The novel mitochondrial iron chelator 5-((methylamino)methyl)-8-hydroxyquinoline protects against mitochondrial-induced oxidative damage and neuronal death. *Biochem. Biophys. Res. Commun.* **2015**, *463*, 787–792. [CrossRef] [PubMed]
308. Lee, D.G.; Park, J.; Lee, H.S.; Lee, S.R.; Lee, D.S. Iron overload-induced calcium signals modulate mitochondrial fragmentation in HT-22 hippocampal neuron cells. *Toxicology* **2016**, *365*, 17–24. [CrossRef] [PubMed]
309. Sanmartín, C.D.; Paula-Lima, A.C.; García, A.; Barattini, P.; Hartel, S.; Núñez, M.T.; Hidalgo, C. Ryanodine receptor-mediated Ca^{2+} release underlies iron-induced mitochondrial fission and stimulates mitochondrial Ca^{2+} uptake in primary hippocampal neurons. *Front. Mol. Neurosci.* **2014**, *7*, 13. [CrossRef] [PubMed]
310. Liu, T.; Liu, W.; Zhang, M.; Yu, W.; Gao, F.; Li, C.; Wang, S.B.; Feng, J.; Zhang, X.Z. Ferrous-Supply-Regeneration Nanoengineering for Cancer-Cell-Specific Ferroptosis in Combination with Imaging-Guided Photodynamic Therapy. *ACS Nano* **2018**, *12*, 12181–12192. [CrossRef] [PubMed]
311. Gao, M.; Yi, J.; Zhu, J.; Minikes, A.M.; Monian, P.; Thompson, C.B.; Jiang, X. Role of Mitochondria in Ferroptosis. *Mol. Cell.* **2019**, *73*, 354–363. [CrossRef] [PubMed]
312. Sumneang, N.; Siri-Angkul, N.; Kumfu, S.; Chattipakorn, S.C.; Chattipakorn, N. The effects of iron overload on mitochondrial function, mitochondrial dynamics, and ferroptosis in cardiomyocytes. *Arch. Biochem. Biophys.* **2020**, *680*, 108241. [CrossRef] [PubMed]
313. Zhou, J.; Jin, Y.; Lei, Y.; Liu, T.; Wan, Z.; Meng, H.; Wang, H. Ferroptosis Is Regulated by Mitochondria in Neurodegenerative Diseases. *Neuro Degener. Dis.* **2020**, *20*, 20–34. [CrossRef] [PubMed]
314. Dong, Y.; Zhang, D.; Yu, Q.; Zhao, Q.; Xiao, C.; Zhang, K.; Jia, C.; Chen, S.; Zhang, B.; Zhang, B.; et al. Loss of Ssq1 leads to mitochondrial dysfunction, activation of autophagy and cell cycle arrest due to iron overload triggered by mitochondrial iron-sulfur cluster assembly defects in Candida albicans. *Int. J. Biochem. Cell Biol.* **2017**, *85*, 44–55. [CrossRef]
315. Maio, N.; Rouault, T.A. Iron-sulfur cluster biogenesis in mammalian cells: New insights into the molecular mechanisms of cluster delivery. *Biochim. Biophys. Acta* **2015**, *1853*, 1493–1512. [CrossRef] [PubMed]
316. Isaya, G. Mitochondrial iron-sulfur cluster dysfunction in neurodegenerative disease. *Front. Pharmacol.* **2014**, *5*, 29. [CrossRef]

317. Maynard, L.S.; Cotzias, G.C. The partition of manganese among organs and intracellular organelles of the rat. *J. Biol. Chem.* **1955**, *214*, 489–495. [CrossRef]
318. Gavin, C.E.; Gunter, K.K.; Gunter, T.E. Manganese and calcium transport in mitochondria: Implications for manganese toxicity. *Neurotoxicology* **1999**, *20*, 445–453. [PubMed]
319. Gunter, T.E.; Puskin, J.S. Manganous ion as a spin label in studies of mitochondrial uptake of manganese. *Biophys. J.* **1972**, *12*, 625–635. [CrossRef]
320. Gavin, C.E.; Gunter, K.K.; Gunter, T.E. Manganese and calcium efflux kinetics in brain mitochondria. Relevance to manganese toxicity. *Biochem. J.* **1990**, *266*, 329–334. [CrossRef]
321. Malthankar, G.V.; White, B.K.; Bhushan, A.; Daniels, C.K.; Rodnick, K.J.; Lai, J.C. Differential lowering by manganese treatment of activities of glycolytic and tricarboxylic acid (TCA) cycle enzymes investigated in neuroblastoma and astrocytoma cells is associated with manganese-induced cell death. *Neurochem. Res.* **2004**, *29*, 709–717. [CrossRef]
322. Malecki, E.A. Manganese toxicity is associated with mitochondrial dysfunction and DNA fragmentation in rat primary striatal neurons. *Brain Res. Bull.* **2001**, *55*, 225–228. [CrossRef]
323. Galvani, P.; Fumagalli, P.; Santagostino, A. Vulnerability of mitochondrial complex I in PC12 cells exposed to manganese. *Eur. J. Pharmacol.* **1995**, *293*, 377–383. [CrossRef]
324. Ahmadi, N.; Ghanbarinejad, V.; Ommati, M.M.; Jamshidzadeh, A.; Heidari, R. Taurine prevents mitochondrial membrane permeabilization and swelling upon interaction with manganese: Implication in the treatment of cirrhosis-associated central nervous system complications. *J. Biochem. Mol. Toxicol.* **2018**, *32*, e22216. [CrossRef] [PubMed]
325. Sarkar, S.; Malovic, E.; Harischandra, D.S.; Ngwa, H.A.; Ghosh, A.; Hogan, C.; Rokad, D.; Zenitsky, G.; Jin, H.; Anantharam, V.; et al. Manganese exposure induces neuroinflammation by impairing mitochondrial dynamics in astrocytes. *Neurotoxicology* **2018**, *64*, 204–218. [CrossRef] [PubMed]
326. Sarkar, S.; Rokad, D.; Malovic, E.; Luo, J.; Harischandra, D.S.; Jin, H.; Anantharam, V.; Huang, X.; Lewis, M.; Kanthasamy, A.; et al. Manganese activates NLRP3 inflammasome signaling and propagates exosomal release of ASC in microglial cells. *Sci. Signal* **2019**, *12*. [CrossRef]
327. Gonzalez, L.E.; Juknat, A.A.; Venosa, A.J.; Verrengia, N.; Kotler, M.L. Manganese activates the mitochondrial apoptotic pathway in rat astrocytes by modulating the expression of proteins of the Bcl-2 family. *Neurochem. Int.* **2008**, *53*, 408–415. [CrossRef] [PubMed]
328. Tamm, C.; Sabri, F.; Ceccatelli, S. Mitochondrial-mediated apoptosis in neural stem cells exposed to manganese. *Toxicol. Sci.* **2008**, *101*, 310–320. [CrossRef] [PubMed]
329. Yin, Z.; Aschner, J.L.; dos Santos, A.P.; Aschner, M. Mitochondrial-dependent manganese neurotoxicity in rat primary astrocyte cultures. *Brain Res.* **2008**, *1203*, 1–11. [CrossRef]
330. Gorojod, R.M.; Alaimo, A.; Porte Alcon, S.; Martinez, J.H.; Cortina, M.E.; Vazquez, E.S.; Kotler, M.L. Heme Oxygenase-1 protects astroglia against manganese-induced oxidative injury by regulating mitochondrial quality control. *Toxicol. Lett.* **2018**, *295*, 357–368. [CrossRef]
331. Rao, K.V.; Norenberg, M.D. Manganese induces the mitochondrial permeability transition in cultured astrocytes. *J. Biol. Chem.* **2004**, *279*, 32333–32338. [CrossRef]
332. Alaimo, A.; Gorojod, R.M.; Miglietta, E.A.; Villarreal, A.; Ramos, A.J.; Kotler, M.L. Manganese induces mitochondrial dynamics impairment and apoptotic cell death: A study in human Gli36 cells. *Neurosci. Lett.* **2013**, *554*, 76–81. [CrossRef] [PubMed]

toxics

MDPI

Review

Molecular Mechanisms of Environmental Metal Neurotoxicity: A Focus on the Interactions of Metals with Synapse Structure and Function

Asuncion Carmona, **Stéphane Roudeau** and **Richard Ortega** *

University of Bordeaux, CNRS, CENBG, UMR-5797, F-33170 Gradignan, France; acarmona@cenbg.in2p3.fr (A.C.); roudeau@cenbg.in2p3.fr (S.R.)
* Correspondence: ortega@cenbg.in2p3.fr

Abstract: Environmental exposure to neurotoxic metals and metalloids such as arsenic, cadmium, lead, mercury, or manganese is a global health concern affecting millions of people worldwide. Depending on the period of exposure over a lifetime, environmental metals can alter neurodevelopment, neurobehavior, and cognition and cause neurodegeneration. There is increasing evidence linking environmental exposure to metal contaminants to the etiology of neurological diseases in early life (e.g., autism spectrum disorder) or late life (e.g., Alzheimer's disease). The known main molecular mechanisms of metal-induced toxicity in cells are the generation of reactive oxygen species, the interaction with sulfhydryl chemical groups in proteins (e.g., cysteine), and the competition of toxic metals with binding sites of essential metals (e.g., Fe, Cu, Zn). In neurons, these molecular interactions can alter the functions of neurotransmitter receptors, the cytoskeleton and scaffolding synaptic proteins, thereby disrupting synaptic structure and function. Loss of synaptic connectivity may precede more drastic alterations such as neurodegeneration. In this article, we will review the molecular mechanisms of metal-induced synaptic neurotoxicity.

Keywords: neurotoxicity; synapse; metal; arsenic; cadmium; lead; manganese; mercury

Citation: Carmona, A.; Roudeau, S.; Ortega, R. Molecular Mechanisms of Environmental Metal Neurotoxicity: A Focus on the Interactions of Metals with Synapse Structure and Function. *Toxics* **2021**, *9*, 198. https://doi.org/10.3390/toxics9090198

Academic Editors: Jason Cannon and Brandon L. Pearson

Received: 21 July 2021
Accepted: 25 August 2021
Published: 27 August 2021

1. Introduction

Three metals, lead (Pb), mercury (Hg), and cadmium (Cd), and the metalloid arsenic (As), have been listed among the ten chemicals of major public health concern by the World Health Organization (WHO) [1]. As, Pb, Hg, and Cd are ranked as first, second, third, and seventh, based on their frequency, toxicity, and potential for human exposure in the priority list of the USA Agency for Toxic Substances and Disease Registry [2]. They all cause neurotoxic effects [3]. Manganese (Mn) is another neurotoxic element of growing concern in terms of environmental overexposures [4]. Sources of human environmental exposure are diverse, from air, food, or drinking water. Environmental exposure to neurotoxic metals is a global health problem affecting millions of people worldwide. For instance, at least 140 million people in 50 countries have been drinking water containing As at levels above the WHO guideline [5]. Depending on the period of exposure over a lifetime, environmental metals affect neurodevelopment [6,7], neurobehavior [8], cognition [9], or are involved in the etiology of neurodegenerative diseases such as Alzheimer's (AD) and Parkinson's (PD) diseases [10–16]. In particular, prenatal and early childhood exposure to environmental metals such as As, Pb, Mn, Cd, or Hg has emerged as strong candidate etiological factors in autism spectrum disorder (ASD) [6,17]. Increased fetal and postnatal uptake of Pb are observed in ASD cases [18]. In adults, high blood levels of As, Hg, and Cd are associated with increased risk of AD [19,20].

The known key molecular pathways of metal-induced toxicity in cells, including neurons, involve the production of reactive oxygen species (ROS), the interaction with sulfhydryl chemical groups (-SH) in proteins, and the competition with binding sites of essential metals (e.g., Fe, Cu, Zn) [21]. Whatever the route of exposure, As, Pb, Hg, and Mn can pass the blood–brain barrier (BBB) and reach the central nervous system (CNS) [11,22]. Cd can barely pass the BBB in adults; however, the BBB is not fully functional in the developing brain [23]. In adults, Cd can be uptaken in the CNS directly through the olfactory pathway and can also alter the BBB contributing to the pathogenesis of neurodegenerative diseases [24]. In neurons, metal-induced molecular damages will affect specific neuronal functions through, for instance, interactions with synaptic vesicles, ion channels, metabolism of neurotransmitters, intracellular signaling pathways, neurotransmitter receptors, transcription machinery regulating synaptic plasticity, or through a combination of these mechanisms [25,26].

In particular, neurotoxic effects will result in altered synaptic transmission and synaptic plasticity. These mechanisms of metal-induced synaptic neurotoxicity are receiving increasing attention. Such mechanisms are consistent with new paradigms on the etiology of major neurological disorders such as ASD, schizophrenia, or AD, suggesting that synaptic alterations may occur early in the pathological process [27], and could involve impairments of the cytoskeletal and mechanical integrity of synaptic structures [28–30]. This review article will focus on the molecular mechanisms of interaction of environmental toxic metals with the synaptic structure and function. We will review the state-of-the-art knowledge on metal interactions with neurotransmitter receptors and with cytoskeletal and scaffolding synaptic proteins.

2. Metal Interactions with Neurotransmitter Receptors

Environmental neurotoxic metals can impair neurotransmitter receptors functions as reviewed in two important articles in this field of research [25,26]. Neurotoxic metals can interact with neurotransmitter receptors by a variety of mechanisms including the modification of their gene and/or protein expression, the indirect disruption of their functions following ROS production, or the direct competition with physiological ions binding sites on the proteins. Here, we will summarize the research works studying molecular interactions between As, Cd, Pb, Mn, and Hg with neurotransmitter receptors, such as the ionotropic glutamate receptors (N-methyl-D-aspartate (NMDA), α-amino-3-hydroxy-5-methyl-4-isoxazolepropionic acid (AMPA), and kainate receptors), the γ-aminobutyric acid (GABA) receptors and the dopamine (DA) receptors.

2.1. Arsenic

Changes in glutamatergic, cholinergic, and dopaminergic systems have been linked to As exposure, resulting in impaired synaptic transmission, as recently reviewed [22]. Arsenic exposure alters the expression of the NMDA receptor (NMDAR) in the hippocampus of animal models, as evidenced in several studies. The NMDAR is a heterotetramer typically composed of GluN1 subunits with GluN2 subunits, or of GluN1 and a mixture of GluN2/GluN3 subunits (Figure 1) [31,32]. Subunit composition of the NMDAR varies during development [31]. Arsenic induces complex modifications in NMDAR subunit expression, with effects of up- or down-regulation at the receptor subunit level, depending on the period of As-exposure during neurodevelopment. For instance, the expression of several NMDAR subunits is downregulated at the protein level following As exposure, as evidenced for GluN1 [33,34], GluN2A [34–36], and GluN2B subunits [34,36,37]. Changes are, however, complex, with concomitant up- and down-regulations following As-exposure, e.g., GluN1 can be down-regulated while GluN2A is up-regulated [33], GluN2B down-regulated while GluN2A is up-regulated [37], or GluN2A down-regulated while GluN1 remains unchanged [35]. This complex picture of As effects on NMDAR subunits expression in function of neurodevelopment stages is exemplified by the comparison of expression at different postnatal periods. On mice exposed to As through drinking water, Western blotting reveals a downregulation of GluN2A 15 days after birth and an upregulation after 90 days, while GluN2B is down-regulated 15 days after birth but no significant changes are observed after 90 days [36].

A similar effect is observed for the AMPA receptor (AMPAR), As-exposure leading to the down-regulation of a specific subunit of the AMPAR [38], such as GluA1, as observed both at the mRNA and protein levels [34,39]. The inhibition of glutamate receptors expression alters synaptic plasticity such as long-term potentiation (LTP), learning, and memory, and is associated with an increase in extracellular glutamate levels [36,38].

Although less investigated than glutamate receptors, changes in the expression of other neurotransmitter receptors following As exposure are also reported. Inorganic As decreases the mRNA expression levels of α7 nicotinic receptors in rats [40], and alters the development of the cholinergic and dopaminergic systems [41]. Chronic exposure to As at a high dose in mice decreases mRNA expression of DA-D2 receptors although there are no changes in DA-D1 receptors in striatum [42].

2.2. Cadmium

Cd exposure affects glutamate, acetylcholine, GABA, and DA neurotransmitter receptors functions in the brain. Cd uptake in neurons is mediated by the NMDAR voltage-dependent calcium channels as indicated by increased Cd influx following stimulation with glutamate or NMDA and glycine [43] (Figure 1). It is also known that extracellular Cd binds directly to the DRPEER sequence of the NMDA channel and inhibits currents in a concentration-dependent manner [44] (Figure 1). The DRPEER sequence is a cluster of charged residues and a proline located extracellularly. This sequence is unique to the GluN1 subunit of the NMDAR and responsible for the high Ca^{2+} flux rate [45]. In addition, Cd impairs the AMPAR-mediated synaptic transmission and short-term plasticity in the rat hippocampus [46].

Cd also interacts with muscarinic acetylcholine receptors. Cd causes cell death in primary cholinergic neurons from the basal forebrain by silencing the muscarinic receptor M1 [47]. In a following study from the same authors, it was shown that Cd effects on muscarinic receptors disruption were induced by oxidative stress [48].

Early studies evidenced that Cd inhibits the neuronal GABAA receptor channel complex through a binding site that was distinct from the recognition sites for GABA and for pharmacological agents [49]. On the other hand, Cd exposure modifies the expression of GABAA receptors in animal studies [50]. In particular, different protein expression levels of GABAARα5 and GABAARδ were observed in the hippocampus of mice offspring after Cd exposure during pregnancy and lactation suggesting that GABAARα5 is more sensitive to the environmental pollutants at puberty and young adulthood, whereas GABAARδ may reflect the accumulation of the environmental pollutants at adulthood.

Finally, Cd neurotoxicity induces motor dysfunctions that have been related to Cd selective effects on DA receptors. Cd exposure decreases mRNA and protein expression of DA-D2 receptors in the striatum of rat brains, whereas levels of expression for DA-D1 receptors were unchanged [51]. Moreover, molecular docking experiments revealed that Cd could directly bind at the competitive site of dopamine on DA-D2 receptors. This selective inhibition of DA-D2 receptors is similar to the one described for As exposure [42].

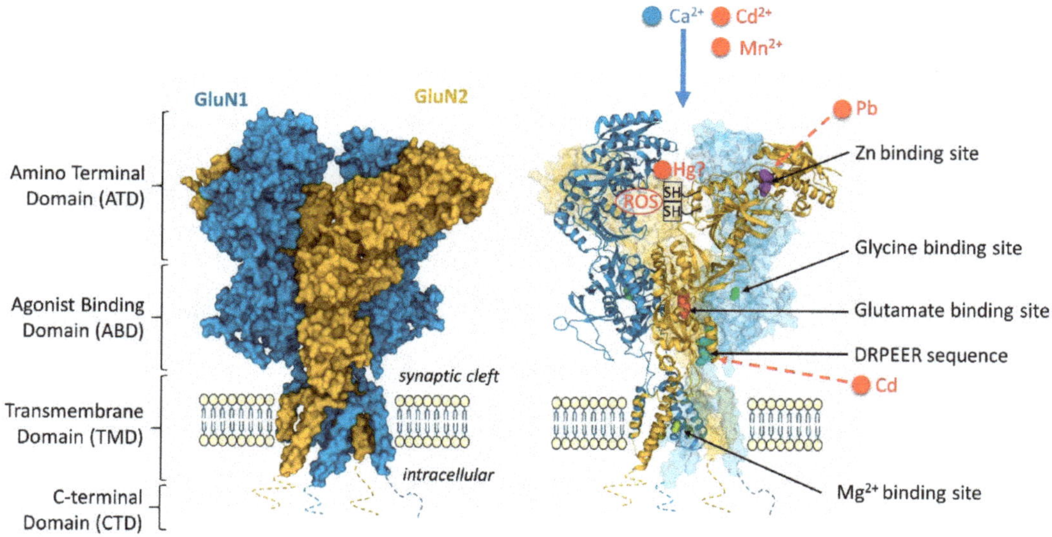

Figure 1. Interaction sites of environmental metal toxicants with the NMDA receptor. Cd^{2+} neuronal uptake is mediated by the NMDAR following its stimulation. Similarly, Mn^{2+} can permeate the plasma membrane through the NMDAR. Cd can bind directly to the DRPEER sequence in the extracellular domain (in the ABD/TMD linker) of the GluN1 subunit and inhibit NMDA-mediated current. Pb competes with zinc to the zinc-binding site of the GluN2 subunit and alters the receptor function. In the case of Hg, there are some indications for interactions with the cysteine -SH groups involved in the control of NMDAR activity, although this suggestion still needs experimental evidence. In most cases, metals could induce ROS that would interact with the –SH groups of the NMDAR. Structure of the GluN1/2B NMDAR (Protein Data Bank accession 4PE5 [32]). Figure inspired from Hansen et al. [52].

2.3. Lead

There is a large body of evidence for the interaction of Pb with glutamatergic signaling affecting more the hippocampus than other brain regions [25]. The adverse effects of Pb-exposure on glutamate neurotransmission have long been identified and involve a reversible inhibition of the NMDA-activated calcium channel current [53,54]. Pb-exposure could block the NMDAR, preventing the influx of Ca^{2+} into the postsynaptic neuron, by interaction with zinc neurophysiology [55] (Figure 1). Zinc is a physiological allosteric inhibitor of NMDAR function [56]. The presence of Pb decreases the inhibitory effect of Zn, suggesting that the two metals compete for binding in the zinc-binding site of the NMDAR [57]. Since Zn binds with high affinity at a regulatory site on the GluN2A subunit, but with lower affinity to the GluN2B subunit, this suggests a preferential sensitivity of GluN2A NMDAR for Pb [57–59].

Pb exposure may also affect the expression of NMDAR. Pb exposure induces a decreased expression of the NMDAR subunit GluN2A in synapses from hippocampal neurons, and an increased targeting of subunit GluN2B to dendritic spines [60]. This effect is particularly critical during neuronal development. Chronic developmental Pb exposure results in decreased levels of GluN2A mRNA and altered levels of GluN1 mRNA in the hippocampus [61–63] and altered expression of GluN1 splice variants [64]. During brain development, there is a shift of GluN2B- to GluN2A-containing NMDAR. These data suggest that Pb delays the normal developmental switch of increased GluN2A incorporation in NMDA receptors during synapse maturation [59].

Pb may prevent the expression of the NMDAR gene by substituting for zinc in the specificity protein-1 (Sp1), a ubiquitously expressed zinc-finger transcription factor which upregulates NMDAR transcription. Pb competition for binding in the zinc finger region of

Sp1 could induce conformational changes that may alter the expression of NMDAR [55]. This hypothesis is based on the observation that Pb exposure alters the binding of Sp1 to DNA, that Zn supplementation has a protective effect and that GluN1 expression globally follows Sp1-to-DNA binding [65].

Electrophysiological studies showed that LTP induction was affected in Pb exposed rats, and this was probably due to excitatory synaptic transmission impairment [66]. This study has evidenced that NMDA and AMPA receptor-mediated current was inhibited, and GluN2A and phosphorylated GluA1 expression were decreased. Moreover, morphological changes were observed with a decline in dendritic spines density and in spine maturation in Pb exposed rats.

The action of Pb on glutamate release, NMDAR function, and structural plasticity can underlie perturbations in synaptic plasticity significantly affecting LTP, and contributing to impairments in hippocampus-mediated learning and memory functions [59,67–69].

NMDA receptors are more sensitive to Pb inhibition than other glutamate channels but Ca^{2+}-permeable AMPAR may also be involved since Pb exposure reduces the expression of the glutamate receptor subunit GluA2 [70]. This result was further confirmed showing that the expression of GluA1, GluA2, GluA3, and GluA4 AMPAR subunits is decreased by Pb exposure in cortical neurons, but with a more pronounced decrease in GluA2 expression [71]. Other results showed that GluA1 and GluA2 expressions are increased in tetrodotoxin-Pb induced synaptic scaling [72].

Although Pb affects mainly the glutamate receptors, the consequences of Pb exposure on DA or GABA systems have also been studied [25]. In the developing rat brain, Pb-exposure can modify the aminergic system, most likely as result of decreased monoamine oxidase (MAO) activity [73]. MAOs are important enzymes in the breakdown of monoamines and in the regulation of monoamine neurotransmission. The expression of MOAs is sensitive to the exposure of various metals, such as Pb [73], but also Mn [74], Hg [13], and uranium [75]. Metal-induced MAO inhibition can thus indirectly impact the aminergic pathways and the expression of dopamine receptors. It has been hypothesized that Pb substitution for Zn may directly target the DA-D1- and -D2 receptors known to contain Zn-binding sites [55]. However, the direct substitution of Zn by Pb in DA receptors still needs to be demonstrated experimentally.

Rats exposed to low levels of Pb have reduced Ca-dependent GABA release in the hippocampus [76]. Similarly, Pb inhibits action potential-dependent GABA release in rat hippocampal slices, possibly through the involvement of voltage-gated calcium channels [77]. However, in other studies of Pb exposure, the GABA synaptic transmission was normal [66].

2.4. Manganese

Mn can interfere with the dopaminergic, cholinergic, glutamatergic, and GABAergic systems [78–80]. Mn disturbs these neurotransmitter systems through multiple mechanisms including the direct interaction with neurotransmitter receptors. Mn is a NMDA calcium channel blocker as evidenced in cultured neurons [81], and in different brain regions (cerebral cortex, hippocampus, striatum, and cerebellum) on Mn exposed rats [82]. On the other hand, Mn inhibits the mRNA and protein expression of NMDAR GluN1, GluN2A, and GluN2B subunits in primary cultured neurons, with a larger decrease of the GluN2A subunit compared to the two others [83]. In vivo exposure to Mn decreased the mRNA and protein expression of GluN1 and GluN2A subunits in the striatum of mice [84], and also decreased the protein levels and mRNA expression of NMDAR GluN1, GluN2A, and GluN2B in the rat hippocampus [85].

However, studies in animal models of Mn-induced brain pathology have shown that the glutamatergic system appear to be mostly unaffected by chronic Mn exposure [86,87]. Although the glutamatergic receptors might not be the primary target of Mn neurotoxicity, they might play a role in Mn accumulation in the central nervous system. Mn can permeate the plasma membrane through the NMDAR [72] (Figure 1). The entry of Mn into the brain is accelerated by the activation of NMDAR (but not AMPAR) in glutamatergic neurons [88]. Another mechanism of Mn neurotoxicity involving glutamatergic receptors is related to Mn induced alpha-synuclein overexpression, which results in the phosphorylation and down-regulation of the GluN2B subunit, and the consequent impaired NMDAR signaling [89].

Interactions of Mn with GABA receptors have also been investigated. Mn exposure decreases GABAA receptor subunit protein expression in the hypothalamus from immature female rats [90]. In vitro experiments on Mn exposed neurons revealed a suppression of GABAA receptors and induction of GABAB receptors, leading to the accumulation of alpha-synuclein [91]. However, similarly to the glutamatergic system, the GABAergic system might not represent the main target of Mn neurotoxicity and more pronounced neurotoxicity is observed towards the dopaminergic system [59,86,92].

Mn effects on DA receptors have been investigated in a variety of animal models. In Mn-treated mice, the mRNA and protein levels expression of the DA-D2 receptor is increased in the striatum, in a dose dependent way, while the expression of DA-D3 or DA-D4 receptors remains unchanged [93]. In a rat experimental model of early life Mn exposure, decreased expression of DA-D1 receptors in nucleus accumbens and dorsal striatum, as well as increased expression of DA-D2 receptors in the prefrontal cortex, were associated with behavioral and learning deficits [94]. In rats exposed to Mn throughout the postnatal days, a persistent increase of DA-D2 receptors protein expression in the dorsal striatum is found [95]. The authors suggest that early Mn exposure depresses presynaptic dopaminergic function, reduces DA levels, causing an up-regulation of DA-D2 receptors expression and a dysregulation of DA-associated signaling pathways. In the striatum of mice exposed to Mn, immunohistochemical activities, protein levels, and mRNA expression of DA-D1 receptors are decreased [84]. The interactions between DA-D1 and NMDAR are inhibited in this animal model resulting in learning and memory dysfunction via injury of striatum. Overall, these studies suggest that Mn exposure induces a decrease of the DA-D1 receptors expression in the frontal cortex as well as an increase of DA-D2 receptors expression in the striatum.

In Mn exposed welders and workers, a PET (positron emission tomography) molecular imaging study of basal ganglia DA-D2 receptors revealed an increased binding of the DA-D2 receptor specific antagonist [11C](N-methyl)benperidol in the substantia nigra, compared to non-exposed workers [96]. This study confirms in humans the link observed in animal models between Mn exposure and DA-D2 receptor increased expression.

2.5. Mercury

Hg interacts with a wide range of neuronal targets including NMDA, GABA, DA, and acetylcholine (ACh) receptors. For example, methyl-Hg (MeHg) induced neuronal toxicity is thought to involve glutamate-mediated excitotoxicity. Activation of NMDAR following MeHg exposure has been reported in developing cortical neurons [97], and in the frontal cortex of adult rats [98]. As indicated in rats, overstimulation of NMDAR following Hg exposure may also contribute to neurotoxicity by inducing excessive dopamine release in the striatum [99]. In hippocampal neurons, $HgCl_2$ and thimerosal (an organomercurial compound used as vaccine preservative) reduce NMDA-evoked currents, with higher toxicity for $HgCl_2$ [100]. It was proposed that Hg could interact with the cysteine -SH groups, crucial for regulating NMDAR function, rendering the NMDAR dysfunctional (Figure 1). This postulate, however, still needs to be confirmed experimentally.

In Purkinje and granule cells of rat cerebellar slices, MeHg alters GABAA receptor-mediated inhibitory synaptic transmission at both presynaptic and postsynaptic sites [101]. In cultured hippocampal neurons, $HgCl_2$ markedly and rapidly potentiates the GABAergic currents while thimerosal works slowly and reduces GABA responses [100]. It was suggested that mercurial compounds most likely interact with cysteine -SH residues at the GABAA receptor complex, critical for its gating properties.

Hg compounds can also alter the brain dopaminergic system. Thimerosal exposure induces a decrease in the density of striatal DA-D2 receptors in exposed rats [102]. In rats exposed to MeHg there is a decrease of DA-D1 receptors in both cortex and striatum, and an increase of DA-D2 receptors in cortex [103]. Here again, the presence of –SH groups in the DA receptors might be the target of Hg binding as shown in biochemical assays exhibiting very stable and non-reversible binding of Hg to the DA-D2 receptor [104].

There are also evidence from animal studies for Hg to affect the global population of ACh receptors. Hg inhibits muscarinic cholinergic ligand binding to the mACh receptor [105], and preferentially affects ACh receptor subtypes M1 and M2 levels in the occipital cortex [106].

3. Metal-Interactions with Proteins of the Synaptic Structure

The disorganization of the synaptic cytoskeletal structure is involved in numerous neurological dysfunctions, including neurodevelopmental disorders [107–109], and neurodegenerative diseases such as AD [110,111]. Recent findings from our team suggest that Cu and Zn physiologically interact with dendro-synaptic tubulin and F-actin to control synapse formation and stability [112,113]. On the other hand, the Shank family of proteins, which are important scaffold proteins involved in the post-synaptic density (PSD) structure, requires Zn to regulate their assembly [114]. As a corollary to these physiological functions, toxic metals could bind directly to cytoskeleton and scaffold proteins (actin, tubulin, SHANK3), possibly by competing with Cu- and/or Zn- binding sites, resulting in the structural disorganization of synapses, thus leading to impairments in synaptic connectivity (Figure 2).

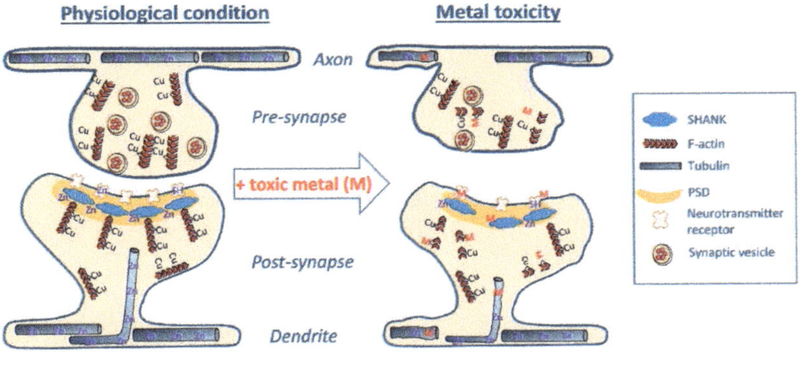

Figure 2. Schematic representation of synaptic toxicity mechanisms by direct competition of toxic metals (M) with physiological metal binding sites (Cu, Zn) within cytoskeletal proteins (e.g., tubulin, F-actin and/or F-actin binding proteins), PSD scaffold proteins (e.g., SHANK), and neurotransmitter receptors (e.g., NMDAR). These metal–protein interactions can damage the synaptic structure and lead to loss of connectivity between neurons.

Actin and tubulin are the most widely expressed cytoskeleton proteins and are involved in the formation, plasticity and stability of synapses [115,116]. Dynamic polymerization or depolymerization of actin filaments (F-actin) serve as driving force for the formation or retraction of dendritic spines. Microtubules are formed by the polymerization of $\alpha\beta$-tubulin heterodimers and are located mainly in axons and dendrites but also in synapses (Figure 2).

On the other hand, SHANK3 protein is a major scaffold protein involved in the multiprotein complex assembly of the PSD. Deletions, mutations, or downregulation of SHANK3 gene are linked to ASD [117]. Zn binds to SHANK3 allowing the PSD assembly [114], and promoting synaptogenesis [118]. Zn deficiency dysregulates the synaptic SHANK3 assembly and may contribute to ASD [119], whereas Zn supplementation could restore neuronal functions [120,121]. Reduced fetal and postnatal uptake of Zn and increased uptake of Pb are observed in ASD cases [18]. In vitro, the exposure to a biometal profile characteristic for ASD (low Zn and high Cd, Pb and Hg) resulted in the reduction of SHANK genes expression along with a reduction of synapse density [122].

3.1. *Arsenic*

The administration of sodium meta-arsenite to Wistar rats results in the destabilization of the cytoskeletal framework observed in sciatic nerves [123]. Protein analysis of the sciatic nerves revealed changes in neurofilament-low (NF-L) protein expression together with the hyper-phosphorylation of NF-L and microtubule-associated protein TAU proteins. Sodium arsenite suppresses neurite outgrowth in mouse neuroblastoma cells, a mechanism induced by the decreased mRNA levels of TAU and tubulin (TUBB5) cytoskeletal genes [124]. In NT2 postmitotic neurons, sodium arsenite exposure significantly reduces neuronal differentiation and the expression of β-tubulin type III protein [125]. These results point out tubulin as a target of As neurotoxicity. Information on molecular mechanisms of toxicity are available from studies of non-neuronal systems. For example, in cultured lung fibroblast, sodium arsenite induces a dose-dependent disassembly of microtubules by targeting the tubulin sulfhydryl groups [126].

3.2. *Cadmium*

$CdCl_2$ induces the disassembly of the cytoskeleton in a variety of cultured neuronal cells, targeting both the actin [127], and the microtubules [128] networks. Cd induces the destruction of microtubules and decreases acetylated tubulin levels in primary rat cortical neurons [129]. Cd down-regulates the gene expression of microtubules dynamics and microtubules motor-based proteins in a neuronal human cellular model [130]. Again, information on molecular interactions with cytoskeleton proteins is available from studies of non-neuronal systems. In cultured lung cells, Cd at concentrations similar to environmental exposures (0.5–1 μM), caused significant oxidation of peptidyl cysteines of proteins regulating actin cytoskeleton and increased filamentous actin polymerization [131].

3.3. *Lead*

Exposure to inorganic Pb reduces the expression of tubulin and induces the disorganization of the cytoskeleton in primary neuronal cells and neonatal brain [132]. Pb toxicity is generally driven by its binding to sulfhydryl proteins, as shown in non-neuronal cells [21]. In porcine brain and in cultured fibroblasts, triethyl lead chloride inhibits microtubule assembly and depolymerizes preformed microtubules by interaction with thiol groups of the tubulin dimer [133]. Moreover, morphological studies showed that density of dendritic spines declined by about 20% in Pb-treated rats, the remaining spines showing an immature form [66].

3.4. *Manganese*

Mn exposure induces dramatic changes in the neuronal cytoskeleton, even at sub-cytotoxic concentrations, in dopaminergic and GABAergic neurons [134]. Environmentally relevant concentrations of Mn impairs cytoskeletal morphology and structure in adult neuronal stem cells by decreasing F-actin polymerization [135]. The molecular mechanisms underlying these effects are not known.

3.5. *Mercury*

In neuroblastoma cells, Hg induces microtubule disruption at concentrations markedly lower than in fibroblasts, indicating a particular sensitivity of nerve cells to Hg [136]. Inorganic Hg exposure results in reduced tubulin expression in primary neuronal cells [124]. A similar reduction in tubulin expression is observed on NT2 postmitotic neurons after MeHg exposure [125]. Using immobilized metal affinity chromatography, thirty-eight Hg-binding proteins have been identified in human neuroblastoma cells, among them tubulin (both α and β subtypes), and various isoforms of actin [137]. In fish muscle, β-actin is one of the main Hg-binding protein together with tropomyosin, a protein also required in the formation of actin filaments [138]. Tubulin sulfhydryl groups have been proposed as the target of Hg toxicity [139]. In vitro exposure of primary cultures of rats to MeHg resulted in neuronal network fragmentation and microtubule depolymerization detected as early as within 1.5 h of MeHg exposure, long before the occurrence of nuclear condensation (6–9 h) [140]. In cortical neurons, Hg exposure compromises the cytoskeleton components, mainly the β-tubulin [141].

More generally, the mechanism of metal induced damage to cytoskeleton proteins may apply not only to neurons but also to other brain cells such as astrocytes. For example, Hg and Mn exposure result in cell swelling in cultured astrocytes, a morphological change also found in AD [142,143]. Mn, Cd, or Hg exposure can alter the cytoskeletal structure of neural stem cells during their differentiation into astrocytes [129,144]. Since astrocytes play an active role in the synaptic organization as conceptualized in the model of the tripartite synapse [145], metal-induced disorganization of the astrocytic cytoskeleton could participate to the impairment of synaptic functions.

4. Conclusions

Neurotoxic environmental metals could directly alter synaptic structure by disrupting the cytoskeleton organization (e.g., F-actin, microtubules), or scaffolding proteins in the postsynaptic compartment (e.g., SHANK3), or indirectly by modifying the levels of expression of neurotransmitters receptors (e.g., glutamate, DA, GABA and ACh receptors) themselves involved in the regulation of synaptic plasticity.

These structural impairments could be due to metal-induced production of ROS, generally targeting sulfhydryl (-SH) groups of proteins involved in synaptic structure. Sulfhydryl groups in neurotransmitter receptors and in cytoskeleton proteins are also potential sites of direct metal interactions. In addition, competition of toxic metal with essential metal (Cu, Zn) binding-sites in synaptic proteins (e.g., NMDAR, tubulin, SHANK3) might also be important molecular targets of metals neurotoxicity.

Overall, these effects would cause the disorganization of the synaptic structure, resulting in synaptic impairments, loss of synaptic connectivity and neurological dysfunctions. This mechanism may apply to a wide variety of neurological disorders, including neurodegenerative diseases such as AD, or developmental disorders (e.g., ASD). The loss of dendritic spines directly correlates with the loss of synaptic function. In AD, early synaptic degeneration could ultimately cause neuronal degeneration through erroneous converting signals derived from structural synaptic changes into the program of cell cycle activation [146].

The extrapolation of data from the literature, in terms of exposure levels, to study molecular mechanisms in cellular and animal models remains complex because the metal concentrations used are often high, although most of the recent data tend to take into account low exposure levels representing environmental exposures. Many questions remain to be solved, such as what are the threshold exposure concentrations showing neurotoxic effects which will depend on each element, its speciation, route and duration of exposure, the combined effects between neurotoxic contaminants, and especially according to the age of the exposed individuals, the periods of neurodevelopment being sensitive to lower metal concentrations than in adults.

A better understanding of the molecular mechanisms involved in cognitive dysfunctions associated with exposure to environmental metals will allow for devising novel prevention strategies, e.g., using supplementation with essential metals (Cu, Zn), when metallic exposure is difficult to prevent. In this respect, the restoration of balanced Cu and Zn homeostasis would be determinant to protect neurons. Organic Cu complexes have been proposed for the therapeutic redistribution of Cu in AD [147], and Zn therapy for the treatment of early AD [148]. Zinc supplementation could also restore neuronal functions in ASD [120,121].

In conclusion, there is robust evidence for toxic effects of metals towards neurotransmitter receptors, synaptic cytoskeleton, and scaffolding proteins suggesting that synapses could be the early targets of environmental metals toxicity. Nevertheless, the precise description of the molecular mechanisms of synaptic toxicity still remains largely unknown and their elucidation will allow the designing of adapted prevention strategies to treat or slow down the progression of devastating neurological diseases such as ASD and AD.

Author Contributions: A.C., S.R., R.O.; writing—review and editing. All authors have read and agreed to the published version of the manuscript.

Funding: This research received no external funding.

Institutional Review Board Statement: Not applicable.

Informed Consent Statement: Not applicable.

Conflicts of Interest: The authors declare no conflict of interest.

References

1. World Health Organization (WHO). Ten Chemicals of Public Health Concern. 2020. Available online: https://www.who.int/news-room/photo-story/photo-story-detail/10-chemicals-of-public-health-concern (accessed on 21 July 2021).
2. ATSDR (Agency for Toxic Substances and Disease Registry). ATSDR's Substance Priority List. 2019. Available online: https://www.atsdr.cdc.gov/spl/#2019spl (accessed on 21 July 2021).
3. ATSDR (Agency for Toxic Substances and Disease Registry). ATDSR's Toxicological Profiles. Available online: https://www.atsdr.cdc.gov/toxprofiledocs/index.html (accessed on 21 July 2021).
4. Miah, M.R.; Ijomone, O.M.; Okoh, C.O.; Ijomone, O.K.; Akingbade, G.T.; Ke, T.; Krum, B.; da Cunha Martins, A., Jr.; Akinyemi, A.; Aranoff, N.; et al. The effects of manganese overexposure on brain health. *Neurochem. Int.* **2020**, *135*, 104688. [CrossRef] [PubMed]
5. WHO. Arsenic—World Health Organizaiton. Fact Sheets. 2018. Available online: https://www.who.int/news-room/fact-sheets/detail/arsenic (accessed on 16 August 2021).
6. Modabbernia, A.; Velthorst, E.; Reichenberg, A. Environmental risk factors for autism: An evidence-based review of systematic reviews and meta-analyses. *Mol. Autism* **2017**, *8*, 1–16. [CrossRef]
7. Chandravanshi, L.; Shiv, K.; Kumar, S. Developmental toxicity of cadmium in infants and children: A review. *Environ. Anal. Health Toxicol.* **2021**, *36*, e2021003. [CrossRef] [PubMed]
8. Weiss, B. Lead, Manganese, and Methylmercury as Risk Factors for Neurobehavioral Impairment in Advanced Age. *Int. J. Alzheimer's Dis.* **2011**, *2011*, 1–11. [CrossRef] [PubMed]
9. Ijomone, O.M.; Ifenatuoha, C.W.; Aluko, O.M.; Ijomone, O.K.; Aschner, M. The aging brain: Impact of heavy metal neurotoxicity. *Crit. Rev. Toxicol.* **2020**, *50*, 801–814. [CrossRef]
10. Monnet-Tschudi, F.; Zurich, M.-G.; Boschat, C.; Corbaz, A.; Honegger, P. Involvement of Environmental Mercury and Lead in the Etiology of Neurodegenerative Diseases. *Rev. Environ. Health* **2006**, *21*, 105–117. [CrossRef]
11. Charlet, L.; Chapron, Y.; Faller, P.; Kirsch, R.; Stone, A.T.; Baveye, P.C. Neurodegenerative diseases and exposure to the environmental metals Mn, Pb, and Hg. *Coord. Chem. Rev.* **2012**, *256*, 2147–2163. [CrossRef]
12. Schofield, K. The Metal Neurotoxins: An Important Role in Current Human Neural Epidemics? *Int. J. Environ. Res. Public Health* **2017**, *14*, 1511. [CrossRef]
13. Siblerud, R.; Mutter, J.; Moore, E.; Naumann, J.; Walach, H. A Hypothesis and Evidence That Mercury May be an Etiological Factor in Alzheimer's Disease. *Int. J. Environ. Res. Public Health* **2019**, *16*, 5152. [CrossRef]
14. Bakulski, K.M.; Seo, Y.A.; Hickman, R.C.; Brandt, D.; Vadari, H.S.; Hu, H.; Park, S.K. Heavy Metals Exposure and Alzheimer's Disease and Related Dementias. *J. Alzheimer's Dis.* **2020**, *76*, 1215–1242. [CrossRef]
15. Raj, K.; Kaur, P.; Gupta, G.; Singh, S. Metals associated neurodegeneration in Parkinson's disease: Insight to physiological, pathological mechanisms and management. *Neurosci. Lett.* **2021**, *753*, 135873. [CrossRef]
16. Rahman, A.; Hannan, A.; Uddin, J.; Rahman, S.; Rashid, M.; Kim, B. Exposure to Environmental Arsenic and Emerging Risk of Alzheimer's Disease: Perspective Mechanisms, Management Strategy, and Future Directions. *Toxics* **2021**, *9*, 188. [CrossRef]
17. Skogheim, T.S.; Weyde, K.V.F.; Engel, S.M.; Aase, H.; Surén, P.; Øie, M.G.; Biele, G.; Reichborn-Kjennerud, T.; Caspersen, I.H.; Hornig, M.; et al. Metal and essential element concentrations during pregnancy and associations with autism spectrum disorder and attention-deficit/hyperactivity disorder in children. *Environ. Int.* **2021**, *152*, 106468. [CrossRef] [PubMed]
18. Arora, M.; Reichenberg, A.; Willfors, C.; Austin, C.; Gennings, C.; Berggren, S.; Lichtenstein, P.; Anckarsäter, H.; Tammimies, K.; Bölte, S. Fetal and postnatal metal dysregulation in autism. *Nat. Commun.* **2017**, *8*, 15493. [CrossRef] [PubMed]
19. Min, J.-Y.; Min, K.-B. Blood cadmium levels and Alzheimer's disease mortality risk in older US adults. *Environ. Health* **2016**, *15*, 69. [CrossRef]
20. Xu, L.; Zhang, W.; Liu, X.; Zhang, C.; Wang, P.; Zhao, X. Circulatory Levels of Toxic Metals (Aluminum, Cadmium, Mercury, Lead) in Patients with Alzheimer's Disease: A Quantitative Meta-Analysis and Systematic Review. *J. Alzheimer's Dis.* **2018**, *62*, 361–372. [CrossRef]
21. Wu, X.; Cobbina, S.J.; Mao, G.; Xu, H.; Zhang, Z.; Yang, L. A review of toxicity and mechanisms of individual and mixtures of heavy metals in the environment. *Environ. Sci. Pollut. Res.* **2016**, *23*, 8244–8259. [CrossRef]

22. Garza-Lombo, C.; Pappa, A.; Panayiotidis, M.I.; Gonsebatt, M.E.; Franco, R. Arsenic-induced neurotoxicity: A mechanistic appraisal. *JBIC J. Biol. Inorg. Chem.* **2019**, *24*, 1305–1316. [CrossRef]
23. Wang, B.; Du, Y. Cadmium and Its Neurotoxic Effects. *Oxidative Med. Cell. Longev.* **2013**, *2013*, 898034. [CrossRef]
24. Pacini, A.; Branca, J.J.V.; Morucci, G. Cadmium-induced neurotoxicity: Still much ado. *Neural Regen. Res.* **2018**, *13*, 1879–1882. [CrossRef]
25. Sadiq, S.; Ghazala, Z.; Chowdhury, A.; Büsselberg, D. Metal Toxicity at the Synapse: Presynaptic, Postsynaptic, and Long-Term Effects. *J. Toxicol.* **2012**, *2012*, 132671. [CrossRef] [PubMed]
26. Marchetti, C. Interaction of metal ions with neurotransmitter receptors and potential role in neurodiseases. *BioMetals* **2014**, *27*, 1097–1113. [CrossRef] [PubMed]
27. Penzes, P.; Cahill, M.; Jones, A.K.; VanLeeuwen, J.-E.; Woolfrey, K.M. Dendritic spine pathology in neuropsychiatric disorders. *Nat. Neurosci.* **2011**, *14*, 285–293. [CrossRef] [PubMed]
28. Wong, G.T.-H.; Chang, R.C.-C.; Law, A.C.-K. A breach in the scaffold: The possible role of cytoskeleton dysfunction in the pathogenesis of major depression. *Ageing Res. Rev.* **2013**, *12*, 67–75. [CrossRef] [PubMed]
29. Kilinc, D. The Emerging Role of Mechanics in Synapse Formation and Plasticity. *Front. Cell. Neurosci.* **2018**, *12*, 483. [CrossRef]
30. Zieger, H.L.; Choquet, D. Nanoscale synapse organization and dysfunction in neurodevelopmental disorders. *Neurobiol. Dis.* **2021**, *158*, 105453. [CrossRef]
31. Paoletti, P.; Bellone, C.; Zhou, Q. NMDA receptor subunit diversity: Impact on receptor properties, synaptic plasticity and disease. *Nat. Rev. Neurosci.* **2013**, *14*, 383–400. [CrossRef]
32. Karakas, E.; Furukawa, H. Crystal structure of a heterotetrameric NMDA receptor ion channel. *Science* **2014**, *344*, 992–997. [CrossRef] [PubMed]
33. Huo, T.-G.; Li, W.-K.; Zhang, Y.-H.; Yuan, J.; Gao, L.-Y.; Yuan, Y.; Yang, H.-L.; Jiang, H.; Sun, G.-F. Excitotoxicity Induced by Realgar in the Rat Hippocampus: The Involvement of Learning Memory Injury, Dysfunction of Glutamate Metabolism and NMDA Receptors. *Mol. Neurobiol.* **2015**, *51*, 980–994. [CrossRef]
34. Zhao, F.; Wang, Z.; Liao, Y.; Wang, G.; Jin, Y. Alterations of NMDA and AMPA receptors and their signaling apparatus in the hippocampus of mouse offspring induced by developmental arsenite exposure. *J. Toxicol. Sci.* **2019**, *44*, 777–788. [CrossRef]
35. Luo, J.-H.; Qiu, Z.-Q.; Zhang, L.; Shu, W.-Q. Arsenite exposure altered the expression of NMDA receptor and postsynaptic signaling proteins in rat hippocampus. *Toxicol. Lett.* **2012**, *211*, 39–44. [CrossRef] [PubMed]
36. Ramos-Chávez, L.A.; Rendón-López, C.R.; Zepeda, A.; Adaya, I.D.S.; Del Razo, L.M.; Gonsebatt, M.E.; Rendón-López, C.R.R. Neurological effects of inorganic arsenic exposure: Altered cysteine/glutamate transport, NMDA expression and spatial memory impairment. *Front. Cell. Neurosci.* **2015**, *9*, 21. [CrossRef] [PubMed]
37. Silva-Adaya, D.; Ramos-Chávez, L.A.; Petrosyan, P.; González-Alfonso, W.L.; Pérez-Acosta, A.; Gonsebatt, M.E. Early Neurotoxic Effects of Inorganic Arsenic Modulate Cortical GSH Levels Associated with the Activation of the Nrf2 and NFκB Pathways, Expression of Amino Acid Transporters and NMDA Receptors and the Production of Hydrogen Sulfide. *Front. Cell. Neurosci.* **2020**, *14*. [CrossRef] [PubMed]
38. Nelson-Mora, J.; Escobar, M.L.; Rodríguez-Durán, L.; Massieu, L.; Montiel, T.; Rodriguez, V.; Hernández-Mercado, K.; Gonsebatt, M.E. Gestational exposure to inorganic arsenic (iAs3+) alters glutamate disposition in the mouse hippocampus and ionotropic glutamate receptor expression leading to memory impairment. *Arch. Toxicol.* **2017**, *92*, 1037–1048. [CrossRef] [PubMed]
39. Maekawa, F.; Tsuboi, T.; Oya, M.; Aung, K.H.; Tsukahara, S.; Pellerin, L.; Nohara, K. Effects of sodium arsenite on neurite outgrowth and glutamate AMPA receptor expression in mouse cortical neurons. *NeuroToxicology* **2013**, *37*, 197–206. [CrossRef]
40. Mónaco, N.M.; Bartos, M.; Dominguez, S.; Gallegos, C.; Bras, C.; Esandi, M.D.C.; Bouzat, C.; Giannuzzi, L.; Minetti, A.; Gumilar, F. Low arsenic concentrations impair memory in rat offpring exposed during pregnancy and lactation: Role of α7 nicotinic receptor, glutamate and oxidative stress. *NeuroToxicology* **2018**, *67*, 37–45. [CrossRef]
41. Chandravanshi, L.P.; Gupta, R.; Shukla, R.K. Arsenic-Induced Neurotoxicity by Dysfunctioning Cholinergic and Dopaminergic System in Brain of Developing Rats. *Biol. Trace Element Res.* **2019**, *189*, 118–133. [CrossRef]
42. Ávila, C.L.M.; Limón-Pacheco, J.H.; Giordano, M.; Rodríguez, V.M. Chronic Exposure to Arsenic in Drinking Water Causes Alterations in Locomotor Activity and Decreases Striatal mRNA for the D2 Dopamine Receptor in CD1 Male Mice. *J. Toxicol.* **2016**, *2016*, 1–10. [CrossRef]
43. Usai, C.; Barberis, A.; Moccagatta, L.; Marchetti, C.; Neurochem, J. Pathways of cadmium influx in mammalian neurons. *J. Neurochem.* **2008**, *72*, 2154–2161. [CrossRef] [PubMed]
44. Tu, Y.-C.; Yang, Y.-C.; Kuo, C.-C. Modulation of NMDA channel gating by Ca^{2+} and Cd^{2+} binding to the external pore mouth. *Sci. Rep.* **2016**, *6*, 37029. [CrossRef]
45. Watanabe, J.; Beck, C.; Kuner, T.; Premkumar, L.S.; Wollmuth, L.P. DRPEER: A Motif in the Extracellular Vestibule Conferring High Ca2+Flux Rates in NMDA Receptor Channels. *J. Neurosci.* **2002**, *22*, 10209–10216. [CrossRef]
46. Wang, S.; Hu, P.; Wang, H.; Wang, M.; Chen, J.; Tang, J.; Ruan, D. Effects of Cd2+ on AMPA receptor-mediated synaptic transmission in rat hippocampal CA1 area. *Toxicol. Lett.* **2008**, *176*, 215–222. [CrossRef] [PubMed]
47. Del Pino, J.; Zeballos, G.; Anadon, M.J.; Díaz, M.J.; Moya, M.T.F.; Díaz, G.G.; García, J.; Lobo, M.; Frejo, M.T. Muscarinic M1 receptor partially modulates higher sensitivity to cadmium-induced cell death in primary basal forebrain cholinergic neurons: A cholinesterase variants dependent mechanism. *Toxicol.* **2016**, *361-362*, 1–11. [CrossRef] [PubMed]

48. Moyano, P.; de Frias, M.; Lobo, M.; Anadon, M.J.; Sola, E.; Pelayo, A.; Díaz, M.J.; Frejo, M.T.; Del Pino, J. Cadmium induced ROS alters M1 and M3 receptors, leading to SN56 cholinergic neuronal loss, through AChE variants disruption. *Toxicol.* **2018**, *394*, 54–62. [CrossRef] [PubMed]
49. Celentano, J.J.; Gyenes, M.; Gibbs, T.; Farb, D. Negative modulation of the gamma-aminobutyric acid response by extracellular zinc. *Mol. Pharmacol.* **1991**, *40*, 766–773. [PubMed]
50. Zhao, Q.; Gao, L.; Liu, Q.; Cao, Y.; He, Y.; Hu, A.; Chen, W.; Cao, J.; Hu, C.; Li, L.; et al. Impairment of learning and memory of mice offspring at puberty, young adulthood, and adulthood by low-dose Cd exposure during pregnancy and lactation via GABAAR α5 and δ subunits. *Ecotoxicol. Environ. Saf.* **2018**, *166*, 336–344. [CrossRef]
51. Gupta, R.; Shukla, R.K.; Pandey, A.; Sharma, T.; Dhuriya, Y.; Srivastava, P.; Singh, M.P.; Siddiqi, M.I.; Pant, A.B.; Khanna, V.K. Involvement of PKA/DARPP-32/PP1α and β- arrestin/Akt/GSK-3β Signaling in Cadmium-Induced DA-D2 Receptor-Mediated Motor Dysfunctions: Protective Role of Quercetin. *Sci. Rep.* **2018**, *8*, 1–18. [CrossRef]
52. Hansen, K.B.; Yi, F.; Perszyk, R.; Furukawa, H.; Wollmuth, L.P.; Gibb, A.; Traynelis, S.F. Structure, function, and allosteric modulation of NMDA receptors. *J. Gen. Physiol.* **2018**, *150*, 1081–1105. [CrossRef]
53. Büsselberg, D.; Michael, D.; Platt, B. Pb2+ reduces voltage- andN-methyl-d-aspartate (NMDA)-activated calcium channel currents. *Cell. Mol. Neurobiol.* **1994**, *14*, 711–722. [CrossRef]
54. Büsselberg, D. Calcium channels as target sites of heavy metals. *Toxicol. Lett.* **1995**, *82*, 255–261. [CrossRef]
55. Ordemann, J.M.; Austin, R.N. Lead neurotoxicity: Exploring the potential impact of lead substitution in zinc-finger proteins on mental health. *Metallomics* **2016**, *8*, 579–588. [CrossRef] [PubMed]
56. Jalali-Yazdi, F.; Chowdhury, S.; Yoshioka, C.; Gouaux, E. Mechanisms for Zinc and Proton Inhibition of the GluN1/GluN2A NMDA Receptor. *Cell* **2018**, *175*, 1520–1532.e15. [CrossRef] [PubMed]
57. Guilarte, T.R.; Miceli, R.C.; Jett, D.A. Biochemical evidence of an interaction of lead at the zinc allosteric sites of the NMDA re-ceptor complex: Effects of neuronal development. *Neurotoxicology* **1995**, *16*, 63–71. [PubMed]
58. Gavazzo, P.; Zanardi, I.; Baranowska-Bosiacka, I.; Marchetti, C. Molecular determinants of Pb2+ interaction with NMDA receptor channels. *Neurochem. Int.* **2008**, *52*, 329–337. [CrossRef] [PubMed]
59. Neal, A.P.; Guilarte, T.R. Mechanisms of lead and manganese neurotoxicity. *Toxicol. Res.* **2013**, *2*, 99–114. [CrossRef]
60. Neal, A.P.; Worley, P.F.; Guilarte, T.R. Lead exposure during synaptogenesis alters NMDA receptor targeting via NMDA receptor inhibition. *NeuroToxicology* **2011**, *32*, 281–289. [CrossRef]
61. Guilarte, T.R.; McGlothan, J.L. Hippocampal NMDA receptor mRNA undergoes subunit specific changes during devel-opmental lead exposure. *Brain Res.* **1998**, *790*, 98–107. [CrossRef]
62. Nihei, M.K.; Desmond, N.L.; McGlothan, J.L.; Kuhlmann, A.C.; Guilarte, T.R. N-methyl-d-aspartate receptor subunit changes are associated with lead-induced deficits of long-term potentiation and spatial learning. *Neuroscience* **2000**, *99*, 233–242. [CrossRef]
63. Zhang, X.-Y.; Liu, A.-P.; Ruan, D.-Y.; Liu, J. Effect of developmental lead exposure on the expression of specific NMDA receptor subunit mRNAs in the hippocampus of neonatal rats by digoxigenin-labeled in situ hybridization histochemistry. *Neurotoxicology Teratol.* **2002**, *24*, 149–160. [CrossRef]
64. Guilarte, T.R.; McGlothan, J.L.; Nihei, M.K. Hippocampal expression of N-methyl-d-aspartate receptor (NMDAR1) subunit splice variant mRNA is altered by developmental exposure to Pb^{2+}. *Mol. Brain Res.* **2000**, *76*, 299–305. [CrossRef]
65. Basha, R.; Wei, W.; Brydie, M.; Razmiafshari, M.; Zawia, N. Lead-induced developmental perturbations in hippocampal Sp1 DNA-binding are prevented by zinc supplementation: In vivo evidence for Pb and Zn competition. *Int. J. Dev. Neurosci.* **2003**, *21*, 1–12. [CrossRef]
66. Wang, T.; Guan, R.-L.; Liu, M.-C.; Shen, X.-F.; Chen, J.Y.; Zhao, M.-G.; Luo, W.-J. Lead Exposure Impairs Hippocampus Related Learning and Memory by Altering Synaptic Plasticity and Morphology During Juvenile Period. *Mol. Neurobiol.* **2015**, *53*, 3740–3752. [CrossRef] [PubMed]
67. Gilbert, M.; Lasley, S. Developmental lead (Pb) exposure reduces the ability of the NMDA antagonist MK-801 to suppress long-term potentiation (LTP) in the rat dentate gyrus, in vivo. *Neurotoxicology Teratol.* **2007**, *29*, 385–393. [CrossRef] [PubMed]
68. Neal, A.P.; Guilarte, T.R. Molecular Neurobiology of Lead (Pb2+): Effects on Synaptic Function. *Mol. Neurobiol.* **2010**, *42*, 151–160. [CrossRef] [PubMed]
69. White, L.; Cory-Slechta, D.; Gilbert, M.; Tiffany-Castiglioni, E.; Zawia, N.; Virgolini, M.; Rossi-George, A.; Lasley, S.; Qian, Y.; Basha, R. New and evolving concepts in the neurotoxicology of lead. *Toxicol. Appl. Pharmacol.* **2007**, *225*, 1–27. [CrossRef]
70. Ishida, K.; Kotake, Y.; Miyara, M.; Aoki, K.; Sanoh, S.; Kanda, Y.; Ohta, S. Involvement of decreased glutamate receptor subunit GluR2 expression in lead-induced neuronal cell death. *J. Toxicol. Sci.* **2013**, *38*, 513–521. [CrossRef]
71. Ishida, K.; Kotake, Y.; Sanoh, S.; Ohta, S. Lead-Induced ERK Activation Is Mediated by GluR2 Non-containing AMPA Receptor in Cortical Neurons. *Biol. Pharm. Bull.* **2017**, *40*, 303–309. [CrossRef] [PubMed]
72. Ding, J.-J.; Zou, R.-X.; He, H.-M.; Tang, Y.-Q.; Wang, H.-L. Effect of Pb Exposure on Synaptic Scaling Through Regulation of AMPA Receptor Surface Trafficking. *Toxicol. Sci.* **2018**, *165*, 224–231. [CrossRef]
73. Devi, C.; Reddy, G.; Prasanthi, R.; Chetty, C.; Reddy, G. Developmental lead exposure alters mitochondrial monoamine oxidase and synaptosomal catecholamine levels in rat brain. *Int. J. Dev. Neurosci.* **2005**, *23*, 375–381. [CrossRef]
74. Zhang, S.; Zhou, Z.; Fu, J. Effect of manganese chloride exposure on liver and brain mitochondria function in rats. *Environ. Res.* **2003**, *93*, 149–157. [CrossRef]

75. Carmona, A.; Malard, V.; Avazeri, E.; Roudeau, S.; Porcaro, F.; Paredes, E.; Vidaud, C.; Bresson, C.; Ortega, R. Uranium exposure of human dopaminergic cells results in low cytotoxicity, accumulation within sub-cytoplasmic regions, and down regulation of MAO-B. *NeuroToxicology* **2018**, *68*, 177–188. [CrossRef] [PubMed]
76. Lasley, S.M.; Gilbert, M.E. Rat Hippocampal Glutamate and GABA Release Exhibit Biphasic Effects as a Function of Chronic Lead Exposure Level. *Toxicol. Sci.* **2002**, *66*, 139–147. [CrossRef]
77. Xiao, C.; Gu, Y.; Zhou, C.-Y.; Wang, L.; Zhang, M.-M.; Ruan, D.-Y. Pb2+ impairs GABAergic synaptic transmission in rat hippocampal slices: A possible involvement of presynaptic calcium channels. *Brain Res.* **2006**, *1088*, 93–100. [CrossRef]
78. Finkelstein, Y.; Milatovic, D.; Aschner, M. Modulation of cholinergic systems by manganese. *NeuroToxicology* **2007**, *28*, 1003–1014. [CrossRef]
79. Goncalves Soares, A.T.; Silva, A.D.C.; Tinkov, A.A.; Khan, H.; Santamaría, A.; Skalnaya, M.G.; Skalny, A.V.; Tsatsakis, A.; Bowman, A.B.; Aschner, M.; et al. The impact of manganese on neurotransmitter systems. *J. Trace Elements Med. Biol.* **2020**, *61*, 126554. [CrossRef]
80. Tinkov, A.; Paoliello, M.; Mazilina, A.; Skalny, A.; Martins, A.; Voskresenskaya, O.; Aaseth, J.; Santamaria, A.; Notova, S.; Tsatsakis, A.; et al. Molecular Targets of Manganese-Induced Neurotoxicity: A Five-Year Update. *Int. J. Mol. Sci.* **2021**, *22*, 4646. [CrossRef]
81. Mayer, M.L.; Westbrook, G.L. Permeation and block of N-methyl-D-aspartic acid receptor channels by divalent cations in mouse cultured central neurones. *J. Physiol.* **1987**, *394*, 501–527. [CrossRef]
82. Guilarte, T.R.; Chen, M.-K. Manganese inhibits NMDA receptor channel function: Implications to psychiatric and cognitive effects. *NeuroToxicology* **2007**, *28*, 1147–1152. [CrossRef] [PubMed]
83. Xu, B.; Xu, Z.-F.; Deng, Y. Effect of manganese exposure on intracellular Ca2+ homeostasis and expression of NMDA receptor subunits in primary cultured neurons. *NeuroToxicology* **2009**, *30*, 941–949. [CrossRef] [PubMed]
84. Song, Q.; Deng, Y.; Yang, X.; Bai, Y.; Xu, B.; Liu, W.; Zheng, W.; Wang, C.; Zhang, M.; Xu, Z. Manganese-Disrupted Interaction of Dopamine D1 and NMDAR in the Striatum to Injury Learning and Memory Ability of Mice. *Mol. Neurobiol.* **2015**, *53*, 6745–6758. [CrossRef]
85. Wang, L.; Fu, H.; Liu, B.; Liu, X.; Chen, W.; Yu, X. The effect of postnatal manganese exposure on the NMDA receptor signaling pathway in rat hippocampus. *J. Biochem. Mol. Toxicol.* **2017**, *31*, e21969. [CrossRef]
86. Calabresiab, P.; Ammassari-Teule, M.; Gubelliniad, P.; Sancesarioa, G.; Morello, M.; Centonze, D.; Marfia, G.A.; Saullea, E.; Passinoc, E.; Picconi, B.; et al. A Synaptic Mechanism Underlying the Behavioral Abnormalities Induced by Manganese Intoxication. *Neurobiol. Dis.* **2001**, *8*, 419–432. [CrossRef] [PubMed]
87. Burton, N.C.; Schneider, J.S.; Syversen, T.; Guilarte, T.R. Effects of Chronic Manganese Exposure on Glutamatergic and GABAergic Neurotransmitter Markers in the Nonhuman Primate Brain. *Toxicol. Sci.* **2009**, *111*, 131–139. [CrossRef] [PubMed]
88. Itoh, K.; Sakata, M.; Watanabe, M.; Aikawa, Y.; Fujii, H. The entry of manganese ions into the brain is accelerated by the activation of N-methyl-d-aspartate receptors. *Neurosci.* **2008**, *154*, 732–740. [CrossRef]
89. Ma, Z.; Liu, K.; Li, X.-R.; Wang, C.; Liu, C.; Yan, D.-Y.; Deng, Y.; Liu, W.; Xu, B. Alpha-synuclein is involved in manganese-induced spatial memory and synaptic plasticity impairments via TrkB/Akt/Fyn-mediated phosphorylation of NMDA receptors. *Cell Death Dis.* **2020**, *11*, 1–15. [CrossRef] [PubMed]
90. Yang, X.; Tan, J.; Xu, X.; Yang, H.; Wu, F.; Xu, B.; Liu, W.; Shi, P.; Xu, Z.; Deng, Y. Prepubertal overexposure to manganese induce precocious puberty through GABAA receptor/nitric oxide pathway in immature female rats. *Ecotoxicol. Environ. Saf.* **2020**, *188*, 109898. [CrossRef]
91. Sun, Y.; He, Y.; Yang, L.; Liang, D.; Shi, W.; Zhu, X.; Jiang, Y.; Ou, C. Manganese induced nervous injury by α-synuclein accumulation via ATP-sensitive K(+) channels and GABA receptors. *Toxicol. Lett.* **2020**, *332*, 164–170. [CrossRef]
92. Stredrick, D.L.; Stokes, A.H.; Worst, T.J.; Freeman, W.; Johnson, E.A.; Lash, L.; Aschner, M.; Vrana, K.E. Manganese-Induced Cytotoxicity in Dopamine-Producing Cells. *NeuroToxicology* **2004**, *25*, 543–553. [CrossRef]
93. Nam, J.; Kim, K. Abnormal Motor Function and the Expression of Striatal Dopamine D2 Receptors in Manganese-Treated Mice. *Biol. Pharm. Bull.* **2008**, *31*, 1894–1897. [CrossRef]
94. Kern, C.H.; Stanwood, G.; Smith, D.R. Preweaning manganese exposure causes hyperactivity, disinhibition, and spatial learning and memory deficits associated with altered dopamine receptor and transporter levels. *Synapse* **2010**, *64*, 363–378. [CrossRef]
95. McDougall, S.A.; Der-Ghazarian, T.; Britt, C.E.; Varela, F.A.; Crawford, C.A. Postnatal manganese exposure alters the expression of D2L and D2S receptor isoforms: Relationship to PKA activity and Akt levels. *Synapse* **2010**, *65*, 583–591. [CrossRef] [PubMed]
96. Criswell, S.R.; Warden, M.N.; Nielsen, S.S.; Perlmutter, J.S.; Moerlein, S.M.; Sheppard, L.; Lenox-Krug, J.; Checkoway, H.; Racette, B.A. Selective D2 receptor PET in manganese-exposed workers. *Neurology* **2018**, *91*, e1022–e1030. [CrossRef] [PubMed]
97. Miyamoto, K.-I.; Nakanishi, H.; Moriguchi, S.; Fukuyama, N.; Eto, K.; Wakamiya, J.; Murao, K.; Arimura, K.; Osame, M. Involvement of enhanced sensitivity of N-methyl-D-aspartate receptors in vulnerability of developing cortical neurons to methylmercury neurotoxicity. *Brain Res.* **2001**, *901*, 252–258. [CrossRef]
98. Juárez, B.I.; Portillo-Salazar, H.; González-Amaro, R.; Mandeville, P.; Aguirre, J.R.; Jiménez, M.E. Participation of N-methyl-d-aspartate receptors on methylmercury-induced DNA damage in rat frontal cortex. *Toxicology* **2005**, *207*, 223–229. [CrossRef]
99. Vidal, L.; Duran, R.; Faro, L.; Campos, F.; Cervantes, R.; Alfonso, M. Protection from inorganic mercury effects on the in vivo dopamine release by ionotropic glutamate receptor antagonists and nitric oxide synthase inhibitors. *Toxicology* **2007**, *238*, 140–146. [CrossRef]

100. Wyrembek, P.; Szczuraszek, K.; Majewska, M.D.; Mozrzymas, J. Intermingled modulatory and neurotoxic effects of thimerosal and mercuric ions on electrophysiological responses to GABA and NMDA in hippocampal neurons. *J. Physiol. Pharmacol.* **2010**, *61*, 753–758.
101. Yuan, Y.; Atchison, W.D. Methylmercury Differentially Affects GABA A Receptor-Mediated Spontaneous IPSCs in Purkinje and Granule Cells of Rat Cerebellar Slices. *J. Physiol.* **2003**, *550 Pt 1*, 191–204. [CrossRef]
102. Olczak, M.; Duszczyk, M.; Mierzejewski, P.; Meyza, K.; Majewska, M.D. Persistent behavioral impairments and alterations of brain dopamine system after early postnatal administration of thimerosal in rats. *Behav. Brain Res.* **2011**, *223*, 107–118. [CrossRef] [PubMed]
103. Coccini, T.; Roda, E.; Castoldi, A.F.; Poli, D.; Goldoni, M.; Vettori, M.V.; Mutti, A.; Manzo, L. Developmental exposure to methylmercury and 2,2′,4,4′,5,5′-hexachlorobiphenyl (PCB153) affects cerebral dopamine D1-like and D2-like receptors of weanling and pubertal rats. *Arch. Toxicol.* **2011**, *85*, 1281–1294. [CrossRef]
104. Scheuhammer, A.M.; Cherian, M. Effects of heavy metal cations, sulfhydryl reagents and other chemical agents on striatal D2 dopamine receptors. *Biochem. Pharmacol.* **1985**, *34*, 3405–3413. [CrossRef]
105. Basu, N.; Kwan, M.; Chan, H.M. Mercury but not Organochlorines Inhibits Muscarinic Cholinergic Receptor Binding in the Cerebrum of Ringed Seals (Phoca hispida). *J. Toxicol. Environ. Health Part A* **2006**, *69*, 1133–1143. [CrossRef]
106. Basu, N.; Scheuhammer, A.M.; Rouvinen-Watt, K.; Evans, R.D.; Grochowina, N.; Chan, L.H. The effects of mercury on muscarinic cholinergic receptor subtypes (M1 and M2) in captive mink. *NeuroToxicology* **2008**, *29*, 328–334. [CrossRef]
107. Sarowar, T.; Grabrucker, A.M. Actin-Dependent Alterations of Dendritic Spine Morphology in Shankopathies. *Neural Plast.* **2016**, *2016*, 1–15. [CrossRef]
108. Joensuu, M.; Lanoue, V.; Hotulainen, P. Dendritic spine actin cytoskeleton in autism spectrum disorder. *Prog. Neuro-Psychopharmacol. Biol. Psychiatry* **2018**, *84 Pt B*, 362–381. [CrossRef]
109. Lasser, M.; Tiber, J.; Lowery, L.A. The Role of the Microtubule Cytoskeleton in Neurodevelopmental Disorders. *Front. Cell. Neurosci.* **2018**, *12*, 165. [CrossRef]
110. Dent, E.W. Of microtubules and memory: Implications for microtubule dynamics in dendrites and spines. *Mol. Biol. Cell* **2017**, *28*, 1–8. [CrossRef] [PubMed]
111. Pelucchi, S.; Stringhi, R.; Marcello, E. Dendritic Spines in Alzheimer's Disease: How the Actin Cytoskeleton Contributes to Synaptic Failure. *Int. J. Mol. Sci.* **2020**, *21*, 908. [CrossRef] [PubMed]
112. Perrin, L.; Roudeau, S.; Carmona, A.; Domart, F.; Petersen, J.D.; Bohic, S.; Yang, Y.; Cloetens, P.; Ortega, R. Zinc and Copper Effects on Stability of Tubulin and Actin Networks in Dendrites and Spines of Hippocampal Neurons. *ACS Chem. Neurosci.* **2017**, *8*, 1490–1499. [CrossRef] [PubMed]
113. Domart, F.; Cloetens, P.; Roudeau, S.; Carmona, A.; Verdier, E.; Choquet, D.; Ortega, R. Correlating STED and synchrotron XRF nano-imaging unveils cosegregation of metals and cytoskeleton proteins in dendrites. *eLife* **2020**, *9*. [CrossRef]
114. Baron, M.K.; Boeckers, T.M.; Vaida, B.; Faham, S.; Gingery, M.; Sawaya, M.R.; Salyer, D.; Gundelfinger, E.D.; Bowie, J.U. An Architectural Framework That May Lie at the Core of the Postsynaptic Density. *Science* **2006**, *311*, 531–535. [CrossRef] [PubMed]
115. Bucher, M.; Fanutza, T.; Mikhaylova, M. Cytoskeletal makeup of the synapse: Shaft versus spine. *Cytoskeleton* **2019**, *77*, 55–64. [CrossRef]
116. Parato, J.; Bartolini, F. The microtubule cytoskeleton at the synapse. *Neurosci. Lett.* **2021**, *753*, 135850. [CrossRef] [PubMed]
117. Durand, C.M.; Betancur, C.; Boeckers, T.M.; Bockmann, J.; Chaste, P.; Fauchereau, F.; Nygren, G.; Rastam, M.; Gillberg, I.C.; Anckarsäter, H.; et al. Mutations in the gene encoding the synaptic scaffolding protein SHANK3 are associated with autism spectrum disorders. *Nat. Genet.* **2006**, *39*, 25–27. [CrossRef] [PubMed]
118. Grabrucker, A.; Knight, M.J.; Proepper, C.; Bockmann, J.; Joubert, M.; Rowan, M.; Nienhaus, G.U.; Garner, C.; Bowie, J.U.; Kreutz, M.R.; et al. Concerted action of zinc and ProSAP/Shank in synaptogenesis and synapse maturation. *EMBO J.* **2011**, *30*, 569–581. [CrossRef]
119. Grabrucker, S.; Jannetti, L.; Eckert, M.; Gaub, S.; Chhabra, R.; Pfaender, S.; Mangus, K.; Reddy, P.P.; Rankovic, V.; Schmeisser, M.J.; et al. Zinc deficiency dysregulates the synaptic ProSAP/Shank scaffold and might contribute to autism spectrum disorders. *Brain* **2014**, *137 Pt 1*, 137–152. [CrossRef]
120. Hagmeyer, S.; Sauer, A.K.; Grabrucker, A.M. Prospects of Zinc Supplementation in Autism Spectrum Disorders and Shankopathies Such as Phelan McDermid Syndrome. *Front. Synaptic Neurosci.* **2018**, *10*. [CrossRef] [PubMed]
121. Fourie, C.; Vyas, Y.; Lee, K.; Jung, Y.; Garner, C.; Montgomery, J.M. Dietary Zinc Supplementation Prevents Autism Related Behaviors and Striatal Synaptic Dysfunction in Shank3 Exon 13–16 Mutant Mice. *Front. Cell. Neurosci.* **2018**, *12*, 374. [CrossRef] [PubMed]
122. Hagmeyer, S.; Mangus, K.; Boeckers, T.M.; Grabrucker, A.M. Effects of Trace Metal Profiles Characteristic for Autism on Synapses in Cultured Neurons. *Neural Plast.* **2015**, *2015*, 1–16. [CrossRef]
123. Vahidnia, A.; Romijn, F.; van der Voet, G.; de Wolff, F. Arsenic-induced neurotoxicity in relation to toxicokinetics: Effects on sciatic nerve proteins. *Chem. Interact.* **2008**, *176*, 188–195. [CrossRef] [PubMed]
124. Aung, K.H.; Kurihara, R.; Nakashima, S.; Maekawa, F.; Nohara, K.; Kobayashi, T.; Tsukahara, S. Inhibition of neurite outgrowth and alteration of cytoskeletal gene expression by sodium arsenite. *NeuroToxicology* **2012**, *34*, 226–235. [CrossRef]
125. Stern, M.; Gierse, A.; Tan, S.; Bicker, G. Human Ntera2 cells as a predictive in vitro test system for developmental neurotoxicity. *Arch. Toxicol.* **2013**, *88*, 127–136. [CrossRef]

126. Zhao, Y.; Toselli, P.; Li, W. Microtubules as a Critical Target for Arsenic Toxicity in Lung Cells in Vitro and in Vivo. *Int. J. Environ. Res. Public Health* **2012**, *9*, 474–495. [CrossRef]
127. Ge, Y.; Song, X.; Chen, L.; Hu, D.; Hua, L.; Cui, Y.; Liu, J.; An, Z.; Yin, Z.; Ning, H. Cadmium induces actin cytoskeleton alterations and dysfunction in Neuro-2a cells. *Environ. Toxicol.* **2019**, *34*, 469–475. [CrossRef] [PubMed]
128. Di Liegro, I.; Gerspacher, C.; Scheuber, U.; Schiera, G.; Proia, P.; Gygax, D. The effect of cadmium on brain cells in culture. *Int. J. Mol. Med.* **2009**, *24*, 311–318. [CrossRef]
129. Wang, T.; Wang, Q.; Song, R.; Zhang, Y.; Yang, J.; Wang, Y.; Yuan, Y.; Bian, J.; Liu, X.; Gu, J.; et al. Cadmium induced inhibition of autophagy is associated with microtubule disruption and mitochondrial dysfunction in primary rat cerebral cortical neurons. *Neurotoxicology Teratol.* **2016**, *53*, 11–18. [CrossRef]
130. Forcella, M.; Lau, P.; Oldani, M.; Melchioretto, P.; Bogni, A.; Gribaldo, L.; Fusi, P.; Urani, C. Neuronal specific and non-specific re-sponses to cadmium possibly involved in neurodegeneration: A toxicogenomics study in a human neuronal cell model. *Neu-rotoxicology* **2020**, *76*, 162–173. [CrossRef]
131. Go, Y.-M.; Orr, M.; Jones, D.P. Actin cytoskeleton redox proteome oxidation by cadmium. *Am. J. Physiol. Cell. Mol. Physiol.* **2013**, *305*, L831–L843. [CrossRef] [PubMed]
132. Choi, W.-S.; Kim, S.-J.; Kim, J.S. Inorganic lead (Pb)- and mercury (Hg)-induced neuronal cell death involves cytoskeletal reorganization. *Lab. Anim. Res.* **2011**, *27*, 219–225. [CrossRef] [PubMed]
133. Zimmermann, H.-P.; Faulstich, H.; Hänsch, G.; Doenges, K.; Stournaras, C. The interaction of triethyl lead with tubulin and microtubules. *Mutat. Res. Mol. Mech. Mutagen.* **1988**, *201*, 293–302. [CrossRef]
134. Stanwood, G.D.; Leitch, D.B.; Savchenko, V.; Wu, J.; Fitsanakis, V.A.; Anderson, D.J.; Stankowski, J.N.; Aschner, M.; McLaughlin, B. Manganese exposure is cytotoxic and alters dopaminergic and GABAergic neurons within the basal ganglia. *J. Neurochem.* **2009**, *110*, 378–389. [CrossRef]
135. Parsons-White, A.B.; Spitzer, N. Environmentally relevant manganese overexposure alters neural cell morphology and differentiation in vitro. *Toxicol. Vitr.* **2018**, *50*, 22–28. [CrossRef]
136. Stoiber, T.; Degen, G.H.; Bolt, H.M.; Unger, E. Interaction of mercury(II) with the microtubule cytoskeleton in IMR-32 neuroblastoma cells. *Toxicol. Lett.* **2004**, *151*, 99–104. [CrossRef] [PubMed]
137. Li, Y.; He, B.; Hu, L.; Huang, X.; Yun, Z.; Liu, R.; Zhou, Q.; Jiang, G. Characterization of mercury-binding proteins in human neuroblastoma SK-N-SH cells with immobilized metal affinity chromatography. *Talanta* **2018**, *178*, 811–817. [CrossRef] [PubMed]
138. Nong, Q.; Dong, H.; Liu, Y.; Liu, L.; He, B.; Huang, Y.; Jiang, J.; Luan, T.; Chen, B.; Hu, L. Characterization of the mercury-binding proteins in tuna and salmon sashimi: Implications for health risk of mercury in food. *Chemosphere* **2021**, *263*, 128110. [CrossRef]
139. Ajsuvakova, O.P.; Tinkov, A.A.; Aschner, M.; Rocha, J.B.; Michalke, B.; Skalnaya, M.G.; Skalny, A.V.; Butnariu, M.; Dadar, M.; Sarac, I.; et al. Sulfhydryl groups as targets of mercury toxicity. *Coord. Chem. Rev.* **2020**, *417*, 213343. [CrossRef]
140. Castoldi, A.F.; Barni, S.; Turin, I.; Gandini, C.; Manzo, L. Early acute necrosis, delayed apoptosis and cytoskeletal breakdown in cultured cerebellar granule neurons exposed to methylmercury. *J. Neurosci. Res.* **2000**, *59*, 775–787. [CrossRef]
141. Xu, F.; Farkas, S.; Kortbeek, S.; Zhang, F.-X.; Chen, L.; Zamponi, G.W.; Syed, N.I. Mercury-induced toxicity of rat cortical neurons is mediated through N-methyl-D-Aspartate receptors. *Mol. Brain* **2012**, *5*, 30. [CrossRef]
142. Moran, J.; Sabanero, M.; Meza, I.; Pasantes-Morales, H. Changes of actin cytoskeleton during swelling and regulatory volume decrease in cultured astrocytes. *Am. J. Physiol. Physiol.* **1996**, *271*, C1901–C1907. [CrossRef] [PubMed]
143. Rao, K.R.; Reddy, P.; Hazell, A.; Norenberg, M. Manganese induces cell swelling in cultured astrocytes. *NeuroToxicology* **2007**, *28*, 807–812. [CrossRef]
144. Mori, H.; Sasaki, G.; Nishikawa, M.; Hara, M. Effects of subcytotoxic cadmium on morphology of glial fibrillary acidic protein network in astrocytes derived from murine neural stem/progenitor cells. *Environ. Toxicol. Pharmacol.* **2015**, *40*, 639–644. [CrossRef] [PubMed]
145. Perea, G.; Navarrete, M.; Araque, A. Tripartite synapses: Astrocytes process and control synaptic information. *Trends Neurosci.* **2009**, *32*, 421–431. [CrossRef] [PubMed]
146. Arendt, T. Synaptic degeneration in Alzheimer's disease. *Acta Neuropathol.* **2009**, *118*, 167–179. [CrossRef] [PubMed]
147. Crouch, P.J.; Barnham, K.J. Therapeutic Redistribution of Metal Ions To Treat Alzheimer's Disease. *Accounts Chem. Res.* **2012**, *45*, 1604–1611. [CrossRef] [PubMed]
148. Squitti, R.; Pal, A.; Picozza, M.; Avan, A.; Ventriglia, M.; Rongioletti, M.C.; Hoogenraad, T. Zinc Therapy in Early Alzheimer's Disease: Safety and Potential Therapeutic Efficacy. *Biomology* **2020**, *10*, 1164. [CrossRef] [PubMed]

toxics

MDPI

Article

Relationship of Blood and Urinary Manganese Levels with Cognitive Function in Elderly Individuals in the United States by Race/Ethnicity, NHANES 2011–2014

Arturo J. Barahona [1], Zoran Bursac [2], Emir Veledar [2,3], Roberto Lucchini [1], Kim Tieu [1] and Jason R. Richardson [1,*]

1 Department of Environmental Health Sciences, Robert Stempel College of Public Health and Social Work, Florida International University, Miami, FL 33199, USA; arbaraho@fiu.edu (A.J.B.); rlucchin@fiu.edu (R.L.); ktieu@fiu.edu (K.T.)
2 Department of Biostatistics, Robert Stempel College of Public Health and Social Work, Florida International University, Miami, FL 33199, USA; zbursac@fiu.edu (Z.B.); emirv@baptisthealth.net (E.V.)
3 Baptist Health South Florida, Miami, FL 33199, USA
* Correspondence: jarichar@fiu.edu

Abstract: Manganese (Mn) is an essential metal with a biphasic relationship with health outcomes. High-level exposure to Mn is associated with manganism, but few data explore the effects of chronic, lower-level Mn on cognitive function in adults. We sought to determine the relationship between blood/urinary manganese levels and cognitive function in elderly individuals using 2011–2014 data from the National Health and Nutrition Examination Survey (NHANES). Weighted multivariate regression models were used to determine correlations, adjusting for several covariates. Blood Mn was inversely associated with the Consortium to Establish a Registry for Alzheimer's Disease (CERAD) immediate learning of new verbal information (p-value = 0.04), but lost significance after adjusting for medical history (p-value = 0.09). In addition, blood Mn was inversely associated with Animal Fluency scores after adjusting for all covariates. Urinary Mn was inversely associated with CERAD immediate learning after adjusting for all covariates (p-value = 0.01) and inversely associated with the Digit Symbol Substitution Test scores (p-value = 0.0002), but lost significance after adjusting for medical history (p-value = 0.13). Upon stratifying by race/ethnicity, other Races and Non-Hispanic (NH)-Blacks had significantly higher blood Mn levels when compared to NH-Whites. Collectively, these findings suggest that increased blood and urinary Mn levels are associated with poorer cognitive function in an elderly US population.

Keywords: manganese; neurotoxicity; environmental exposure; cognitive function; race; ethnicity; Alzheimer's disease; CERAD; animal fluency; DSST

Citation: Barahona, A.J.; Bursac, Z.; Veledar, E.; Lucchini, R.; Tieu, K.; Richardson, J.R. Relationship of Blood and Urinary Manganese Levels with Cognitive Function in Elderly Individuals in the United States by Race/Ethnicity, NHANES 2011–2014. *Toxics* **2022**, *10*, 191. https://doi.org/10.3390/toxics10040191

Academic Editors: Richard Ortega and Asuncion Carmona

Received: 4 March 2022
Accepted: 7 April 2022
Published: 14 April 2022

1. Introduction

Manganese (Mn) is a trace metal with various essential functions in the human body and plays a vital role in the metabolism of glucose and lipids and the synthesis of various proteins [1]. Mn also plays a protective role against oxidative stress; it is a critical component of Mn superoxide dismutase (MnSOD), a reactive oxygen species scavenging enzyme [1]. Furthermore, Mn has been previously linked to sex-related metabolic differences and is essential in reproduction [2]. While there is an average of 0.012 g of Mn in a 70 kg human body, high-level Mn exposure has been linked to neurotoxicity despite the critical functions it plays [3,4]. Acute, high exposure to Mn has been demonstrated to result in manganism, which is clinically manifested as slow and clumsy movements, tremors, difficulty in walking, and facial muscle spasms [5]. Alternatively, chronic exposure has been linked to an extrapyramidal syndrome similar to Parkinson's disease (PD) [5,6].

Epidemiological studies indicate a possible association between environmental exposure to Mn and cognitive function. It is suggested that the main exposure to Mn is occupational, where it is often inhaled by welders in the construction and agricultural industries [6]. Flynn and Susi found that occupational exposure to Mn is highly correlated to manganism across multiple studies [7]. A cross-sectional study conducted among foundry workers found that more chronic levels of metals such as aluminum, lead, and Mn were correlated with lower levels of cognition and higher stress levels [8]. Additionally, increased blood Mn levels have been associated with decreased cognitive function, as defined by the Auditory Verbal Learning Test in ferroalloy workers [9]. Pathologically, welders with increased chronic Mn exposure have been found to have decreased neural activation when compared to healthy controls, which also correlated with lower executive function, as defined by sorting and word-color tests [10].

Several studies have also highlighted associations between chronic environmental exposure to Mn and cognitive function in the general population. Some alternative routes of exposure to Mn may include contaminated water and diet [11,12]. The Environmental Protection Agency (EPA) has set a regulatory standard of 0.05 mg/L in drinking water, while the World Health Organization's (WHO) standard is 0.08 mg/L [13]. While the WHO standard was set based on the tolerable daily intake for infants, it can be generalized to the whole population, as infants represent the most susceptible subgroup [14]. Nevertheless, experts estimate that over 100 million people remain at risk for Mn toxicity worldwide. This risk is understated as most individuals are also at risk for co-exposure to heavy metals including Pb and As in drinking water [15]. A systemic review conducted by Zoni and Lucchini evaluated data from five countries and found six out of ten articles suggesting adverse effects of Mn on cognitive function, as defined by cognitive, motor, or behavioral changes [16]. Computer modeling data of measured air monitoring paired with cognitive testing suggests chronic environmental exposure to Mn found in the air may result in cognitive deficits in adults [17]. These findings have been validated by a recent study that found an association between high Mn emission and cognitive dysfunction in residential communities in South Africa [18].

The effects of Mn on developmental stages in children have also been widely reported. A cross-sectional study in Brazil found negative associations between Mn hair levels and cognitive performance in children and caregivers as measured by Full-Scale IQ and Raven's Progressive Matrix, respectively [19]. The Mn concentrations observed in this population were considered above average; nevertheless, a different study based in Canada found similar results with lower concentrations. Bouchard and co-workers found significant positive associations between Mn hair levels and hyperactive and oppositional behavior in children exposed to naturally high Mn levels in tap water [20]. Alternatively, Mn levels have been found to have a biphasic effect, where prenatal Mn is beneficial for adolescent cognition but has toxic effects at later time windows [21,22]. While these examples suggest an association between Mn exposure and cognitive function in adults and children, few data explore the effects of lower-level Mn on cognitive function in elderly individuals.

This study investigated possible associations between Mn exposure and cognitive function using a representative U.S. elderly population obtained from the National Health and Examination Survey (NHANES). NHANES uses a complex sampling design and over samples persons 60 and older, as well as minority groups, to obtain an accurate representation of the United States population. Covariates such as race/ethnicity are important to consider as research suggests disadvantages in education and socioeconomic status across groups may influence cognitive function [23,24]. These same factors may increase one's risk of exposure to other toxicants that may exacerbate the effects of Mn exposure. We used exposure biomarkers for Mn, including blood and urine, obtained from NHANES 2011–2012 and 2013–2014 cycles to conduct linear regression analyses, adjusting for several covariates to determine their relationship to cognitive function.

2. Materials and Methods

2.1. Study Deseign and Data Sources

A cross-sectional study design was used to assess the effects of Mn exposure on cognitive function. The data were obtained from NHANES, a program for the National Center for Health Statistics (NCHS), a division of the Center for Disease Control and Prevention (CDC). NHANES is designed to assess the health and nutritional status of different populations across the United States by collecting data through a combination of health interviews and physical examinations. The health interviews include demographic, socioeconomic, and health-related questions, while the physical examinations include cognitive and physiological measurements performed by medical professionals. Two study samples for blood and urine were used of 2068 and 950 participants, respectively. Participants were selected using a complex, multistage probability design with appropriate weights to obtain a representative sample of the US population. Data from the 2011–2012 and 2013–2014 cycles were merged to conduct the analyses. The inclusion criteria consisted of individuals 60 years of age or older living in the US who had completed the Consortium to Establish a Registry for Alzheimer's Disease (CERAD), Animal Fluency and Digital Symbol Substitution Test (DSST) test scores and also had measurements of blood and urine Mn levels.

2.2. Cognitive Function Assessments

Measurements of cognitive function were collected through at-home patient interviews using three different tests of cognitive performance: (1) word learning and recall modules from the CERAD, (2) the Animal Fluency test, and (3) the DSST. The CERAD assessed immediate and delayed learning ability for new verbal information [25]. It consisted of three learning trials and a delayed recall. Participants were instructed to read aloud 10 unrelated words and were immediately asked to recall as many words as possible. The order of the 10 words was changed in each trial. The delayed recall was assessed after the Animal Fluency and DSST tests were finished and about 8–10 min after the start of the CERAD's learning trials. The final results included two scores—one for immediate word learning and another for delayed recall. The Animal Fluency test assessed categorical verbal fluency [26]. Participants were asked to name as many animals as possible in one minute, and a point was given for each animal named. The DSST assessed processing speed, sustained attention, and working memory [27]. Participants were asked to fill out a form that had a key with nine numbers paired with symbols. Participants were given two minutes to copy the corresponding symbols, and a point was given for each correct match. While higher test scores indicate better cognitive performance, these tests were not meant to replace a medical diagnosis but to examine the association of cognitive function with Mn exposure.

2.3. Manganese Exposure Assessments

Single blood and urine samples used for analysis were collected at mobile examination centers (MECs) and shipped to the CDC's Division of Laboratory Sciences in Atlanta, Georgia, for analysis. Once in the lab, manganese concentrations were determined using inductively coupled plasma mass spectrometry, and the resulting electrical signals were used to determine the concentration of the element [28]. The lower detection limit for blood Mn was 0.99 μg/L, and a fill value was used in the cases where the result was below the limit of detection; this value was the lower limit of detection divided by the square root of 2. The limit of detection for urinary Mn was 0.13 μg/L, and, likewise, a fill value was used in the cases where the result was below the limit of detection; this value was the lower limit of detection divided by the square root of 2.

2.4. Covariates

Multiple covariates were included in the analysis to rule out any possible confounding factors or variables related to cognitive function or Mn exposure. These variables include age (60–64, 65–69, 70–74, 75–79, and 80 and older), gender (male and female), race/ethnicity (Hispanic, Non-Hispanic White, Non-Hispanic and Other Race), level of education (<high school, high school, >high school), poverty index ratio (PIR), marital status (Married/Living with a partner, Widowed/Divorced/Separated and Never Married), alcohol consumption (<12 drinks per year and >12 drinks per year), hypertension (yes and no), diabetes (yes and no), coronary heart disease (yes and no) and stroke (yes and no).

2.5. Statistical Analysis

SAS/STATv14.2 was used for statistical analyses. NHANES data includes sampling weights incorporated into the analyses to adjust for possible biases such as unequal distribution poststratification and nonresponse. A new sampling weight was constructed for the merged 2011–2012 and 2013–2014 data cycles, following the publicly available protocol on the CDC's website (CDC, 2021). The measurements of cognitive function are subject to floor and ceiling effects due to the wide range in function among the elderly population [1]. A composite cognitive z-score was created by averaging the standardized measurements of cognitive function to control for these effects, and the Kolmogorov–Smirnov test was used to assess normality. Stratified analyses between measurements of cognitive function were used to determine differences between tests. A descriptive analysis was performed to determine the distributions of demographic characteristics for individuals who completed all cognitive function tests. Univariate linear regression analyses were conducted to determine associations between the composite outcome z-score and individual covariates. Linear regression models were used to assess the relationship between Mn and the composite outcome z-score while adjusting for covariates. Any individuals with missing data were excluded from the respective models. Results were considered statistically significant at an alpha level of 0.05 and reported per 10 unit change in Mn level.

3. Results

Blood and urinary Mn data were analyzed as two separate cohorts to increase power. From the blood Mn cohort, 2068 participants had available data on cognitive function out of 3110 participants evaluated. These participants were on average 69.1 years of age. There were 1011 males and 1057 females in the cohort (46% vs. 54%, respectively). The composite outcome z-score was approximately normally distributed based on the Kolmogorov–Smirnov test (p-value = 0.1273), had a minimum and maximum of −2.5 and 2.5, respectively, and had a mean of 0.24. The average concentration of blood Mn levels was 9.4 μg/L per participant; about 22% of participants fell below the limit of detection. Table 1 depicts the demographic composition of the participants who had data on all four tests of cognitive function as well as a blood Mn measurement. Covariates including gender, age, race/ethnicity, education, PIR, marital status, alcohol consumption, hypertension, diabetes, coronary artery disease, and stroke were all significantly associated with the composite outcome z-score (Table S1). Univariate analyses suggested the need for their adjustment in the regression models.

Upon stratification of cognitive function by cognitive task, a 10 unit increase in blood Mn was associated with a 0.2-point decrease in CERAD (immediate) in model 2, adjusted for gender, age, race/ethnicity, PIR, and marital status (β = −0.2, 95% CI −0.3 to −0.01) (Table 2). This association was no longer significant in models 3 and 4 after adjusting for alcohol consumption and comorbidities (Table 2). A 10 unit increase in blood Mn was also associated with a 0.8-point decrease in AF scores in all models, after adjusting for all covariates (β = −0.8, 95% CI −1.4 to −0.1) (Table 3). Blood Mn was not significantly associated with CERAD (delayed) or DSST scores (Tables S2 and S3). Blood Mn was not significantly associated with composite outcome z-score in unadjusted or adjusted linear regression models (Table S4).

Table 1. Characteristics of the Mn-Blood study population (n = 2068).

Variables	N	Weighted Sample	(%) *	Variables	N	Weighted Sample	Mean (SE)	Median (IQR)
Gender								
Male	1011	1,7508,127	46	Age (year)	2068	38,452,145	69.1 (0.24)	67.7 (11)
Female	1057	20,944,017	54	Blood Mn (µg/L)	2068	38,452,145	9.4 (0.08)	8.8 (3.8)
Race/Ethnicity				CERAD learning	2068	38,452,145	6.5 (0.1)	6.5 (2)
Hispanic	385	2,618,094	7	CERAD recall	2068	38,452,145	6.1 (0.1)	5.8 (3.2)
NH White	988	30,935,957	80	Animal Fluency	2068	38,452,145	18 (0.2)	17 (7.5)
NH Black	491	2,980,219	8	DSST	2068	38,452,145	52 (0.7)	53 (24)
Other Race	204	1,917,874	5	z-score	2068	38,452,145	0.24 (0.04)	0.28 (1.1)
Education								
<High School	531	6,133,440	16					
High School	490	8,783,404	23					
>High School	1046	23,530,632	61					
Missing	1							
PIR								
≤0.99	333	3,298,332	9					
≥1	1550	32,672,780	91					
Missing	185							
Marital Status								
Married/Living with partner	1201	25,353,317	69					
Widowed/Divorced/Separated	742	11,403,761	31					
Missing	125							
Alcohol Consumption								
>12 drinks/year	1397	27,847,784	73					
<12 drinks/year	640	10,098,855	27					
Missing	31							
HTN								
Yes	1262	21,869,127	57					
No	804	16,514,849	43					
Missing	2							
DM								
Yes	472	7,293,920	20					
No	1508	29,747,613	80					
Missing	88							
CAD								
Yes	172	3,086,811	8					
No	1886	35,263,550	92					
Missing	10							
Stroke								
Yes	138	2,341,705	6					
No	1926	36,057,144	94					
Missing	4							

* Weighted percentage, mean, SE, median, and IQR. CERAD, Consortium to Establish a Registry for Alzheimer's Disease; DSST, Digital Symbol Substitution Test; HTN, hypertension; DM, diabetes mellitus; CAD, coronary artery disease.

Table 2. Mn-Blood levels (continuous) in relation to CERAD (immediate).

	Model 1 (n = 2068) *		Model 2 (n = 1772) †		Model 3 (n = 1744) ‡		Model 4 (n = 1650) §	
Variables	**β (95% CI) ¶**	***p*-Value**	**β (95% CI) ¶**	***p*-Value**	**β (95% CI) ¶**	***p*-Value**	**β (95% CI) ¶**	***p*-Value**
Mn	−0.03 (−0.2 to 0.2)	0.714	−0.2 (−0.3 to −0.01)	0.04	−0.2 (−0.3 to 0.01)	0.07	−0.2 (−0.3 to 0.03)	0.09
Gender								
Male			referent	referent	referent	referent	referent	referent
Female			6 (4 to 7)	<0.0001	6 (4 to 8)	<0.0001	6 (4 to 8)	<0.0001
Age (year)			−0.8 (−0.9 to −0.7)	<0.0001	−0.8 (−0.9 to −0.6)	<0.0001	−0.8 (−0.9 to −0.6)	<0.0001
Race/Ethnicity								
NH White			referent	referent	referent	referent	referent	referent
Hispanic			−5 (−8 to −2)	0.005	−4 (−8 to −1)	0.01	−4 (−8 to −1)	0.01
NH Black			−2 (−4 to 1)	0.18	−1 (−4 to 1)	0.31	−0.5 (−3 to 2)	0.65
Other Race			−2 (−5 to 1)	0.21	−2 (−5 to 1)	0.23	−1 (−4 to 2)	0.32
Education								
>High School			referent	referent	referent	referent	referent	referent
High School			−4 (−6 to −2)	0.0004	−4 (−6 to −2)	0.001	−4 (−6 to −2)	0.001
<High School			−6 (−10 to −2)	0.004	−6 (−10 to −2)	0.01	−5 (−10 to −1)	0.01
PIR								
≤0.99			referent	referent	referent	referent	referent	referent
≥1			5 (3 to 7)	<0.0001	5 (3 to 7)	<0.0001	5 (3 to 7)	<0.0001
Marital Status								
Married/Living with partner			referent	referent	referent	referent	referent	referent
Widowed/Divorced/ Separated			−1 (−4 to 1)	0.31	−1 (−4 to 1)	0.34	0.8 (−3 to 2)	0.48
Alcohol Consumption								
<12 drinks/year					referent	referent	referent	referent
>12 drinks/year					1 (−1 to 4)	0.17	1 (−1 to 4)	0.19
HTN								
No							referent	referent
Yes							−1 (−3 to 1)	0.33
DM								
No							referent	referent
Yes							−2 (−4 to −1)	0.002
Stroke								
No							referent	referent
Yes							−0.1 (−3 to 3)	0.97
CAD								
No							referent	referent
Yes							−2 (−6 to 3)	0.43

* Unadjusted. † Adjusted for age, gender, ethnicity, education, PIR, and marital status. ‡ Adjusted for age, gender, ethnicity, education, PIR, marital status, and alcohol consumption. § Adjusted for age, gender, ethnicity, education, PIR, marital status, alcohol consumption, HTN, DM, stroke, and CAD. ¶ Weighted β and 95% Confidence Interval.

Table 3. Mn-Blood levels (continuous) in relation to AF.

	Model 1 (n = 2068) *		Model 2 (n = 1772) †		Model 3 (n = 1744) ‡		Model 4 (n = 1650) §	
Variables	**β (95% CI) ¶**	**p-Value**	**β (95% CI) ¶**	**p-Value**	**β (95% CI) ¶**	**p-Value**	**β (95% CI) ¶**	**p-Value**
Mn	0.4 (−1 to 0.2)	0.21	−0.8 (−1 to −0.1)	0.02	−0.7 (−1 to −0.1)	0.02	−0.8 (−1 to −0.1)	0.03
Gender								
Male			referent	referent	referent	referent	referent	referent
Female			−0.5 (−9 to 8)	0.91	0.9 (−8 to 10)	0.83	2 (−7 to 10)	0.64
Age (year)			−3 (−3 to −2)	<0.0001	−3 (−3 to −2)	<0.0001	−2 (−3 to −2)	<0.0001
Race/Ethnicity								
NH White			referent	referent	referent	referent	referent	referent
Hispanic			−23 (−30 to −20)	<0.0001	−23 (−30 to −20)	<0.0001	−23 (−30 to −20)	<0.0001
NH Black			−33 (−40 to −30)	<0.0001	−32 (−40 to −20)	<0.0001	−31 (−40 to −20)	<0.0001
Other Race			−24 (−30 to −10)	0.0005	−23 (−40 to −10)	0.0007	−21 (−30 to −10)	0.003
Education								
>High School			referent	referent	referent	referent	referent	referent
High School			−29 (−40 to −20)	<0.0001	−28 (−40 to −20)	<0.0001	−28 (−40 to −20)	<0.0001
<High School			−33 (−40 to −30)	<0.0001	−32 (−40 to −30)	<0.0001	−31 (−40 to −20)	<0.0001
PIR								
≤0.99			referent	referent	referent	referent	referent	referent
≥1			12 (2 to 20)	0.02	11 (2 to 20)	0.02	11 (1 to 20)	0.03
Marital Status								
Married/Living with partner			referent	referent	referent	referent	referent	referent
Widowed/Divorced/ Separated			−0.4 (−9 to 8)	0.93	−0.7 (−10 to 8)	0.88	−1 (−10 to 8)	0.82
Alcohol Consumption								
<12 drinks/year					referent	referent	referent	referent
>12 drinks/year					7 (0.8 to 10)	0.03	7 (0.8 to 10)	0.03
HTN								
No							referent	referent
Yes							−6 (−10 to 0.2)	0.06
DM								
No							referent	referent
Yes							−8 (−10 to −1)	0.03
Stroke								
No							referent	referent
Yes							−6 (−20 to 5)	0.27
CAD								
No							referent	referent
Yes							−9 (−20 to 4)	0.16

* Unadjusted. † Adjusted for age, gender, ethnicity, education, PIR, and marital status. ‡ Adjusted for age, gender, ethnicity, education, PIR, marital status, and alcohol consumption. § Adjusted for age, gender, ethnicity, education, PIR, marital status, alcohol consumption, HTN, DM, stroke, and CAD. ¶ Weighted β and 95% Confidence Interval.

From the urinary Mn cohort, 950 participants had available data on cognitive function out of 3110 participants evaluated. These participants were on average 69.3 years of age. There were 464 males and 486 females in the cohort (46% vs. 54%, respectively). The average concentration of urinary Mn levels was 0.19 µg/L per participant; about 64% of participants fell below the limit of detection. The high percentage of measurements in the urine cohort explains why the average concentration was so close to the limit of detection. This may further be explained by the fact that more than 90% of Mn is excreted into feces [29,30]. Table 4 depicts the demographic composition of the participants who had data on all four tests of cognitive function as well as a blood Mn measurement. Similar to the blood Mn cohort, all covariates in the urinary Mn cohort were significantly associated with the composite z-score and, therefore, adjusted for in the models (Table S5).

Upon stratification of cognitive task, a 10 unit increase in urinary Mn was associated with a 2.4-point decrease in CERAD (immediate) after adjusting for all covariates in model 4 ($\beta = -2.4$, 95% CI -4.1 to -0.7) (Table 5). A 10 unit increase in urinary Mn was also associated with a 30-point decrease in DSST scores in models 1 to 3 ($\beta = -30$, 95% CI -40 to -10); however, significance diminished when adjusting for all covariates in model 4 (Table 6). Urinary Mn was not significantly associated with CERAD (delayed) or AF scores (Tables S6 and S7). Urinary Mn was also not significantly associated with composite outcome z-score in unadjusted or adjusted linear regression models (Table S8).

Univariate analyses between all covariates and z-score showed that race/ethnicity had the most substantial effect on cognitive function (Tables S1 and S5). To further investigate this, the mean blood and urinary Mn levels were compared across each race/ethnicity. Figure 1 shows Other Races had significantly higher blood Mn levels when compared to NH-Whites, while it was significantly lower in NH-Blacks. Hispanics did not have significant differences in mean blood Mn levels compared to NH-Whites (Figure 1). These racial and ethnic differences are similar to those previously found and may be explained by dietary differences [2]. Consumption of tea, which has a considerable amount of Mn, is believed to play a significant role in Mn intake [31]. Additionally, there were no significant differences observed in mean urinary Mn levels across all four groups (Figure 2).

Lastly, linear regression analysis showed no significant associations between blood Mn levels and urinary Mn levels in unadjusted and adjusted models (Table S9).

Table 4. Characteristics of the Mn-Urine study population (n = 950).

Variables	N	Weighted Sample	(%) *	Variables	N	Weighted Sample	Mean (SE)	Median (IQR)
Gender								
Male	464	7,702,844	46	Age (year)	950	16,795,626	69.3 (0.3)	67.9 (10.7)
Female	486	9,092,783	54	Urinary Mn (μg/L)	950	16,795,626	0.19 (0.03)	0.09 (0.07)
Race/Ethnicity				CERAD learning	950	16,795,626	6.6 (0.01)	6.6 (2.1)
Hispanic	176	1,216,571	7	CERAD recall	950	16,795,626	6.3 (0.1)	6.1 (3.3)
NH White	451	13,351,140	79	Animal Fluency	950	16,795,626	18 (0.3)	16 (8)
NH Black	225	1,438,019	9	DSST	950	16,795,626	52 (0.8)	51 (25)
Other Race	98	789,895	5	z-score	950	16,795,626	0.23 (0.04)	0.29 (1.1)
Education								
<High School	259	2,847,794	17					
High School	209	3,422,549	20					
>High School	480	10,513,695	63					
Missing	2							
PIR								
≤0.99	166	1,539,528	10					
≥1	693	13,934,209	90					
Missing	91							

Table 4. *Cont.*

Variables	N	Weighted Sample	(%) *	Variables	N	Weighted Sample	Mean (SE)	Median (IQR)
Marital Status								
Married/Living with partner	557	10,964,908	65					
Widowed/Divorced/Separated	336	5,008,167	30					
Never married	55	810,401	5					
Missing	2							
Alcohol Consumption								
>12 drinks/year	644	12,105,591	73					
<12 drinks/year	288	4,409,341	27					
Missing	18							
HTN								
Yes	589	96,29,710	57					
No	361	7,165,916	43					
DM								
Yes	214	3,081,562	19					
No	694	12,955,019	81					
Missing	42							
CAD								
Yes	76	1,380,980	8					
No	869	15,348,084	92					
Missing	5							
Stroke								
Yes	49	760,890	5					
No	900	16,002,161	95					
Missing	1							

* Weighted percentage, mean, SE, median, and IQR. CERAD, Consortium to Establish a Registry for Alzheimer's Disease; DSST, Digital Symbol Substitution Test; HTN, hypertension; DM, diabetes mellitus; CAD, coronary artery disease.

Table 5. Mn-Urine levels (continuous) in relation to CERAD (immediate).

	Model 1 (*n* = 2068) *		Model 2 (*n* = 1772) †		Model 3 (*n* = 1744) ‡		Model 4 (*n* = 1650) §	
Variables	**β (95% CI) ¶**	***p*-Value**	**β (95% CI) ¶**	***p*-Value**	**β (95% CI) ¶**	***p*-Value**	**β (95% CI) ¶**	***p*-Value**
Mn	−0.7 (−2 to 0.2)	0.16	−1 (−2 to 0.3)	0.12	−1 (−2 to 0.2)	0.09	−2 (−4 to −0.7)	0.01
Gender								
Male			referent	referent	referent	referent	referent	referent
Female			7 (5 to 9)	<0.0001	7 (5 to 9)	<0.0001	7 (5 to 9)	<0.0001
Age (year)			−0.8 (−1 to −0.6)	<0.0001	−0.8 (−1 to −0.5)	<0.0001	−0.7 (−1 to −0.5)	<0.0001
Race/Ethnicity								
NH White			referent	referent	referent	referent	referent	referent
Hispanic			−7 (−10 to −4)	<0.0001	−7 (−10 to −4)	<0.0001	−6 (−10 to −4)	0.0001
NH Black			0.2 (−2 to 3)	0.89	0.5 (−2 to 3)	0.67	0.7 (−2 to 4)	0.6
Other Race			−5 (−10 to 1)	0.1	−4 (−10 to 11)	0.12	−4 (−10 to 1)	0.13

Table 5. *Cont.*

	Model 1 (n = 2068) *		Model 2 (n = 1772) †		Model 3 (n = 1744) ‡		Model 4 (n = 1650) §	
Variables	β (95% CI) ¶	p-Value	β (95% CI) ¶	p-Value	β (95% CI) ¶	p-Value	β (95% CI) ¶	p-Value
Education								
>High School			referent	referent	referent	referent	referent	referent
High School			−4 (−10 to −4)	0.0003	−4 (−6 to −2)	0.001	−3 (−6 to −1)	0.003
<High School			−7 (−10 to −4)	0.0002	−7 (−10 to 3)	0.0004	−6 (−10 to −3)	0.001
PIR								
≤0.99			referent	referent	referent	referent	referent	referent
≥1			1 (−2 to 5)	0.4	1 (−2 to −5)	0.51	0.9 (−3 to 5)	0.59
Marital Status								
Married/Living with partner			referent	referent	referent	referent	referent	referent
Widowed/Divorced/ Separated			−2 (−5 to 0.6)	0.12	−2 (−5 to 0.5)	0.11	−2 (−5 to 0.6)	0.13
Alcohol Consumption								
<12 drinks/year					referent	referent	referent	referent
>12 drinks/year					2 (−1 to 5)	0.19	2 (−1 to 5)	0.24
HTN								
No							referent	referent
Yes							−2 (−5 to 0.5)	0.1
DM								
No							referent	referent
Yes							−1 (−4 to 1)	0.27
Stroke								
No							referent	referent
Yes							−3 (−7 to 2)	0.27
CAD								
No							referent	referent
Yes							0.3 (−6 to 7)	0.93

* Unadjusted. † Adjusted for age, gender, ethnicity, education, PIR, and marital status. ‡ Adjusted for age, gender, ethnicity, education, PIR, marital status, and alcohol consumption. § Adjusted for age, gender, ethnicity, education, PIR, marital status, alcohol consumption, HTN, DM, stroke, and CAD. ¶ Weighted β and 95% Confidence Interval.

Table 6. Mn-Urine levels (continuous) in relation to DSST.

	Model 1 (n = 2068) *		Model 2 (n = 1772) †		Model 3 (n = 1744) ‡		Model 4 (n = 1650) §	
Variables	β (95% CI) ¶	p-Value	β (95% CI) ¶	p-Value	β (95% CI) ¶	p-Value	β (95% CI) ¶	p-Value
Mn	−20 (−40 to −10)	0.003	−30 (−40 to −10)	0.002	−30 (−40 to −10)	0.0002	−30 (−70 to 10)	0.13
Gender								
Male			referent	referent	referent	referent	referent	referent
Female			70 (40 to 100)	<0.0001	80 (50 to 100)	<0.0001	70 (50 to 100)	<0.0001
Age (year)			−10 (−12 to −9)	<0.0001	−10 (−12 to −9)	<0.0001	−10 (−2 to −9)	<0.0001
Race/Ethnicity								
NH White			referent	referent	referent	referent	referent	referent
Hispanic			−120 (−150 to −90)	<0.0001	−120 (−150 to −90)	<0.0001	−120 (−150 to −90)	<0.0001
NH Black			−110 (−130 to −80)	<0.0001	−100 (−120 to −80)	<0.0001	−100 (−130 to −70)	<0.0001
Other Race			−20 (−60 to 20)	0.29	−10 (−50 to 20)	0.45	−20 (−60 to 20)	0.38

Table 6. *Cont.*

	Model 1 (n = 2068) *		Model 2 (n = 1772) †		Model 3 (n = 1744) ‡		Model 4 (n = 1650) §	
Variables	**β (95% CI) ¶**	***p*-Value**	**β (95% CI) ¶**	***p*-Value**	**β (95% CI) ¶**	***p*-Value**	**β (95% CI) ¶**	***p*-Value**
Education								
>High School			referent	referent	referent	referent	referent	referent
High School			−80 (−110 to −40)	<0.0001	−70 (−110 to −40)	0.0003	−60 (−100 to −80)	0.002
<High School			−130 (−160 to −100)	<0.0001	−120 (−150 to −90)	<0.0001	−110 (−140 to −80)	<0.0001
PIR								
≤0.99			referent	referent	referent	referent	referent	referent
≥1			80 (40 to 110)	<0.0001	70 (40 to 110)	0.0002	70 (40 to 110)	0.0002
Marital Status								
Married/Living with partner			referent	referent	referent	referent	referent	referent
Widowed/Divorced/Separated			−20 (−50 to 10)	0.09	−30 (−50 to 10)	0.06	−30 (−60 to 10)	0.04
Alcohol Consumption								
<12 drinks/year					referent	referent	referent	referent
>12 drinks/year					40 (10 to 70)	0.01	30 (10 to 60)	0.03
HTN								
No							referent	referent
Yes							−2 (−30 to 20)	0.88
DM								
No							referent	referent
Yes							−60 (−100 to −30)	0.0003
Stroke								
No							referent	referent
Yes							−20 (−70 to 40)	0.5
CAD								
No							referent	referent
Yes							−50 (−100 to 10)	0.1

* Unadjusted. † Adjusted for age, gender, ethnicity, education, PIR, and marital status. ‡ Adjusted for age, gender, ethnicity, education, PIR, marital status, and alcohol consumption. § Adjusted for age, gender, ethnicity, education, PIR, marital status, alcohol consumption, HTN, DM, stroke, and CAD. ¶ Weighted β and 95% Confidence Interval.

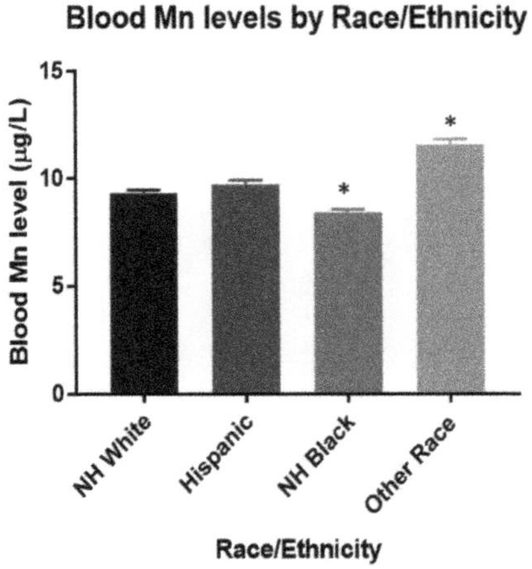

Figure 1. Estimated mean blood Mn levels were 9.3, 9.7, 8.4, and 11.6 µg/L for NH-White (n = 988), Hispanic (n = 385), NH-Black (n = 491), and Other Race (n = 204), respectively. * $p < 0.0001$. Data were analyzed by One-way ANOVA with Tukey's *post-hoc* test holding NH White as the control. The limit of detection used was 0.99 µg/L.

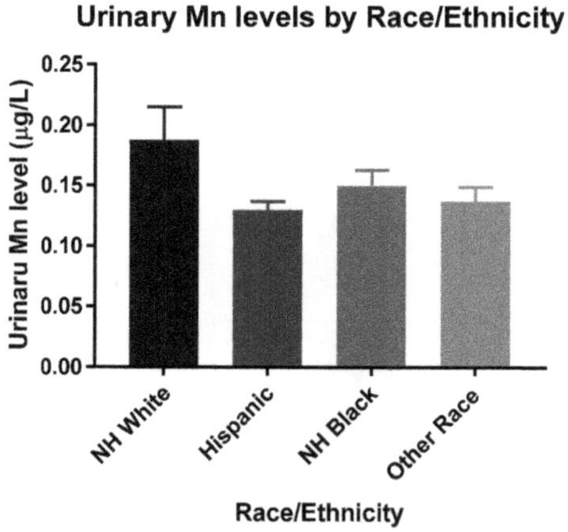

Figure 2. Estimated mean urinary Mn levels were 0.19, 0.13, 0.15, and 0.14 µg/L for NH-White (n = 451), Hispanic (n = 176), NH-Black (n = 225), and Other Race (n = 98), respectively. Data were analyzed by One-way ANOVA with Tukey's *post-hoc* test holding NH White as the control. The limit of detection used was 0.13 µg/L.

4. Discussion

To our knowledge, this is the first study to investigate possible associations between blood and urinary Mn levels and cognitive function in an elderly US-based population. We found significant inverse associations between blood and urinary Mn levels with cognitive function when stratified by four cognitive function tests, while there was no significant association between blood and urinary Mn levels and composite outcome z-score. Specifically, blood Mn was inversely associated with CERAD (immediate) after adjusting for gender, age, race/ethnicity, PIR, and marital status; this association is no longer significant after adjusting for medical history. In addition, blood Mn was inversely associated with AF scores after adjusting for all covariates. Finally, urinary Mn was inversely associated with CERAD (immediate) after adjusting for all covariates and inversely associated with DSST scores in models 1 to 3, while no longer significant after adjusting for medical history. Each cognitive function test measures different cognitive domains that are influenced by various brain regions and interactions among these regions, and these results may suggest a region-specific effect of Mn [25–27].

A similar NHANES study has previously reported an inverse association between urinary Mn levels and high blood pressure in adults [32]. Oulhote and co-workers reported similar differences in blood Mn levels across different race/ethnicities, indicating the importance of considering race/ethnicity as a confounding factor when conducting future studies [2]. Studies in children have reported high Mn exposure associated with poor childhood development characterized by motor skills and behavioral performance [33]. Furthermore, elevated Mn levels in private wells are associated with increased congenital disabilities [34]. Together, these data suggest that Mn may be implicated in a broad range of adverse effects that may influence cognitive function and highlight the need for further research on the mechanisms through which it induces toxic effects.

The exact molecular mechanisms of Mn-induced neurotoxicity have not been fully established. It is believed that Mn exposure has a biphasic relationship with health outcomes, suggesting Mn serves as an essential nutrient at an optimal concentration but has toxic effects at low and high levels [35]. Mn is transported across the blood-brain barrier (BBB) via several pathways, such as the divalent metal transporter 1 (DMT1) and the transferrin receptor system and accumulates in Fe-rich regions of the basal ganglia [11,35]. Mutations in the Mn efflux and influx transporter genes SLC30A10 and SLC30A8 can alter Mn levels in the Golgi apparatus and induce neurotoxicity via aberrant vesicular trafficking [35,36].

Evidence suggests Mn may increase autophagy at higher levels, implicating a role in the degradation of protein aggregates associated with neurodegenerative diseases such as PD or Alzheimer's disease AD [37]. Its implication in neurodegenerative diseases is further evidenced by its ability to alter mitochondrial function and induce dopaminergic cell loss [35]. Furthermore, Mn accumulation affects the metabolism of other metals, and Mn toxicity may indirectly lead to toxic increases of other metals as well [1]. Epidemiological studies investigating possible environmental associations between Mn and adverse effects can help provide further insight into prevention methods.

A major strength of this study is the utility of a representative U.S.-based elderly population. NHANES is designed to oversample minority populations and apply adjusted weights to ensure representation. Nevertheless, the data reported in this study should be interpreted with caution based on several limitations. First, there were data excluded from both cohorts due to missing values. The final cohorts represent 66% and 31% of the participants considered for blood and urine analyses, respectively. Second, blood and urine Mn levels are generally thought to reflect acute exposure to Mn. The half-life of urinary Mn is believed to be about 30 h, while it is only about 2 h in the blood [38,39]. This difference may explain why there was no significant association found between blood Mn and urinary Mn levels as the renal route of excretion is only responsible for up to about 5% of Mn excretion [38,40]. The biologic half-life of Mn in the brain has been previously reported to be between 51–74 days, also suggesting that blood and serum levels may not be reflective of the total concentration of Mn in the body [41]. The average blood and urine Mn levels seen in our sample population also fall below the average Mn levels reported in the Agency for Toxic Substances and Disease Registry of 4–15 μg/L in blood and 1–8 μg/L in urine [42]. Moreover, previous studies have found urine to be a poor biomarker for Mn exposure [43]. Ellingsen and co-workers found weak associations between urinary Mn and inhalable aerosol fractions in a group of workers in Mn-alloy producing plants [44]. Future research should focus on using biomarkers that accurately depict long-term exposure. For example, Laohaudomchok and co-workers found toenails to be a measure of Mn levels up to 12 months after exposure [45]. Ideally, an integrated approach that considers multiple biomarkers for Mn exposure provides a more accurate representation of the total body burden of exposure [46].

Lastly, only biomarkers of Mn were used, and it is possible that exposure to other neurotoxic agents may have confounded the results. For example, other metals such as Al, Se, Pb, Zn, Cd, and Hg have all been implicated in a broad range of neurodegenerative disorders and may share similar routes of exposure to Mn [47–49]. Zn transporters, ZIP8 and ZIP14, have been suggested to also be involved in Mn transport pathways, indicating that co-exposure of these two metals may have combinatorial effects [50]. Cognitive dysfunction is a common clinical presentation of many neurodegenerative diseases. These diseases are multifaceted and co-exposure of a variety of environmental factors such as pesticides or particles found in air pollution may further affect the role of Mn in cognitive function.

5. Conclusions

Collectively, these findings suggest that increased blood and urinary Mn levels are associated with poorer cognitive function in an elderly US population. These data further support the need for preventative approaches to reduce excess Mn exposure. Future studies incorporating biomarker matrices that are more stable and using cohorts with repeated samples available will be important in determining the impact of chronic, lower-level Mn exposure on cognitive function.

Supplementary Materials: The following supporting information can be downloaded at: https://www.mdpi.com/article/10.3390/toxics10040191/s1, Table S1: Mn-Blood univariate analyses of the association between z-score and covariates (n = 2068), Table S2: Mn-Blood levels (continuous) in relation to CERAD (delayed), Table S3: Mn-Blood levels (continuous) in relation to DSST, Table S4: Mn-Blood levels (continuous) in relation to z-score, Table S5: Mn-Urine univariate analyses of the association between z-score and covariates (n = 950), Table S6: Mn-Urine levels (continuous) in relation to CERAD (delayed), Table S7: Mn-Urine levels (continuous) in relation to AF, Table S8: Mn-Urine levels (continuous) in relation to cognitive function, Table S9: Mn-Blood levels (continuous) in relation to Mn-Urine.

Author Contributions: Conceptualization, A.J.B. and J.R.R.; data curation A.J.B.; formal analysis, A.J.B., Z.B. and E.V.; methodology, A.J.B., Z.B., E.V., R.L. and J.R.R.; writing—original draft preparation, A.J.B.; writing—review and editing, Z.B., E.V., R.L., K.T. and J.R.R.; supervision, J.R.R. All authors have read and agreed to the published version of the manuscript.

Funding: This research was funded by A.J.B., C.V. Starr Fellowship; Z.B., grant number NIMHD U54 MD012393; R.L., grant number R01ES019222; K.T., grant number R35ES030523; and J.R.R., grant number, R01ES026057 and R13AG069380.

Institutional Review Board Statement: The NHANES study was approved by National Center for Health Statistics Research Ethics Review Board.

Informed Consent Statement: Informed consent was obtained from all subjects involved in the study, as per the National Center for Health Statistics Research guidelines. More information can be found at https://www.cdc.gov/nchs/nhanes/genetics/genetic_participants.htm, accessed on 3 March 2022.

Data Availability Statement: The NHANES data is publicly available at https://www.cdc.gov/nchs/nhanes/genetics/genetic_participants.htm, accessed on 3 March 2022.

Conflicts of Interest: The authors declare no conflict of interest.

References

1. Li, L.; Yang, X. The Essential Element Manganese, Oxidative Stress, and Metabolic Diseases: Links and Interactions. *Oxidative Med. Cell. Longev.* **2018**, *2018*, 7580707. [CrossRef] [PubMed]
2. Oulhote, Y.; Mergler, D.; Bouchard, M.F. Sex-and age-differences in blood manganese levels in the US general population: National health and nutrition examination survey 2011–2012. *Environ. Health* **2014**, *13*, 87. [CrossRef] [PubMed]
3. Lide, D.R.; Fredrikse, H.P.R. *CRC Handbook of Chemistry and Physics*; Chemical Composition of the Human Body; CRC Press: Boca Raton, FL, USA, 1994; pp. 7–11.
4. Dobson, A.W.; Erikson, K.M.; Aschner, M. Manganese neurotoxicity. *Ann. N. Y. Acad. Sci.* **2004**, *1012*, 115–128. [CrossRef] [PubMed]
5. Lucchini, R.G.; Martin, C.J.; Doney, B.C. From Manganism to Manganese-Induced Parkinsonism: A Conceptual Model Based on the Evolution of Exposure. *Neuromol. Med.* **2009**, *11*, 311–321. [CrossRef] [PubMed]
6. Martinez-Finley, E.J.; Gavin, C.E.; Aschner, M.; Gunter, T.E. Manganese neurotoxicity and the role of reactive oxygen species. *Free Radic. Biol. Med.* **2013**, *62*, 65–75. [CrossRef] [PubMed]
7. Flynn, M.R.; Susi, P. Neurological risks associated with manganese exposure from welding operations—A literature review. *Int. J. Hyg. Environ. Health* **2009**, *212*, 459–469. [CrossRef]
8. Mohammed, R.S.; Ibrahim, W.; Sabry, D.; El-Jaafary, S.I. Occupational metals exposure and cognitive performance among foundry workers using tau protein as a biomarker. *Neurotoxicology* **2020**, *76*, 10–16. [CrossRef]
9. Rolle-McFarland, D.; Liu, Y.; Mostafaei, F.; Zauber, S.E.; Zhou, Y.; Li, Y.; Fan, Q.; Zheng, W.; Nie, L.H.; Wells, E.M. The association of bone, fingernail and blood manganese with cognitive and olfactory function in Chinese workers. *Sci. Total Environ.* **2019**, *666*, 1003–1010. [CrossRef]
10. Seo, J.; Chang, Y.; Jang, K.E.; Park, J.W.; Kim, Y.-T.; Park, S.-J.; Jeong, K.S.; Kim, A.; Kim, S.H.; Kim, Y. Altered executive function in the welders: A functional magnetic resonance imaging study. *Neurotoxicol. Teratol.* **2016**, *56*, 26–34. [CrossRef]
11. Aschner, M.; Erikson, K.M.; Hernández, E.H.; Tjalkens, R. Manganese and its role in Parkinson's disease: From transport to neuropathology. *Neuromol. Med.* **2009**, *11*, 252–266. [CrossRef]
12. Chen, P.; Culbreth, M.; Aschner, M. Exposure, epidemiology, and mechanism of the environmental toxicant manganese. *Environ. Sci. Pollut. Res.* **2016**, *23*, 13802–13810. [CrossRef] [PubMed]
13. World Health Organization. *No. WHO/HEP/ECH/WSH/2021.5*; Manganese in Drinking Water: Background Document for Development of WHO Guidelines for Drinking-Water Quality; World Health Organization: Geneva, Switzerland, 2021.

14. Mitchell, E.J.; Frisbie, S.H.; Roudeau, S.; Carmona, A.; Ortega, R. How much manganese is safe for infants? A review of the scientific basis of intake guidelines and regulations relevant to the manganese content of infant formulas. *J. Trace Elem. Med. Biol.* **2021**, *65*, 126710. [CrossRef] [PubMed]
15. Amrose, S.E.; Cherukumilli, K.; Wright, N.C. Chemical Contamination of Drinking Water in Resource-Constrained Settings: Global Prevalence and Piloted Mitigation Strategies. *Annu. Rev. Environ. Resour.* **2020**, *45*, 195–226. [CrossRef]
16. Zoni, S.; Lucchini, R.G. Manganese exposure: Cognitive, motor and behavioral effects on children: A review of recent findings. *Curr. Opin. Pediatrics* **2013**, *25*, 255–260. [CrossRef]
17. Bowler, R.M.; Kornblith, E.S.; Gocheva, V.V.; Colledge, M.; Bollweg, G.; Kim, Y.; Beseler, C.L.; Wright, C.W.; Adams, S.W.; Lobdell, D. Environmental exposure to manganese in air: Associations with cognitive functions. *NeuroToxicology* **2015**, *49*, 139–148. [CrossRef]
18. Racette, B.A.; Nelson, G.; Dlamini, W.W.; Hershey, T.; Prathibha, P.; Turner, J.R.; Checkoway, H.; Sheppard, L.; Nielsen, S.S. Environmental Manganese Exposure and Cognitive Control in a South African Population. *NeuroToxicology* **2022**, *89*, 31–40. [CrossRef]
19. Menezes-Filho, J.A.; Novaes, C.D.O.; Moreira, J.C.; Sarcinelli, P.N.; Mergler, D. Elevated manganese and cognitive performance in school-aged children and their mothers. *Environ. Res.* **2011**, *111*, 156–163. [CrossRef]
20. Bouchard, M.; Laforest, F.; Vandelac, L.; Bellinger, D.; Mergler, D. Hair Manganese and Hyperactive Behaviors: Pilot Study of School-Age Children Exposed through Tap Water. *Environ. Health Perspect.* **2007**, *115*, 122–127. [CrossRef]
21. Claus Henn, B.; Ettinger, A.S.; Schwartz, J.; Téllez-Rojo, M.M.; Lamadrid-Figueroa, H.; Hernández-Avila, M.; Schnaas, L.; Amarasiriwardena, C.; Bellinger, D.C.; Hu, H.; et al. Early postnatal blood manganese levels and children's neurodevelopment. *Epidemiology* **2010**, *21*, 433–439. [CrossRef]
22. Bauer, J.A.; White, R.F.; Coull, B.A.; Austin, C.; Oppini, M.; Zoni, S.; Fedrighi, C.; Cagna, G.; Placidi, D.; Guazzetti, S.; et al. Critical windows of susceptibility in the association between manganese and neurocognition in Italian adolescents living near ferro-manganese industry. *NeuroToxicology* **2021**, *87*, 51–61. [CrossRef]
23. Díaz-Venegas, C.; Downer, B.; Langa, K.; Wong, R. Racial and ethnic differences in cognitive function among older adults in the USA. *Int. J. Geriatr. Psychiatry* **2016**, *31*, 1004–1012. [CrossRef] [PubMed]
24. Avila, J.F.; Vonk, J.M.; Verney, S.P.; Witkiewitz, K.; Rentería, M.A.; Schupf, N.; Mayeux, R.; Manly, J.J. Sex/gender differences in cognitive trajectories vary as a function of race/ethnicity. *Alzheimer's Dement.* **2019**, *15*, 1516–1523. [CrossRef] [PubMed]
25. Morris, J.C.; Heyman, A.; Mohs, R.C.; Hughes, J.P.; van Belle, G.; Fillenbaum, G.; Mellits, E.D.; Clark, C. The Consortium to Establish a Registry for Alzheimer's Disease (CERAD). Part I. Clinical and neuropsychological assessment of Alzheimer's disease. *Neurology* **1989**, *39*, 1159–1165. [PubMed]
26. Strauss, E.; Sherman, E.M.S.; Spreen, O. A compendium of neuropsychological tests: Administration, norms, and commentary. *Am. Chem. Soc.* **2006**, *3*, 500–501.
27. Wechsler, D. *WAIS Manual*; Psychological Corporation: New York, NY, USA, 1997.
28. Laboratory Procedure Manual. In *Organic Analytical Toxicology Branch*; Division of Laboratory Sciences, National Center for Environmental Health: Atlanta, GA, USA, 2013.
29. Trumbo, P.; Yates, A.A.; Schlicker, S.; Poos, M. Dietary reference intakes: Vitamin A, vitamin K, arsenic, boron, chromium, copper, iodine, iron, manganese, molybdenum, nickel, silicon, vanadium, and zinc. *J. Am. Diet Assoc.* **2001**, *101*, 294–301. [CrossRef]
30. Aschner, J.L.; Aschner, M. Nutritional aspects of manganese homeostasis. *Mol. Asp. Med.* **2005**, *26*, 353–362. [CrossRef]
31. Hope, S.J.; Daniel, K.; Gleason, K.L.; Comber, S.; Nelson, M.; Powell, J.J. Influence of tea drinking on manganese intake, manganese status and leucocyte expression of MnSOD and cytosolic aminopeptidase P. *Eur. J. Clin. Nutr.* **2006**, *60*, 1–8. [CrossRef]
32. Wu, C.; Woo, J.G.; Zhang, N. Association between urinary manganese and blood pressure: Results from National Health and Nutrition Examination Survey (NHANES), 2011–2014. *PLoS ONE* **2017**, *12*, e0188145. [CrossRef]
33. Liu, W.; Xin, Y.; Li, Q.; Shang, Y.; Ping, Z.; Min, J.; Cahill, C.M.; Rogers, J.T.; Wang, F. Biomarkers of environmental manganese exposure and associations with childhood neurodevelopment: A systematic review and meta-analysis. *Environ. Health* **2020**, *19*, 104. [CrossRef]
34. Sanders, A.P.; Desrosiers, T.A.; Warren, J.L.; Herring, A.H.; Enright, D.; Olshan, A.F.; Meyer, R.E.; Fry, R.C. Association between arsenic, cadmium, manganese, and lead levels in private wells and birth defects prevalence in North Carolina: A semi-ecologic study. *BMC Public Health* **2014**, *14*, 955. [CrossRef]
35. Balachandran, R.C.; Mukhopadhyay, S.; McBride, D.; Veevers, J.; Harrison, F.E.; Aschner, M.; Haynes, E.N.; Bowman, A.B. Brain manganese and the balance between essential roles and neurotoxicity. *J. Biol. Chem.* **2020**, *295*, 6312–6329. [CrossRef] [PubMed]
36. Carmona, A.; Zogzas, C.E.; Roudeau, S.; Porcaro, F.; Garrevoet, J.; Spiers, K.M.; Salomé, M.; Cloetens, P.; Mukhopadhyay, S.; Ortega, R. SLC30A10 Mutation Involved in Parkinsonism Results in Manganese Accumulation within Nanovesicles of the Golgi Apparatus. *ACS Chem. Neurosci.* **2019**, *10*, 599–609. [CrossRef] [PubMed]
37. Pfalzer, A.C.; Bowman, A.B. Relationships Between Essential Manganese Biology and Manganese Toxicity in Neurological Disease. *Curr. Environ. Health Rep.* **2017**, *4*, 223–228. [CrossRef] [PubMed]
38. Roels, H.; Lauwerys, R.; Genet, P.; Sarhan, M.J.; de Fays, M.; Hanotiau, I.; Buchet, J.-P. Relationship between external and internal parameters of exposure to manganese in workers from a manganese oxide and salt producing plant. *Am. J. Ind. Med.* **1987**, *11*, 297–305. [CrossRef]

39. Zheng, W.; Kim, H.; Zhao, Q. Comparative toxicokinetics of manganese chloride and methylcyclopentadienyl manganese tricarbonyl (MMT) in Sprague-Dawley rats. *Toxicol. Sci.* **2000**, *54*, 295–301. [CrossRef]
40. Gandhi, D.; Rudrashetti, A.P.; Rajasekaran, S. The impact of environmental and occupational exposures of manganese on pulmonary, hepatic, and renal functions. *J. Appl. Toxicol.* **2022**, *42*, 103–129. [CrossRef]
41. Takeda, A.; Sawashita, J.; Okada, S. Biological half-lives of zinc and manganese in rat brain. *Brain Res.* **1995**, *695*, 53–58. [CrossRef]
42. Toxicological Profile for Manganese. In *Agency for Toxic Substances and Disease Registry (ATSDR)*; US Department of Health and Human Services: Atlanta, GA, USA, 2012.
43. Smith, D.; Gwiazda, R.; Bowler, R.; Roels, H.; Park, R.; Bs, C.T.; Lucchini, R. Biomarkers of Mn exposure in humans. *Am. J. Ind. Med.* **2007**, *50*, 801–811. [CrossRef]
44. Ellingsen, D.G.; Hetland, S.M.; Thomassen, Y. Manganese air exposure assessment and biological monitoring in the manganese alloy production industry. *J. Environ. Monit.* **2003**, *5*, 84–90. [CrossRef]
45. Laohaudomchok, W.; Lin, X.; Herrick, R.F.; Fang, S.C.; Cavallari, J.; Christiani, D.C.; Weisskopf, M.G. Toenail, Blood, and Urine as Biomarkers of Manganese Exposure. *J. Occup. Environ. Med.* **2011**, *53*, 506–510. [CrossRef]
46. Levin-Schwartz, Y.; Gennings, C.; Henn, B.C.; Coull, B.A.; Placidi, D.; Lucchini, R.; Smith, D.R.; Wright, R.O. Multi-media biomarkers: Integrating information to improve lead exposure assessment. *Environ. Res.* **2020**, *183*, 109148. [CrossRef] [PubMed]
47. Brown, R.C.; Lockwood, A.H.; Sonawane, B.R. Neurodegenerative Diseases: An Overview of Environmental Risk Factors. *Environ. Health Perspect.* **2005**, *113*, 1250–1256. [CrossRef] [PubMed]
48. Cannon, J.; Greenamyre, J.T. The Role of Environmental Exposures in Neurodegeneration and Neurodegenerative Diseases. *Toxicol. Sci.* **2011**, *124*, 225–250. [CrossRef]
49. Charlet, L.; Chapron, Y.; Faller, P.; Kirsch, R.; Stone, A.T.; Baveye, P.C. Neurodegenerative diseases and exposure to the environmental metals Mn, Pb, and Hg. *Coord. Chem. Rev.* **2012**, *256*, 2147–2163. [CrossRef]
50. Fujishiro, H.; Kambe, T. Manganese transport in mammals by zinc transporter family proteins, ZNT and ZIP. *J. Pharmacol. Sci.* **2022**, *148*, 125–133. [CrossRef]

toxics

MDPI

Article

Concentrations of Lead, Mercury, Selenium, and Manganese in Blood and Hand Grip Strength among Adults Living in the United States (NHANES 2011–2014)

M. Corinaud J. Gbemavo [1,2] and Maryse F. Bouchard [1,2,*]

[1] Department Environmental and Occupational Health, School of Public Health, Université de Montréal, Montreal, QC H3C 3J7, Canada; mahude.corinaud.josue.gbemavo@umontreal.ca
[2] CHU Sainte-Justine Hospital Research Center, Montreal, QC H3T 1C5, Canada
* Correspondence: maryse.bouchard@umontreal.ca

Abstract: Exposure to lead and mercury can cause deficits in neuromotor function. Selenium and manganese are essential elements, hence both deficiency and excess could result in decreased neuromotor function. We aimed to examine hand grip strength, a marker of neuromotor function, and blood concentrations of lead, mercury, selenium, and manganese in the general U.S. population. We used data from the National Health and Nutrition Examination Survey (NHANES, 2011–2014) on 6199 participants ages 20–79 years. We assessed associations of blood concentration for these elements and grip strength with generalized regression models, and cubic splines to detect possible nonlinear relations, adjusting for confounders. The results showed that mercury and manganese were not associated with grip strength. Lead was associated with weaker grip strength in women (for 10-fold increase in lead, −2.4 kg; 95% CI: −4.2, −0.5), but not in men. Higher selenium was associated with stronger grip strength in women (8.5 kg; 95% CI: 1.9, 15.1) and men (4.6; 95% CI: −11.9, 21.0), although the association was not significant in the latter. In conclusion, lead exposure was associated with weaker grip strength in women, even at the low exposure levels in the population. Furthermore, low blood selenium level was associated with weaker grip strength, suggesting that some individuals might have selenium deficiency manifesting with poorer neuromotor function.

Keywords: grip strength; metals; neuromotor system; neurotoxicity; NHANES

Citation: Gbemavo, M.C.J.; Bouchard, M.F. Concentrations of Lead, Mercury, Selenium, and Manganese in Blood and Hand Grip Strength among Adults Living in the United States (NHANES 2011–2014). *Toxics* **2021**, *9*, 189. https://doi.org/10.3390/toxics9080189

Academic Editors: Jaymie R. Meliker and Gunnar Toft

Received: 22 June 2021
Accepted: 16 August 2021
Published: 17 August 2021

1. Introduction

Environmental exposure to some metals may lead to neurotoxic effects that can manifest by alterations in different functions of the nervous system, such as cognitive impairments, mental health or mood disorders, and impairment in neuromotor function. For instance, the adverse effects of lead exposure on the nervous system have been very well documented, especially during development [1] and in adults with high levels of exposure in the workplace [2,3]. Fewer studies have addressed potential adverse effects on neuromotor function at the lower exposure levels found in adults from the general population. Blood lead concentration was associated with weaker hand grip strength in a community-based study among men in unadjusted models, but not after adjustment for covariates [4]. However, higher blood lead was also associated with other neuromotor function indicators, such as poor fine motor skills, assessed with a handwriting test, and with slower walking speed in older individuals from the general population [5,6].

Mercury is a global pollutant with known neurotoxic properties. The greatest risks for the integrity of the nervous system are thought to arise from exposure to methylmercury, especially when exposure occurs during the perinatal period [7]. Some studies also reported that high exposure during adulthood, such as in residents of Minamata, could cause severe neuromotor deficits such as cerebellar ataxia and poor motor strength [8]. Another study reported poorer motor strength and motor dexterity with higher hair methylmercury levels in adults consuming mercury-contaminated fish from the Amazonian basin [9]. However, much fewer data are available on the potential effects on neuromotor function arising from low-level exposure during adulthood.

Selenium and manganese are essential nutrients, but intake of these elements at levels exceeding the homeostatic capacity may cause toxic effects. Both elements are constituents of enzymes protecting against oxidative damage in cells [10,11], and thus could be important for protecting brain and muscle cells implicated in normal neuromotor function. Selenium is quite abundant in a normal diet, being found in a diversity of foods, including meat, seafood, and grains [11]. A few studies in elderly populations reported that lower levels of blood selenium were associated with poorer coordination, slower motor speed [12], and weaker grip strength [13]. Elevated exposure to selenium can also exert toxic effects, and it has been estimated that a narrow range of intake separates deficiency from toxicity [14]. Selenium acute toxicity was reported in individuals who consumed misformulated supplement products [14], including symptoms of fatigue and muscle pain lasting several years after the consumption of the faulty supplements. However, high intake of selenium has not been linked to overt toxicity in populations living in areas with selenium-rich soils in the United States [15] and China [16].

An adequate intake of manganese is also necessary for normal function of the nervous system, with both deficiency and excess possibly related to adverse nervous system effects [10]. The main source of exposure to the general population is through the diet, with this element being present in many foods. The neurotoxicity of this metal has been well-described in workers exposed to airborne manganese particles, which can induce a neurodegenerative syndrome similar to Parkinson's disease [17]. Evidence of a neurotoxic effect in adults exposed to environmental sources (non-occupational) is sparse, but could occur in some populations, such as those living in the vicinity of metallurgical industries [18] and mines [19], and in those consuming manganese-contaminated drinking water [20]. However, no study has yet investigated potential nervous system deficits in relation to exposure to manganese in the general adult population with low, background exposure levels.

In summary, there are few reports on potential neurotoxicity associated with exposure to lead, mercury, selenium, and manganese in adults from the general population, as most studies focused on children or special populations with elevated exposure levels. Hence, the aim of our study was to examine the association between hand grip strength, a marker of neuromotor function, and blood concentrations of lead, mercury, selenium, and manganese in the general U.S. population. We relied on data from the U.S. National Health and Nutrition Examination Survey (NHANES), which provided measures of hand grip strength. Motor weakness is one of the most common effects of exposure to several neurotoxicants. Hence, the grip strength test is often used in the assessment of neurotoxicity in humans as well as in animal models [21].

We hypothesized that higher blood levels of lead and mercury might be associated with weaker hand grip strength. As selenium and manganese are essential nutrients at low levels, but can also have neurotoxic effects at higher levels, we hypothesized that the relation between blood concentrations and grip strength might display an inverse U-shaped curve.

2. Materials and Methods

2.1. Study Population

We used data collected during the cycles 2011–2012 and 2013–2014 of the NHANES, a nationally representative of the non-institutionalized U.S. population. Data were gathered from respondents through questionnaires, a medical examination in mobile examination, and laboratory analysis of biological samples. Among the 19,931 respondents of NHANES cycles 2011–2012 and 2013–2014, 6199 met the following inclusion criteria: (i) age between 20 and 79 years; (ii) data on grip strength of both hands; and (iii) data on blood concentrations of lead, mercury, manganese, and selenium. We excluded participants without valid hand grip measurements ($n = 180$ with incorrect arm/hand position during the test) and those who reported having had a hand surgery ($n = 203$). The study was approved by the National Center for Health Statistics Research Ethics Review Board and participants gave signed informed consent [22].

2.2. Measurement of Blood Metals

A sample of venous blood was collected from each survey participant during the physical examination; frozen at −20 °C; and shipped to the Division of Laboratory Sciences, National Center for Environmental Health, CDC (Atlanta, GA, USA) for analysis. The concentrations of lead, mercury, selenium, and manganese were measured using an inductively coupled plasma-mass spectrometer dynamic reaction cell (Elan ICP-DRC-MS instrument). State-of-the-art quality controls were applied, including standard reference materials for external calibration and spiked pools for internal quality control. NHANES reports blood concentrations for inorganic mercury and total mercury (the difference representing organic mercury). Although organic and inorganic forms of mercury can have a differential impact on the nervous system, both have neurotoxic properties. Hence, we only used total mercury in our analysis. Further details on the laboratory procedures are available elsewhere [23].

2.3. Measurement of Hand Grip Strength

Grip strength was evaluated using the Hand Dynamometer (Takei Digital gripper force gauge model T.K.K.5401) to obtain a measurement in kg of the maximum force exerted by hands. Before measurement of handgrip strength, the dynamometer was adjusted to participants' hand size. Participants were instructed to squeeze the dynamometer as hard as possible using one hand. The test was repeated three times for each hand, alternating hands with a 60 s rest between measurements on the same hand. The measures were considered valid when the participant was able to perform the test while standing and managed to form a 90° angle with the index on the dynamometer handle. In our analysis, we used the combined hands' grip strength, representing the sum of the largest reading from each hand. Further details on the grip strength test procedures are available elsewhere [24].

2.4. Potential Confounders

The potential confounding factors considered were as follows: age (in years), education (less than 9th grade; 9–12th grade (no diploma); high school graduate/GED equivalent; some college or associate degree; college graduate or above), race/ethnicity (non-Hispanic White; non-Hispanic Black; non-Hispanic Asian and other non-Hispanic groups; Mexican American and other Hispanics), family income to poverty ratio grouped into quartiles (<0.87, 0.87–1.65, 1.65–3.58, >3.58), body mass index (BMI, in kg/m^2), smoking (never, occasional, regular smoker), and alcohol consumption (no, yes). Participants who smoked less than 100 cigarettes in their lifetime were labelled as 'never smokers'; those who smoked cigarettes, pipes, cigars, little cigars, or electronic cigarettes during the past 5 days were 'regular smokers'; those who did not smoke in the last 5 days were 'occasional smokers'. For alcohol consumption, participants were grouped into 'yes' if they reported consuming >12 alcoholic beverages in the past year, and into 'no' otherwise.

2.5. Statistical Analyses

All analysis were stratified by sex because there is a large difference in hand grip strength between men and women and because previous studies have reported differential susceptibility to metal neurotoxicity in men and women (e.g., [25]). Given the large sample size, these sex-stratified analyses could be performed with sufficient statistical power. For descriptive purposes, we analyzed whether participant characteristics were associated with grip strength, using ANOVA or Kruskal Wallis test (for between-groups unequal and equal variance, respectively).

The main outcome was the combined hands' grip strength, representing the sum of the largest dynamometer reading from each hand. We used generalized regression models (GLMs) for complex survey samples to analyze the association between blood concentrations of elements (lead, mercury, selenium, and manganese) and grip strength, adjusting for confounding factors. Each element concentration in blood was modelled separately and a sensitivity analysis was carried out by including all four elements in the models to produce association estimates for mutually-adjusted elements. Blood concentrations of all elements were log10-transformed to normalize the distribution and entered as continuous values in models. Age, education, BMI, and race/ethnicity were including a priori in all models because they are strong predictors of grip strength [26]. Additional covariates were included in models when associated in univariate analysis with both grip strength and at least one blood concentration of the following elements (at $p < 0.20$): family income-to-poverty ratio, smoking status, and alcohol consumption. All models were adjusted for the same covariates. We also ran restricted cubic splines models to generate plots showing the shape of the associations, setting 5 knots, placed at the 5th, 25th, 50th, 75th, and 95th percentiles. This allows for a visual exploration of the data to detect thresholds in the associations, or inverse 'U-shaped' curve that may characterize associations for essential nutrients (i.e., manganese and selenium).

In addition to the GLMs on the outcome for combined hands' grip strength (analyzed as continuous values), we also ran analyses on grip strength categorized into more clinically relevant groups. In order to do this, we standardized grip strength measures for age and sex by calculating z-scores, and then we created two groups: 'low' and 'normal' grip strength, defined as a z-score ≤20th percentile and >20th percentile, respectively. This cut-off to define motor weakness based on hand dynamometer measures of grip strength is similar to previous studies (e.g., [27]). Binary logistic models for complex surveys were used to assess whether element blood concentrations were associated with the risk of low grip strength score (i.e., age- and sex-specific z-score below the 20th percentile). The same covariates were included as in GLMs (i.e., age, BMI, education, poverty to income ratio, alcohol, and smoking).

The sampling weights, strata, and primary sampling units created by the U.S. National Center for Health Statistics (NCHS) were applied to all statistical analyses according to NCHS guidelines to account for the complex, stratified multistage probability sample design of NHANES. All analyses were performed using R (version 3.5.0), using the packages survey and rms. p-values below 0.05 were considered to be statistically significant.

3. Results

In total, 6199 patients were included in our study: 3091 women (49.7%) and 3108 men (50.1%). Table 1 presents the characteristics of the study population by sex. The mean age was 45.9 (SD, 15.9) years for women and 45.8 (SD, 16.3) years for men. The average grip strength in men (mean, 88.9; SD, 17.8 kg) was much higher than that of women (57.0; 11.0 kg). Most participants reported never having smoked (66.5% of women and 49.1% of men). However, the majority of women and men reported drinking alcohol (58.8% and 80.4%, respectively).

Table 2 shows descriptive statistics on the concentrations of elements in blood of the women and men included in this study. Median concentrations were higher in men than women for lead, mercury, and selenium, but not for manganese. Blood lead median concentrations were 1.12 μg/dL and 1.68 μg/dL for women and men, respectively. Blood mercury median concentrations were 1.58 μg/L in women and 1.74 μg/L in men. The median blood selenium level was 193.4 μg/L in women and 198.2 μg/L in men. For manganese, women presented slightly higher blood levels than men, with median concentrations of 10.74 μg/L in women and 9.22 μg/L in men.

Table 1. Description of the study population stratified by sex.

Characteristic	Women (n = 3091)	Men (n = 3108)
Age (years); mean (SD)	45.9 (15.9)	45.8 (16.3)
BMI (kg/m^2); mean (SD)	29.3 (7.6)	28.4 (6.2)
Grip strength (kg); mean (SD)	57.0 (11)	88.9 (17.8)
Education level; n (%)		
Less than 9th grade	195 (6.3)	263 (8.5)
9–12th grade (no diploma)	374 (12.1)	424 (13.6)
High school graduate/GED equivalent	622 (20.1)	707 (22.8)
Some college or associate degree	1055 (34.1)	908 (29.2)
College graduate or above	845 (27.3)	806 (25.9)
Race/ethnic group; n (%)		
Non-Hispanic White	1192 (38.6)	1174 (37.8)
Non-Hispanic Black	744 (24.1)	758 (24.4)
Non-Hispanic Asian/other non-Hispanic	496 (16.1)	521 (16.8)
Mexican American and other Hispanic	659 (21.3)	655 (21.1)
Family income to poverty ratio; n (%)		
0.00–0.87	602 (19.5)	519 (16.7)
0.87–1.65	649 (20.9)	667 (21.5)
1.65–3.58	749 (24.2)	746 (24)
3.58–5.00	857 (27.7)	935 (30.1)
Missing data	234 (7.6)	241 (7.7)
Smoking status; n (%)		
Never	2057 (66.6)	1525 (49.1)
Occasional smoker	459 (14.9)	689 (22.2)
Regular smoker	493 (15.9)	806 (25.9)
Missing data	82 (2.6)	88 (2.8)
Alcohol consumption; n (%)		
No	1008 (32.6)	427 (13.7)
Yes	1816 (58.8)	2499 (80.4)
Missing data	267 (8.6)	182 (5.9)

Table 2. Descriptive statistics for element concentrations in blood for women (n = 3091) and men (n = 3108).

Blood Concentration	Min	Percentile 25	Median	Percentile 75	Max	SD	Geometric Mean
Lead (μg/dL)							
Women	0.11	0.55	0.88	1.36	19.4	1.06	0.89
Men	0.17	0.80	1.22	1.89	61.29	2.19	1.26
Mercury (μg/L)							
Women	0.11	0.42	0.82	1.76	36.99	2.29	0.89
Men	0.11	0.43	0.84	1.87	50.81	2.85	0.93
Selenium (μg/L)							
Women	105.4	177.2	191.0	206.3	734.8	28.88	191.6
Men	105.8	181.3	196.6	212.1	635.8	26.43	196.6
Manganese (μg/L)							
Women	1.88	7.82	9.85	12.69	62.51	4.40	10.02
Men	2.69	7.02	8.73	10.65	54.92	3.37	8.73

Table 3 presents the mean grip strength with respect to individual characteristics for men and women, respectively, and from univariate analysis. For both men and women, grip strength decreased significantly with age ($r = -0.38$ in men and -0.35 in women, indicating a decrease of 0.35 kg of grip strength per increase of 1 year of age in women for instance). When examining grip strength by age groups to allow for nonlinear relations, mean grip strength increased with age, peaking in the 30–39 years group, and then decreased in older age groups, similarly in men and women. Grip strength increased significantly with BMI in both sexes. Grip strength also varied significantly with education and race/ethnicity. Men and women in the lower education group (i.e., less than 9th grade) had poorer grip strength than those in the other groups. In addition, a stronger grip strength was observed in non-Hispanic black men and women compared with other race/ethnicity groups. In men, higher family income to poverty ratio was associated with stronger grip strength, but there was no such relation in women. Men and women who consumed alcohol had a lower grip strength than those who did not. Men and women who reported being regular smokers had stronger grip strength than the others.

Table 3. Grip strength with respect to individual characteristics in women and men.

Characteristic	Women (n = 3091)	Men (n = 3108)
	Pearson's r (IC95%)	
Age (years) [1]	−0.35 (−0.38, −0.32)	−0.38 (−0.41, −0.35)
BMI (kg/m^2) [1]	0.20 (0.16, 0.23)	0.18 (0.15, 0.22)
	Mean (SD)	
Age groups (years) [2]		
20–29	59.8 (10.3)	93.5 (17.3)
30–39	61.1 (10.4)	97.4 (17.4)
40–49	59.8 (11.1)	92.6 (15.6)
50–59	55.9 (9.8)	86.8 (15.3)
60–69	52.3 (9.1)	79.8 (16.1)
70–79	46.6 (9.5)	73.1 (13.4)
Education level [2]		
Less than 9th grade	50.8 (10.4)	78.9 (16.5)
9–12th grade (no diploma)	55.5 (11.1)	88.0 (19.3)
High school graduate/GED equivalent	56.4 (11.3)	90.4 (17.1)
Some college or associate degree	58.2 (11.2)	91.1 (17.7)
College graduate or above	58.1 (10.2)	88.9 (17.1)
Race/ethnicity [2]		
Non-Hispanic White	57.1 (10.7)	90.8 (17.5)
Non-Hispanic Black	61.9 (11.6)	93.2 (18.7)
Non-Hispanic Asian/other non-Hispanic	53.6 (9.8)	83.5 (16.1)
Mexican American and other Hispanic	53.9 (9.8)	84.8 (16.6)
Family income to poverty ratio [3]		
0.00–0.87	57.3 (11.7)	86.7 (18.1)
0.87–1.65	56.5 (11.5)	88.8 (18.4)
1.65–3.58	57.2 (11.1)	89.6 (18.1)
3.58–5.00	57.8 (10.3)	90.7 (16.8)
Missing	54.5 (10.0)	84.6 (17.9)
Smoking status [2]		
Never	56.6 (11.0)	89.9 (17.9)
Occasional smoker	56.6 (10.8)	84.8 (16.8)
Regular smoker	58.7 (11.3)	91.2 (17.7)
Missing	60.3 (10.8)	84.6 (17.9)
Alcohol consumption [2]		
No	58.1 (10.8)	89.9 (17.5)
Yes	55.1 (11.4)	84.8 (18.5)
Missing	60.3 (10.8)	83.0 (19.9)
All participants	57.0 (11.1)	88.9 (17.8)

[1] Significant association between age and BMI and grip strength for both sexes ($p < 0.001$). [2] Significant difference between characteristic's groups and grip strength for both sexes ($p < 0.001$). [3] In women, there was no significant difference in mean grip strength with respect to family income to poverty ratio ($p = 0.28$), but the difference was significant in men ($p < 0.001$).

The results of the GLMs for complex survey samples used to investigate associations between blood concentration of elements and grip strength are summarized in Table 4. A higher concentration of lead in blood was associated with significantly weaker grip strength in women ($p < 0.05$), with a 10-fold increase in lead being associated with lower grip strength by 2.37 kg (95% CI: −4.24, −0.50). Using the results showing that an increase in 1 year of age was associated with a 0.35 kg decrease in grip strength in women (Table 3), we could estimate that the −2.37 kg difference in grip strength is equivalent to about 6.8 years of aging (2.37/0.35). Blood lead was not associated with grip strength in men (β = 1.46 kg; 95% CI: −2.18, 5.10). Blood mercury concentration was not significantly associated with grip strength in either sex. The direction of the association estimate was positive for both men and women, indicating higher grip strength with higher blood mercury levels, but the confidence intervals were wide and far from significant. Higher blood selenium was significantly associated with stronger grip strength in women (β = 8.49 kg; 95% CI: 1.89, 15.10) and in men (β = 4.57 kg; 95% CI: −11.89, 21.03), but the association did not reach significance in the latter. Blood concentration of manganese was not significantly associated with grip strength in men or women.

Table 4. Change in grip strength for a 10-fold increase in the concentration of each blood element in men and women.

Blood Concentration	Women (n = 2609)		Men (n = 2706)	
	β (CI95%)	***p*-Value**	**β (CI95%)**	***p*-Value**
Lead (μg/dL)	−2.37 (−4.24, −0.50)	0.03	1.87 (−1.69, 5.43)	0.32
Mercury (μg/L)	0.18 (−0.89, 1.26)	0.75	1.38 (−0.22, 2.98)	0.11
Selenium (μg/L)	8.49 (1.89, 15.10)	0.02	4.57 (−11.89, 21.03)	0.59
Manganese (μg/L)	−2.08 (−4.76, 059)	0.15	−1.49 (−5.87, 2.88)	0.51

Note: Estimates are from GLMs for complex survey design for the association between blood concentrations of elements separately and grip strength, adjusting for age, BMI, education, race/ethnicity, income to poverty ratio, smoking, and alcohol consumption.

Figures 1 and 2 show the shape of the association between blood element concentrations and grip strength with cubic splines for women and men, respectively. It shows that the association between blood lead and grip strength in women appears linear, with a steady decrease in strength with increased blood lead levels (Figure 1a). For selenium in women, there is a steep increase in grip strength up until approximately 200 μg/L, then grip strength continues to increase with higher selenium, but the slope is less pronounced. The spline for blood mercury and grip strength was flat, indicating no association. In men, splines for blood lead and selenium suggest stronger grip strength with higher blood levels, whereas it was the reverse for manganese, but none of the association were very strong, consistent with the lack of significant associations in the GLM analysis. Similar to the results in women, the spline for blood mercury and grip strength was flat, indicating no association.

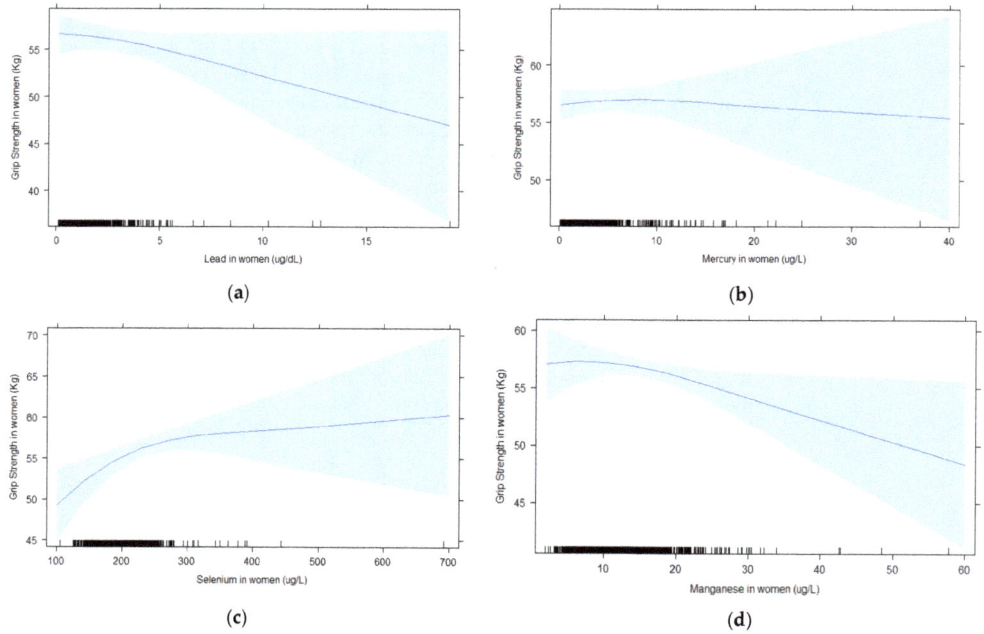

Figure 1. For women, associations between blood element concentrations and grip strength analyzed with splines (5 knots) for (**a**) lead, (**b**) mercury, (**c**) selenium, and (**d**) manganese. All models were adjusted for age, BMI, education, race/ethnicity, income to poverty ratio, smoking, and alcohol consumption.

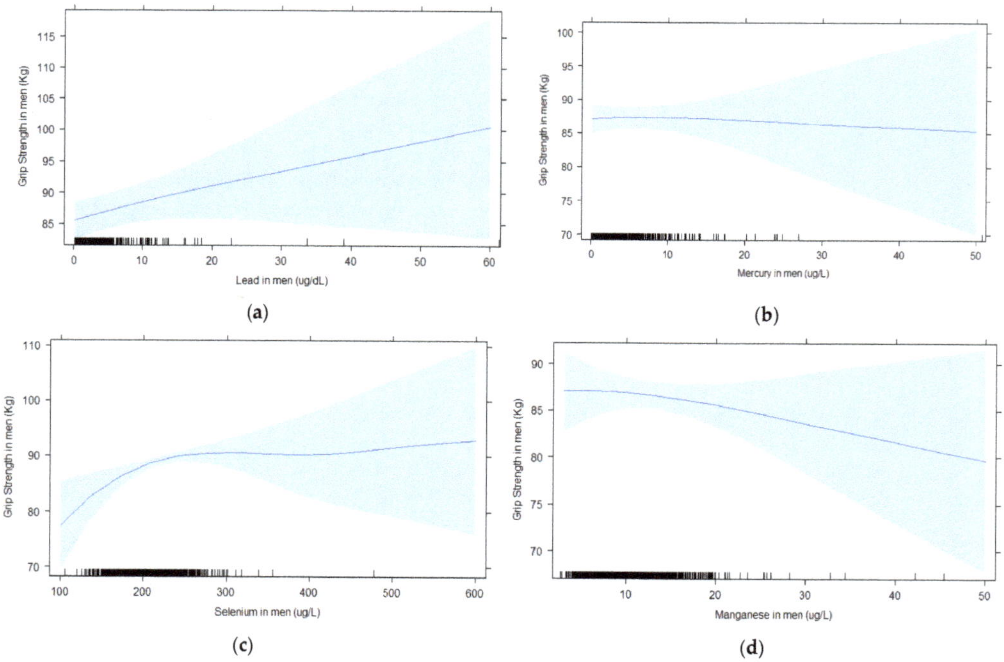

Figure 2. For men, associations between blood element concentrations and grip strength analyzed with splines (5 knots) for (**a**) lead, (**b**) mercury, (**c**) selenium, and (**d**) manganese. All models were adjusted for age, BMI, education, race/ethnicity, income to poverty ratio, smoking, and alcohol consumption.

We also analyzed grip strength as a binary variable, categorizing participants as having 'low' or 'normal' grip strength (Table 5). Consistent with the GLM results, a 10-fold increase in blood lead was significantly associated with an elevated risk of having low grip strength among women (OR = 1.76; 95% CI: 1.09, 2.84). For selenium, a 10-fold increase in blood levels was significantly associated with a decreased risk of having a low grip strength among women (OR = 0.03; 95% CI: 0.003, 0.4). Blood concentrations of mercury and manganese showed no association in these analyses.

Table 5. Odds of low grip strength for a 10-fold increase in blood concentration of elements.

Blood Concentration	Women (n = 2609)	Men (n = 2706)
	OR (95% CI)	OR (95% CI)
Lead (μg/dL)	1.76 (1.09, 2.84)	0.87 (0.58, 1.30)
Mercury (μg/L)	0.96 (0.69, 1.32)	0.77 (0.55, 1.06)
Selenium (μg/L)	0.03 (0.003, 0.4)	3.39 (0.27, 41.38)
Manganese (μg/L)	0.83 (0.38, 1.78)	1.53 (0.64, 3.67)

Note: Results are from logistic regressions for complex survey design for the association between blood concentration of elements and low grip strength (i.e., grip strength ≤ 20th percentile), adjusting for age, BMI, education, race/ethnicity, income to poverty ratio, smoking, and alcohol consumption.

Finally, we ran GLMs including all four elements and the results were similar to those from the single-exposure models (Table 6). The association estimates between blood lead and weaker grip strength and that between blood selenium and stronger grip strength in women remained of similar magnitude and statistically significant. Likewise, manganese and mercury were not associated with grip strength in these models for mutually-adjusted elements.

Table 6. Change in grip strength for a 10-fold increase in the concentration of blood elements, from models including all four elements.

Blood Concentration	Women (n = 2609)		Men (n = 2706)	
	β (CI95%)	p-Value	β (CI95%)	p-Value
Lead (μg/dL)	−2.50 (−4.40, −0.61)	0.02	1.46 (−2.18, 5.10)	0.45
Mercury (μg/L)	0.34 (−0.72, 1.39)	0.54	1.31 (−0.36, 2.97)	0.15
Selenium (μg/L)	8.19 (1.69, 14.69)	0.03	3.67 (−12.79, 20.14)	0.67
Manganese (μg/L)	−1.84 (−4.48, 0.79)	0.20	−1.87 (−6.37, 2.62)	0.43

Estimates are from GLMs for complex survey design for the association between blood concentrations of elements and grip strength, adjusting for age, BMI, education, race/ethnicity, income to poverty ratio, smoking, and alcohol consumption.

4. Discussion

In the present study among adults from the general U.S. population, we observed that higher concentrations of lead were associated with weaker grip strength in women, and this association appeared to be approximately linear. We observed no association between blood lead and grip strength in men, similar to a previous study conducted on older men (i.e., >65 years) from the general population [4]. Other studies also assessed neuromotor function in relation to lead exposure, but relied on different tests for assessing this outcome. For instance, an analysis of older adults (i.e., >58 years, NHANES 1999–2002) reported that walking speed decreased significantly with increasing blood lead [6]. Similar to our study, this association was observed in women, whereas null findings were observed in men. The adverse effects of high occupational lead exposure on neuromotor function are well-documented [2,3,28], and the present study suggests that this might also apply to women from the general U.S. population. The magnitude of the association estimate for a ten-fold increase in blood lead levels was equivalent to about 6.8 years of aging. These results were obtained after controlling for several important potential confounders, including age and socioeconomic status.

With respect to mercury, we did not observe a significant association between blood levels of this metal and grip strength. Moreover, further exploration of the data for a potential threshold effect (e.g., an association appearing at the most extreme values of blood mercury concentrations) did not reveal any indication of an effect of mercury in this population. Previous studies linking adverse effects of mercury on neuromotor function among adults were conducted in populations with much higher levels of exposure owing to the frequent consumption of mercury-contaminated fish [29] or occupational exposure [30]. We did not find any other study investigating mercury exposure in relation to neuromotor function in the adult general population, hence there is no comparable investigation to place our findings into perspective. The lack of association between blood mercury and grip strength is likely due to the low levels of exposure experienced in this population and/or to the well-documented confusion bias introduced by the good nutrients present in fish such as omega-3 fatty acids and selenium. Fish and seafoods are the most important source of exposure to mercury in the U.S. population [31].

Our study suggests an association between low blood selenium levels and weaker grip strength in both men and women, although the results were significant only for the latter. Our results are consistent with those from previous studies showing associations between low blood selenium levels and weaker grip strength in older women living in Baltimore, Maryland [13] and in older adults living in Chianti, Italy [32]. The same results were observed in another study that evaluated other indicators of neuromotor function such as upper and lower limb coordination [12]. Furthermore, our findings showing that performance on motor function increases more steeply at lower than higher blood concentrations was also observed in another study conducted among Spanish and American adults [27]. It is noteworthy that all previous studies on selenium and neuromotor function were conducted on older populations, hence the present study extends these findings to individuals of younger age groups. The stronger hand grip observed in individuals with higher blood selenium could be due to the role of selenoproteins in muscular function. Skeletal muscles are important sites of selenium storage, and selenoproteins are known to be involved in muscle function. Hence, mutations in the gene causing deficiency in selenoprotein N are known to cause inherited neuromuscular disorders characterized by generalized muscle atrophy and muscle weakness [33]. Our findings showing that individuals with low blood selenium had weaker grip strength suggest that selenium deficiency might be prevalent in the general U.S. population, manifesting with poorer neuromotor function.

Like selenium, manganese is also an essential element for human health. However, we did not observe a significant association between blood manganese and grip strength in the present study. We have not been able to identify any other studies that have investigated grip strength in relation to blood manganese in the general population. However, populations with high environmental exposure to manganese because of their proximity to polluting industries might display neuromotor deficits [18–20]. Furthermore, our exploration of the shape of the exposure–response relation did not reveal other potential association, such as an inverse U-shaped curve, that may characterize essential nutrients. The lack of association between low blood manganese and grip strength is not surprising, as adverse health effects due to manganese deficiency in humans have only been reported under experimental conditions [34]. This element is essential, being a constituent of several metalloproteins [10], but dietary sources are common and provide sufficient intake for most, if not all individuals in the population.

An important strength of the present study is the large sample, being representativeness of the general, non-institutionalized population. In our analysis, we applied adequate methods for complex surveys to ensure that the findings are generalizable to the general U.S. population. All the analyses were adjusted for important confounders, and the results were robust when all four elements were included in the same model (i.e., results for mutually-adjusted elements). Further studies might be useful to explore whether interactions between these different elements might occur. Our study has some limitations, including the cross-sectional study design, which does not allow to assess the temporality of associations. Another limitation is our reliance on blood as a biomarker for manganese and mercury, as other biological matrices might be better biomarkers of exposure. For instance, a review and meta-analysis has concluded that hair was a better biomarker of exposure to manganese than blood [35]. Likewise, to the extent that the largest and most concerning form of mercury exposure is methylmercury, measurements made in hair might be more useful in future studies aimed at detecting neurofunctional effects resulting from overexposure to this metal [36]. Finally, the present study focused on hand grip strength because no other measure was available in the survey to assess neuromotor function. However, grip strength is a measure of particular interest because compelling evidence from longitudinal studies indicates that it is predictive of disability [37,38] and of mortality [39].

5. Conclusions

The findings from the present study indicate that higher blood lead was associated with weaker grip strength in women, even at the low exposure levels encountered in this population. Furthermore, low blood selenium levels were associated with weaker grip strength, suggesting that some individuals might have selenium deficiency manifesting with poorer neuromotor function. There was no indication that circulating concentrations of manganese and mercury were associated with neuromotor strength at the levels of exposure experienced by this population.

Author Contributions: M.F.B. had the original study idea, supervised the statistical analysis, made critical edits to the manuscripts, provided funding for the study, and approved the final version of the manuscript. M.C.J.G. drafted the manuscript, carried out the statistical analysis of the data, and approved the final version of the manuscript. Both authors have read and agreed to the published version of the manuscript.

Funding: This research was supported by the Canada Research Chair Program (M.F.B. holds the Canada Research Chair on Environmental Contaminants and Populations' Health).

Institutional Review Board Statement: The NHANES study was approved by the U.S. National Center for Health Statistics (NCHS) Research Ethics Review Board (ERB approval #2011-17).

Informed Consent Statement: Informed consent was obtained from all subjects involved in the study.

Data Availability Statement: The data used in the present study are available at https://wwwn.cdc.gov/nchs/nhanes/Default.aspx (accessed on 3 November 2018).

Conflicts of Interest: The authors declare they have no actual or potential competing financial interests.

References

1. Hanna-Attisha, M.; Lanphear, B.; Landrigan, P. Lead Poisoning in the 21st Century: The Silent Epidemic Continues. *Am. J. Public Health* **2018**, *108*, 1430. [CrossRef]
2. Blond, M.; Netterstrom, B. Neuromotor function in a cohort of Danish steel workers. *NeuroToxicology* **2007**, *28*, 336–344. [CrossRef]
3. Iwata, T.; Yano, E.; Karita, K.; Dakeishi, M.; Murata, K. Critical dose of lead affecting postural balance in workers. *Am. J. Ind. Med.* **2005**, *48*, 319–325. [CrossRef]
4. Khalil, N.; Faulker, K.A.; Greenspan, G.L.; Cauley, J.A. Associations between bone mineral density, grip strength, and lead body burden in older men. *J. Am. Geriatr. Soc.* **2014**, *62*, 141–146. [CrossRef] [PubMed]
5. Grashow, R.; Spiro, A.; Taylor, K.M.; Newton, K.; Shrairman, R.; Landau, A.; Sparrow, D.; Hu, H.; Weisskopf, M. Cumulative lead exposure in community-dwelling adults and fine motor function: Comparing standard and novel tasks in the VA Normative Aging Study. *NeuroToxicology* **2013**, *35*, 154–161. [CrossRef]
6. Ji, J.S.; Elbaz, A.; Weisskopf, M.G. Association between Blood Lead and Walking Speed in the National Health and Nutrition Examination Survey (NHANES 1999–2002). *Environ. Health Perspect.* **2013**, *121*, 711–716. [CrossRef]
7. Grandjean, P.; Landrigan, P. Neurobehavioural effects of developmental toxicity. *Lancet Neurol.* **2014**, *13*, 330–338. [CrossRef]
8. Igata, A. Epidemiological and clinical features of Minamata disease. *Environ. Res.* **1993**, *63*, 157–169. [CrossRef] [PubMed]
9. Lebel, J.; Mergler, D.; Branches, F.; Lucotte, M.; Amorim, M.; Larribe, F.; Dolbec, J. Neurotoxic Effects of Low-Level Methylmercury Contamination in the Amazonian Basin. *Environ. Res.* **1998**, *79*, 20–32. [CrossRef]
10. Erikson, K.M.; Aschner, M. Manganese: Its Role in Disease and Health. *Met Ions Life Sci.* **2019**, *19*, 253–266. [CrossRef]

11. Rayman, M.P. Selenium and human health. *Lancet* **2012**, *379*, 1256–1268. [CrossRef]
12. Shahar, A.; Patel, K.V.; Semba, R.D.; Bandinelli, S.; Shahar, D.; Ferrucci, L.; Guralnik, J.M. Plasma selenium is positively related to performance in neurological tasks assessing coordination and motor speed. *Mov. Disord.* **2010**, *25*, 1909–1915. [CrossRef]
13. Beck, J.; Ferrucci, L.; Sun, K.; Walston, J.; Fried, L.P.; Varadhan, R.; Guralnik, J.M.; Semba, R.D. Low serum selenium concentrations are associated with poor grip strength among older women living in the community. *BioFactors* **2007**, *29*, 37–44. [CrossRef] [PubMed]
14. Morris, J.S.; Crane, S.B. Selenium Toxicity from a Misformulated Dietary Supplement, Adverse Health Effects, and the Temporal Response in the Nail Biologic Monitor. *Nutrients* **2013**, *5*, 1024–1057. [CrossRef] [PubMed]
15. Longnecker, M.; Taylor, P.R.; Levander, O.A.; Howe, M.; Veillon, C.; McAdam, P.A.; Patterson, K.Y.; Holden, J.M.; Stampfer, M.J.; Morris, J.S.; et al. Selenium in diet, blood, and toenails in relation to human health in a seleniferous area. *Am. J. Clin. Nutr.* **1991**, *53*, 1288–1294. [CrossRef] [PubMed]
16. Yang, G.Q.; Wang, S.Z.; Zhou, R.H.; Sun, S.Z. Endemic selenium intoxication of humans in China. *Am. J. Clin. Nutr.* **1983**, *37*, 872–881. [CrossRef]
17. Rodier, J. Manganese Poisoning in Moroccan Miners. *Occup. Environ. Med.* **1955**, *12*, 21–35. [CrossRef]
18. Butler, L.; Gennings, C.; Peli, M.; Borgese, L.; Placidi, D.; Zimmerman, N.; Hsu, H.-H.L.; Coull, B.A.; Wright, R.; Smith, D.R.; et al. Assessing the contributions of metals in environmental media to exposure biomarkers in a region of ferroalloy industry. *J. Expo. Sci. Environ. Epidemiol.* **2018**, *29*, 674–687. [CrossRef] [PubMed]
19. Rodríguez-Agudelo, Y.; Riojas-Rodríguez, H.; Ríos, C.; Rosas, I.; Pedraza, E.S.; Miranda, J.; Siebe, C.; Texcalac, J.L.; Santos-Burgoa, C. Motor alterations associated with exposure to manganese in the environment in Mexico. *Sci. Total Environ.* **2006**, *368*, 542–556. [CrossRef]
20. Kondakis, X.G.; Makris, N.; Leotsinidis, M.; Prinou, M.; Papapetropoulos, T. Possible health effects of high manganese concentration in drinking water. *Arch. Environ. Health* **1989**, *44*, 175–178. [CrossRef]
21. Maurissen, J.P.; Marable, B.R.; Andrus, A.K.; Stebbins, K.E. Factors affecting grip strength testing. *Neurotoxicol. Teratol.* **2003**, *25*, 543–553. [CrossRef]
22. CDC. *National Health and Nutrition Examination Survey. 2013–2014 Interviewer Procedures Manual 2013*; National Center for Health Statistics: Hyattsville, MD, USA, 2013.
23. CDC. *Lab Protocole for Blood Metals in Whole Blood. NHANES 2013–2014*; National Center for Health Statistics: Hyattsville, MD, USA, 2014.
24. CDC. *Muscle Strength Procedure Manual. NHANES 2013–2014*; National Center for Health Statistics: Hyattsville, MD, USA, 2013.
25. Mergler, D.; Baldwin, M.; Bélanger, S.; Larribe, F.; Beuter, A.; Bowler, R.; Panisset, M.; Edwards, R.; De Geoffroy, A.; Sassine, M.P.; et al. Manganese neurotoxicity, a continuum of dysfunction: Results from a community based study. *NeuroToxicology* **1999**, *20*, 327–342.
26. Lang, I.A.; Scarlett, A.; Guralnik, J.M.; Depledge, M.H.; Melzer, D.; Galloway, T.S. Age-Related Impairments of Mobility Associated with Cobalt and Other Heavy Metals: Data from NHANES 1999–2004. *J. Toxicol. Environ. Health Part A* **2009**, *72*, 402–409. [CrossRef] [PubMed]
27. García-Esquinas, E.; Carrasco-Rios, M.; Ortolá, R.; Prieto, M.S.; Pérez-Gómez, B.; Gutiérrez-González, E.; Banegas, J.; Queipo, R.; Olmedo, P.; Gil, F.; et al. Selenium and impaired physical function in US and Spanish older adults. *Redox Biol.* **2020**, *38*, 101819. [CrossRef]
28. Schwartz, B.S.; Lee, B.-K.; Bandeen-Roche, K.; Stewart, W.; Bolla, K.; Links, J.; Weaver, V.; Todd, A. Occupational Lead Exposure and Longitudinal Decline in Neurobehavioral Test Scores. *Epidemiology* **2005**, *16*, 106–113. [CrossRef] [PubMed]
29. Dolbec, J.; Mergler, D.; Passos, C.-J.S.; De Morais, S.S.; Lebel, J. Methylmercury exposure affects motor performance of a riverine population of the Tapajós river, Brazilian Amazon. *Int. Arch. Occup. Environ. Health* **2000**, *73*, 195–203. [CrossRef]
30. Harari, R.; Harari, F.; Gerhardsson, L.; Lundh, T.; Skerfving, S.; Strömberg, U.; Broberg, K. Exposure and toxic effects of elemental mercury in gold-mining activities in Ecuador. *Toxicol. Lett.* **2011**, *213*, 75–82. [CrossRef]
31. Wells, E.M.; Kpylev, L.; Nachman, R.; Radke, E.G.; Segal, D. Seafood, wine, rice, vegetables, and other food items associated with mercury biomarkers among seafood and non-seafood consumers: NHANES 2011–2012. *J. Expo. Sci. Environ. Epidemiol.* **2020**, *30*, 504–514. [CrossRef] [PubMed]
32. Lauretani, F.; Semba, R.D.; Bandinelli, S.; Ray, A.L.; Guralnik, J.M.; Ferrucci, L. Association of low plasma selenium concentrations with poor muscle strength in older community-dwelling adults: The InCHIANTI Study123. *Am. J. Clin. Nutr.* **2007**, *86*, 347–352. [CrossRef]
33. Castets, P.; Lescure, A.; Guicheney, P.; Allamand, V. Selenoprotein N in skeletal muscle: From diseases to function. *J. Mol. Med.* **2012**, *90*, 1095–1107. [CrossRef]
34. ATSDR. *Toxicological Profile for Manganese 2012*; US Department of Health and Human Services, Agency for Toxic Substances and Disease Registry: Atlanta, GA, USA, 2012.
35. Liu, W.; Xin, Y.; Li, Q.; Shang, Y.; Ping, Z.; Min, J.; Cahill, C.M.; Rogers, J.T.; Wang, F. Biomarkers of environmental manganese exposure and associations with childhood neurodevelopment: A systematic review and meta-analysis. *Environ. Health* **2020**, *19*, 1–22. [CrossRef] [PubMed]
36. Branco, V.; Caito, S.; Farina, M.; Da Rocha, J.B.T.; Aschner, M.; Carvalho, C. Biomarkers of mercury toxicity: Past, present, and future trends. *J. Toxicol. Environ. Health Part B* **2017**, *20*, 119–154. [CrossRef]

37. Rantanen, T.; Guralnik, J.M.; Foley, D.; Masaki, K.; Leveille, S.; Curb, J.D.; White, L. Midlife Hand Grip Strength as a Predictor of Old Age Disability. *JAMA* **1999**, *281*, 558–560. [CrossRef]
38. Cooper, R.; Kuh, D.; Cooper, C.; Gale, C.R.; Lawlor, D.A.; Matthews, F.; Hardy, R.; The FALCon and HALCyon Study Teams. Objective measures of physical capability and subsequent health: A systematic review. *Age Ageing* **2010**, *40*, 14–23. [CrossRef] [PubMed]
39. Cooper, R.; Kuh, D.; Hardy, R.; Mortality Review Group; on behalf of the FALCon and HALCyon Study Teams. Objectively measured physical capability levels and mortality: Systematic review and meta-analysis. *BMJ* **2010**, *341*, c4467. [CrossRef] [PubMed]

toxics

MDPI

Article

Methylmercury-Induced Metabolic Alterations in *Caenorhabditis elegans* Are Diet-Dependent

Nicole Crawford [1], Megan Martell [1], Tyson Nielsen [1], Belal Khalil [1], Farooq Imtiaz [1], Etienne Nguidjo [1], Jennifer L. Newell-Caito [2], Julia Bornhorst [3], Tanja Schwerdtle [4] and Samuel W. Caito [1,*]

1 Department of Pharmaceutical Sciences, Husson University School of Pharmacy, Bangor, ME 04401, USA; crawfordn@husson.edu (N.C.); martellm@husson.edu (M.M.); nielsent@husson.edu (T.N.); khalilb@husson.edu (B.K.); imtiazf@husson.edu (F.I.); etiennenguidjo08@yahoo.ca (E.N.)
2 Department of Molecular and Biomedical Sciences, University of Maine, Orono, ME 04469, USA; jennifer.newellcaito@maine.edu
3 Food Chemistry, Faculty of Mathematics and Natural Sciences, University of Wuppertal, 42119 Wuppertal, Germany; bornhorst@uni-wuppertal.de
4 Department of Food Chemistry, Institute of Nutritional Science, University of Potsdam, 14558 Potsdam, Germany; tanja.schwerdtle@bfr.bund.de
* Correspondence: caitos@husson.edu

Abstract: Methylmercury (MeHg) is a well-known neurotoxicant; however, its role in metabolic diseases has been gaining wider attention. Chronic exposure to MeHg in human populations shows an association with diabetes mellitus and metabolic syndrome (MS). As the incidences of both obesity and MS are on the rise globally, it is important to understand the potential role of MeHg in the development of the disease. There is a dearth of information on dietary interactions between MeHg and lipids, which play an important role in developing MS. We have previously shown that MeHg increases food seeking behaviors, lipid levels, fat storage, and pro-adipogenic gene expression in *C. elegans* fed the standard OP50 *Escherichia coli* diet. However, we hypothesized that these metabolic changes could be prevented if the worms were fed a bacterial diet lower in lipid content. We tested whether *C. elegans* developed metabolic alterations in response to MeHg if they were fed two alternative *E. coli* strains (HT115 and HB101) that are known absorb significantly less lipids from their media. Additionally, to explore the effect of a high-lipid and high-cholesterol diet on MeHg-induced metabolic dysfunction, we supplemented the OP50 strain with twice the standard concentration of cholesterol in the nematode growth media. Wild-type worms fed either the HB101 or HT115 diet were more resistant to MeHg than the worms fed the OP50 diet, showing a significant right-hand shift in the dose–response survival curve. Worms fed the OP50 diet supplemented with cholesterol were more sensitive to MeHg, showing a significant left-hand shift in the dose–response survival curve. Changes in sensitivity to MeHg by differential diet were not due to altered MeHg intake in the worms as measured by inductively coupled mass spectrometry. Worms fed the low-fat diets showed protection from MeHg-induced metabolic changes, including decreased food consumption, lower triglyceride content, and lower fat storage than the worms fed either of the higher-fat diets. Oxidative stress is a common characteristic of both MeHg exposure and high-fat diets. Worms fed either OP50 or OP50 supplemented with cholesterol and treated with MeHg had significantly higher levels of reactive oxygen species, carbonylated proteins, and loss of glutathione than the worms fed the HT115 or HB101 low-lipid diets. Taken together, our data suggest a synergistic effect of MeHg and dietary lipid levels on MeHg toxicity and fat metabolism in *C. elegans*, which may affect the ability of MeHg to cause metabolic dysfunction.

Keywords: methylmercury; diet; cholesterol; high fat; low fat

Citation: Crawford, N.; Martell, M.; Nielsen, T.; Khalil, B.; Imtiaz, F.; Nguidjo, E.; Newell-Caito, J.L.; Bornhorst, J.; Schwerdtle, T.; Caito, S.W. Methylmercury-Induced Metabolic Alterations in *Caenorhabditis elegans* Are Diet-Dependent. *Toxics* **2021**, *9*, 287. https://doi.org/10.3390/toxics9110287

Academic Editors: Richard Ortega and Asuncion Carmona

Received: 1 September 2021
Accepted: 29 October 2021
Published: 2 November 2021

1. Introduction

Metabolic syndrome (MS) and obesity are major health concerns with increasing prevalence worldwide. MS is defined as a multifactorial condition characterized by insulin resistance, diabetes mellitus (DM), dyslipidemia, and obesity. It has become increasingly evident that many factors influence the prevalence of MS, including environmental factors [1,2]. One such environmental agent emerging as a potential obesogen is methylmercury (MeHg). MeHg is a well-known neurotoxin, which, in developmental exposures, causes cognitive and behavioral dysfunction in children and is linked to the development of neurodegenerative diseases, such as Parkinson's disease [3,4]. While currently being debated, there is growing evidence for a link between MeHg exposure and the development of MS. The National Health and Nutrition Examination Survey (NHANES) and Korean NHANES (KNHANES) data from 2003–2014 and 2011–2013, respectively, support an association between blood heavy metal levels (which include Hg) with MS, obesity, and lipid dysregulation [5–7]. These studies highlight the effect of heavy metals on the development of MS; however, it is unclear as to whether the observed metabolic effects were due to a single metal or a synergism of multiple metals in the mixture. However it has been shown that elevated blood mercury levels are also associated with increased visceral adipose tissue [8], and that toenail mercury levels (a marker for chronic Hg exposure) are associated with the development of MS [9]. We have recently shown that MeHg significantly increased lipid storage, altered feeding behavior, and increased the transcription of MS-related genes in *Caenorhabditis elegans* (*C. elegans*) [10]. Mechanisms that lead to MeHg-induced dyslipidemia are not known.

As a toxicant, MeHg primarily enters the human body through our diet. MeHg is a major contaminant of our fish supply, with greater Hg concentrations present in fish higher up the food chain, such as tuna, shark, swordfish, and mackerel [11]. While there are many dietary benefits from regular fish consumption, such as increased polyunsaturated fatty acids (PUFA) and selenium intake, the level of Hg ingested is an important consideration, especially for children and pregnant women. Studies have shown that dietary factors can affect how much Hg enters the body and its toxicity. MeHg enters cells through a molecular mimicry mechanism. MeHg readily binds to thiol groups. When bound to the amino acid cysteine, the MeHg–cysteine molecule resembles the amino acid methionine, and is able to enter cells through the large amino acid transporter 1 and 2 (LAT1 and LAT2) [12]. In worms, the amino acid transporters 1–3 are homologs to LAT1 and LAT2, which transport MeHg into cells [13]. If worms are fed a diet enriched in methionine, MeHg transport into cells is significantly decreased, as well as its toxic effects [13]. Similar effects have been observed in mammalian systems [14,15]. In addition to binding thiols, MeHg has high affinity for selenium. In diets enriched in selenium, MeHg will bind selenium and selenoproteins rather than thiols, preventing glutathione depletion and MeHg toxicity [16,17].

The relationship between dietary fats and methylmercury exposure has gained much attention. Polyunsaturated n-3 fatty acids, such as eicosapentaenoic acid (20:5n-3, EPA) and docosahexaenoic acid (22:6n-3, DHA), have multiple health benefits, from lowering serum low-density lipoprotein levels to being cardio-protective and preventing metabolic diseases [18,19]. These fatty acids are high in fish species that also contain significant Hg levels. Therefore, understanding the relationship between PUFA and Hg consumption is important. Longitudinal studies performed in the Seychelle Islands have shown beneficial effects on cognition in children from maternal exposure to PUFAs found in fish contaminated with MeHg [20–22]. n-3 PUFAS may protect against MeHg toxicity, either by decreasing apoptosis or by reducing MeHg uptake [23]. Interestingly, a meta-analysis has shown that high circulating n-3 PUFA levels or fish consumption correlate with a lower risk of developing metabolic syndrome [24]. While PUFAs are an important type of fatty acid, our human diets can be varied and contain multiple other types of lipids. Little is known about the effects of other dietary lipids on MeHg toxicity. As high-total-fat, high-saturated-fat, and high-cholesterol diets are all implicated in the development of metabolic syndrome, we were interested in whether changing the bacterial strain fed to *C. elegans* would affect the worm's response to MeHg. We hypothesized that high-fat diets would synergize the metabolic dysfunction caused by MeHg exposure. To test this hypothesis, we exposed worms to MeHg and fed them one of four test *E. coli* diets: the standard diet strain (OP50), HB101 or HT115 (two diets previously shown to cause lower triglyceride accumulation than OP50), or a high-cholesterol diet (OP50 grown on plates containing twice the standard concentration of cholesterol). We then tested for MeHg lethality and Hg accumulation. We determined that the low-fat diets were more protective against MeHg lethality than OP50 and the high-cholesterol diet despite equivalent Hg accumulation. We then examined intracellular triglyceride content and lipid accumulation, as well as pro-adipogenic gene expression in worms exposed to MeHg and fed the test diets. As feeding in *C. elegans* is linked to specific neurobehavior, we assessed feeding and locomotor behaviors controlled by the dopaminergic, serotonergic, and glutamatergic neurotransmitter systems in worms exposed to MeHg and fed the test diets. Finally, as oxidative stress is an important determinant in neurotoxicity and metabolic toxicity, we measured reactive oxygen species (ROS) levels, protein carbonyl content, glutathione content, and antioxidant response element activation following MeHg exposure and test diet feeding.

2. Materials and Methods

2.1. Reagents

Unless otherwise stated, all reagents were obtained from Sigma-Aldrich (St. Louis, MO, USA). Primers used in this study included *tba-1* (F: AGACCAACAAGCCGATGGAG, R: TCCAGTGCGGATCTCATCAAC), *cebp-1* (F: CACTGACATGCCGAACAACG, R: AGAGAGTCTTGTCTTGCGAAGG, *sbp-1* (F: GGCGGCGAAGATTGTGATTC, R: CACTGACATGCCGAACAACG), *fat-6* (F: AGAGGAGAGCAAGAAGATCCCA, R: TCACGGTTTGCCATTTTGCC), and *vit-2* (F: TGATGAGTCCACCAACGAGTTC, R: TTGCTCCTCGTCTCTCTCGT).

2.2. C. elegans Strains and Worm Maintenance

C. elegans strains were maintained at 20 °C on Nematode Growth Medium (NGM) plates seeded with either *Escherichia coli* strains OP50, HT115, or HB101, as previously described [25]. Additionally, worms were maintained on a 2x cholesterol NGM plate (10 mg/mL cholesterol) seeded with OP50. *C. elegans* are cholesterol auxotrophs; studies have shown that above 5 mg/mL cholesterol levels are high in the nematodes [26]. For the majority of the study, wild-type N2 worms were used. We also used the VP596 (dvls19[pAF15(gst-4::GFP [green fluorescent protein]::NLS)];vsls33[dop-3::RFP (red fluorescent protein)] strain to measure antioxidant response element activity. Both strains were obtained from the Caenorhabditis Genetic Center (CGC; University of Minnesota). The bleaching method was used to harvest eggs for synchronous L1 populations, as previously described [27]. Briefly, embryos were isolated from gravid worms using a bleaching solution (1% NaOCl and 0.25 M NaOH) followed by a sucrose gradient to segregate eggs from worm and bacterial debris. Synchronized L1 worms were treated with MeHg for 30 min in M9 liquid buffer at 25 °C on a tube rotator, and then plated on the NGM plates seeded with the differing *E. coli* diets. We have previously shown that these concentrations are below the LD_{50} for MeHg in *C. elegans* and correlate to concentrations of MeHg in the worm that are below the US EPA reference dose of 0.1 μg/kg/d [28,29].

2.3. Dose–Response Survival Curves

The lethal dose 50% (LD_{50}) of MeHg for N2 *C. elegans* strains fed differing diets was determined by treating 5000 synchronized L1 worms with doses ranging from 1 to 200 μM MeHg for 30 min in M9 liquid buffer at 25 °C on a tube rotator. All exposures were carried out in triplicate and repeated 5 times. After treatment, worms were washed 3 times with M9 buffer, transferred to OP50-, HT115-, or HB101-seeded NGM plates or OP50-seeded 2x cholesterol NGM plates, and manually counted for lethality 24 h after MeHg treatment.

2.4. Mercury Quantification

Inductively coupled mass spectrometry (ICP-MS, Agilent 8800 ICP-QQQ) was used to measure intraworm concentrations of Hg. A total of 50,000 worms per sample were treated with MeHg and then fed for 48 h on one of the four test diets before washing with M9 and flash-freezing in liquid nitrogen. The samples were then sonicated. After centrifugation, an aliquot of the supernatant was used to measure protein concentration via the BCA assay. The rest of the sample was digested in the microwave with 1.6 mL bidest H_2O, 250 µL HNO_3 suprapur® and 250 µL HCl suprapur®. Hg content was measured with No gas mode ICP-MS. Rhodium (0.01 µg/L) was used as internal standard. The calibration was prepared in 10% HNO_3 suprapur® and 10% HCl suprapur® using a concentration range of 1–300 ng/L. The washout solution contained 1 ppm gold in 5% HNO_3 and 5% HCl. The content of Hg was calculated by dividing total Hg by total protein (ng Hg/mg protein).

2.5. Triglyceride Quantification

The Enzychrom™ triglyceride quantification kit (BioAssay Systems, Hayward, CA, USA) was used to measure total intracellular triglycerides. Following the MeHg treatment, 200,000 worms were fed the test diets for 48 h and were homogenized in triglyceride assay buffer. Extracts were incubated for 30 min at room temperature with the triglyceride assay reagent mix, and absorbency (optical density: 570 nm) was read. Data are expressed as mmol triglycerides/µg protein.

2.6. Nile Red Staining

Previously, we have shown that fat storage sites are increased by MeHg through two methods, BODIPY 493/503 and Nile Red [10]. As the Nile Red method amends itself to screening multiple treatment groups, we chose to quantify fat storage sites using this method. Twenty thousand L1 worms were incubated with MeHg, washed, and were transferred to OP50-, HT115-, or HB101-seeded NGM plates or OP50-seeded 2x cholesterol NGM plates. Seventy-two h after treatment, worms were washed off the plates and were fixed for Nile Red staining, as previously described [30]. One thousand worms were first washed with 0.1% triton in PBS, and then fixed in 40% isopropanol for 3 min. Fixed worms were next incubated with 3 µg/mL Nile Red in 40% isopropanol for 30 min followed by another M9 wash step to remove excess dye. Worms were loaded onto a 96-well plate and Nile Red fluorescence was read at excitation 560 nm, emission 590 nm. Data were normalized to worm number and protein levels.

2.7. RNA Isolation and Real-Time qPCR Gene Expression

RNA from 20,000 worms per treatment was isolated using Trizol solution followed by chloroform extraction. cDNA was then synthesized from 1 mg of total RNA using the Appled Biosystems' High-Capacity cDNA Reverse Transcription Kit (Thermo Fisher Scientific). Real-time PCR analysis was then performed using PerfeCTa SYBR Green FastMix (QuantaBio, Beverly, MA, USA).

2.8. Feeding Behavioral Analysis

L1 worms were seeded on OP50-, HT115-, or HB101-spread NGM plates, or OP50-spread 2x cholesterol NGM plates following MeHg treatment and were assessed 72 h post-treatment for behaviors associated with nematode feeding; these include pharyngeal pumping, locomotion, and the basal slowing response. For pharyngeal pumping, 10 worms were transferred to fresh NGM/2x cholesterol NGM plates spread with the corresponding *E. coli* diet and the number of pharynx pumps was counted for 30 s. Locomotion was assessed by the body-bend assay: worms were plated on an unseeded NGM/2x cholesterol NGM plate and scored for the number of forward-directed body-bends during a 30 s timespan. The basal slowing response assay was used to measure dopamine-dependent behavior that mediates the worm's slowing movement to consume food. The basal slowing response assay was performed as previously described [31]. The number of forward-directed body-bends was scored for worms placed either on NGM or 2x cholesterol NGM plates seeded or unseeded with OP50, HB101, or HT115 *E. coli*. For all behavioral assays, 2x cholesterol plates were only used for experiments where worms were fed OP50 grown on 2x cholesterol plates. These data are presented as the change in body-bends, calculated by subtracting the number of body-bends of worms plated on *E. coli*-seeded plates from the number of body-bends of worms plated on unseeded plates.

2.9. Glutathione Quantification

The 5,5′-dithiobis-2-nitrobenzoic acid–GSH disulfide reductase recycling method was used to measure total intracellular glutathione (GSH) levels, as previously described [32] in whole worm extracts from 30,000 worms.

2.10. Intracellular Reactive Oxygen Species Determination

Intracellular reactive oxygen species (ROS) were measured using 2,7-dichlorodihydrofluorescein diacetate (DCFD), as previously described [33]. Briefly, 20 worms treated with MeHg and fed on the test diet for 72 h were loaded onto a black 96-well plate and treated with 25 μM DCFDA. Green fluorescence (excitation 490 nm, emission 520 nm) was read immediately and subsequently every 30 min for 6 h.

2.11. Protein Oxidation Quantification

In this study, 2,4-dinitrophenylhydrazine (DNPH) labeling was used to measure protein carbonylation, a type of protein oxidation, as previously described. Using Yasuda et al.'s method [34], 50,000 treated and test-diet-fed worms were sonicated in 5 mM EDTA with protease inhibitors. Protein was precipitated out of solution using 20% trichloroacetic acid and then incubated with 10 mM DNPH for 1 h. After excess DNPH was washed off, samples were suspended in 6 M guanidine hydrochloride, loaded onto a 96-well plate, and absorbance was read at 380 nm. Concentration of oxidized protein was calculated using Beer–Lambert's law (molar absorptivity coefficient of DNPH is 21 $mM^{-1}cm^{-1}$). Data were normalized to total protein concentration.

2.12. Oxidative Stress Reporter Assay

Activation of the antioxidant response element was measured using the VP596 strain, which expresses GFP under the control of the promoter for the SKN-1 target GSH S transferase 4 (*gst-4*). SKN-1, the worm homolog of nuclear factor (erythroid-derived-2)-like 2 (Nrf2), is a transcription factor that binds the antioxidant response element and transcribes antioxidant genes in response to environmental insults. VP596 worms also express RFP under the *dop-3* promoter. L1 VP596 worms were treated with MeHg (10 or 20 μM) for 30 min, washed, and transferred to agar plates to be maintained for 72 h on the different diets. Worms were then washed off the plates, loaded onto a 96-well plate, and levels of RFP and GFP florescence were measured (RFP: excitation 544 nm, emission 590 nm; and GFP: excitation 485 nm, emission 520 nm). Antioxidant response element activity was represented as GFP florescence divided by RFP florescence (normalization to worm number).

2.13. Statistics

Statistical analyses were performed using Prism 8 software (Graphpad, San Diego, CA, USA). Statistical analysis of significance was carried out either by Student's *t*-test of LD_{50} (Figure 1) or two-way analysis of variance (ANOVA) followed by Tukey's multiple comparisons test. Values of $p < 0.05$ were considered statistically significant.

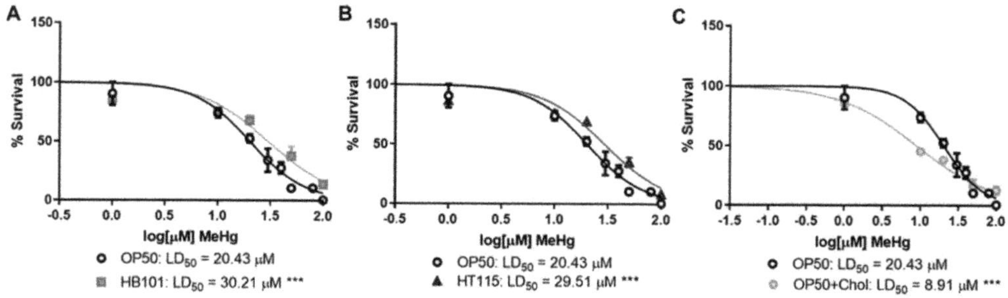

Figure 1. Diet affects MeHg toxicity. N2 worms were treated with increasing concentrations of MeHg for 30 min and then transferred to NGM plates seeded with either OP50 or (**A**) HB101, (**B**) HT115, or (**C**) OP50 supplemented with cholesterol. Dose–response survival curves were generated and LD_{50} values were calculated from five independent experiments. *** $p < 0.001$ as compared with N2 MeHg-treated worms fed OP50.

3. Results

3.1. Bacterial Diet Affects MeHg Toxicity

As diet is a major environmental factor in health and disease development, we examined whether altering the strain of *E. coli* fed to wild-type N2 worms would affect MeHg toxicity. The standard *E. coli* diet used in nematode culture is OP50. This strain was originally selected due to its ability to form a thin, transparent monolayer, allowing for ease of visualization of the worms under a microscope [35]. How well the OP50 strain emulates the nutrition that a *C. elegans* worm would receive in the wild has not been accurately determined. In addition to the standard OP50 diet, we selected two diets previously shown to be lower in dietary lipids: HB101 and HT115 [36]. In comparison to OP50, worms fed HB101 had ~20% fewer free fatty acids and 50% fewer triglycerides [36]. Furthermore, the fatty acid content of triglycerides in worms fed HB101 contained ~50% fewer branched fatty acids and significantly increased the monounsaturated fatty acid percentage in total fatty acids than worms fed OP50 [36]. In comparison to worms fed OP50, worms fed HT115 had ~50% less triglycerides but did not have significant differences in total free fatty acid content or fatty acid content [36]. Similar to HB101, worms fed HT115 had 50% fewer branched-chain fatty acids in their triglycerides as compared to OP50 worms. Feeding worms either HT115 or HB101 has no effect on the mean lifespan of *C. elegans*, but does result in decreased basal fat storage of dietary lipids [36]. Lastly, we created a high-cholesterol diet by feeding OP50 *E. coli* twice the standard cholesterol concentration in NGM plates. We exposed N2 worms to increasing concentrations of MeHg, plated them on the four different diets, and generated dose–response survival curves (Figure 1). Worms fed either HB101 or HT115 were more resistant to the toxic effects of MeHg, exhibiting a right-hand shift in their dose–response curves as compared to N2 fed OP50 (LD_{50}s of 30.21 and 29.51 μM for HB101 and HT115, respectively, as compared to 20.43 μM for OP50). In contrast, worms fed the 2x cholesterol OP50 diet were more sensitive to MeHg, showing a left-hand shift in their dose–response curves, as compared to N2 fed OP50 (LD_{50} of 8.91 μM). These data suggest that a bacterial diet affects the toxicity of MeHg in nematodes.

3.2. Diet Did Not Alter Mercury Accumulation

Dietary components have been shown to affect the accumulation of MeHg; for example, the amino acid methionine competes with MeHg for passage through the large amino acid transporter into cells [12]. We therefore were interested in whether the differential toxicity to MeHg among the worms fed the four diets was due to differential accumulation of MeHg in the worms. N2 worms were treated with either 10 or 20 μM MeHg and plated for 48 h on NGM plates that contained one of the four test diets. Levels of Hg in the worms were quantified by inductively coupled mass spectrometry (ICPMS), as previously described [37]. N2 worms fed either the HB101, HT115, or 2x cholesterol OP50 diet accumulated similar levels of Hg following the 10 or 20 μM MeHg treatments as compared to the worms fed OP50 and treated with 10 or 20 μM MeHg (Figure 2). Worms treated with 10 μM MeHg and fed HT115 appeared to have lower levels of Hg accumulation than worms fed OP50 and treated with MeHg; however, this trend was not statistically significant. This suggests that there was no difference in transport, accumulation, or elimination between the worms fed the four diets that could account for the variance in the dose–response curves seen in Figure 1.

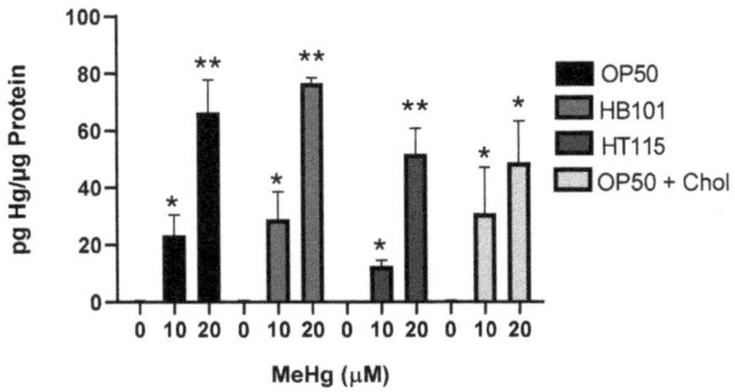

Figure 2. Bacterial diet does not affect MeHg content in worms. Hg content was measured by ICP-MS in worms fed OP50, HT115, HB101, or OP50+ cholesterol diets 48 h after MeHg treatment. Hg levels are expressed as pg Hg/μg protein. Data represent four independent experiments. * $p < 0.05$, ** $p < 0.01$ as compared to untreated, OP50-fed control.

3.3. *Bacterial Diet Altered Lipid Accumulation in Response to MeHg*

We have previously shown that MeHg increases the triglyceride content of N2 worms fed OP50 [10]. As HB101, HT115, and 2x cholesterol OP50 contained different lipid profiles to OP50, we were interested in whether the worms would accumulate lipids in response to MeHg at similar levels. L1 N2 were treated with 10 or 20 μM MeHg for 30 min, and were allowed to feed on one of the four test diets and mature for 48 h before extraction of triglycerides. MeHg dose-dependently increased the total triglyceride content in N2 worms (Figure 3A–C). Triglyceride levels in the worms fed HB101 or HT115 were significantly lower than those of the N2 worms treated with 20 μM (Figure 3A,B), while worms fed 2x cholesterol OP50 contained significantly more triglycerides in response to 10 and 20 μM MeHg than the worms fed OP50 (Figure 3C).

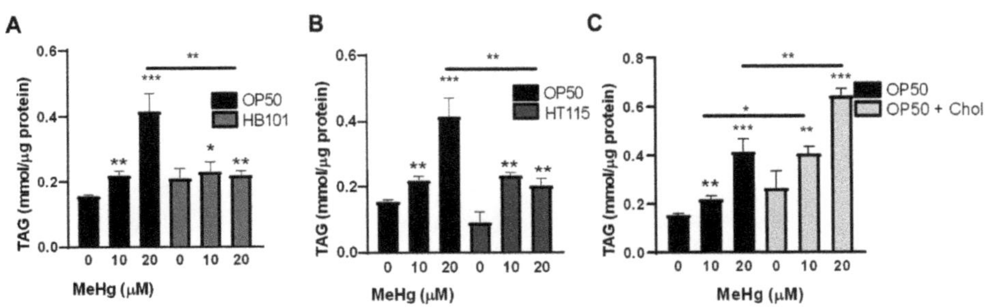

Figure 3. MeHg increases triglyceride content in worms fed a high-fat, but not low-fat, diet. Total triglycerides were measured in lysates from N2 worms fed OP50 or (**A**) HB101, (**B**) HT115, or (**C**) OP50 supplemented with cholesterol 48 h after MeHg treatment. Data are expressed as mean triglycerides mmol/μg protein ± SEM. All data are representative of five independent experiments. * $p < 0.05$, ** $p < 0.01$, *** $p < 0.001$ as compared to untreated, OP50-fed control. Horizontal bars represent comparisons between Hg-treated worms fed different diets.

We next assessed whether a bacterial diet would affect intracellular lipid storage sites. N2 worms were treated with MeHg, fed one of the four test diets, and, 72 h after exposure, lipid storage sites were stained with Nile Red, as previously described [30]. Both 10 and 20 μM MeHg treatments increased lipid storage in N2 worms (Figure 4A–C). Worms that were fed either the HB101 or HT115 diets showed decreased levels of lipid storage sites in response to MeHg than worms fed OP50 (Figure 4A,B). However, worms fed OP50 with 2x cholesterol showed increased lipid storage sites in response to MeHg compared to worms fed OP50 with standard cholesterol levels (Figure 4C). Taken together these data suggest that lipid accumulation in response to MeHg is altered by the lipid content of the diet.

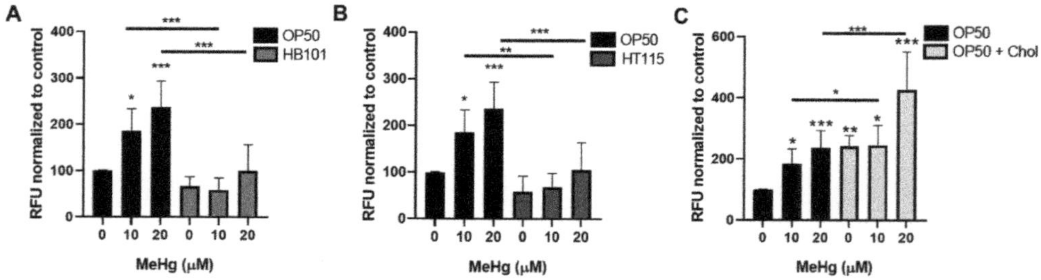

Figure 4. MeHg-induced fat accumulation in worms fed high-fat, but not low-fat, diets. Worms were treated with MeHg for 30 min and placed on NGM plates containing OP50 or (**A**) HB101, (**B**) HT115, or (**C**) OP50 supplemented with cholesterol. Then, 72 h after treatment, worms were fixed, stained with Nile Red, and fluorescence was measured. Data represent mean Nile Red fluorescence normalized to worm number and protein content ± SEM from five independent experiments. * $p < 0.05$, ** $p < 0.01$, *** $p < 0.001$ as compared with untreated, OP50-fed control. Horizontal bars represent comparisons between Hg-treated worms fed different diets.

3.4. MeHg-Induced Pro-Adipogenic Gene Transcription Is Diet-Dependent

We have previously shown that MeHg induces the expression of several genes involved in lipid accumulation and metabolic disease [10]. MeHg increases the expression of *cebp-1* (worm homolog of C/EBP) in worms fed OP50. C/EBP is a regulator of adipocyte differentiation, hyperplasia (increase in adipocyte cell numbers), and hypertrophy (adipocyte cell size) [38,39]. MeHg exposure increased the expression of *cebp-1* in N2 worms fed OP50 72 h post-treatment as compared to untreated controls (Figure 5A). Worms fed with either HB101 or HT115 showed significantly lower expression of *cebp-1* than worms fed OP50. However, *cebp-1* expression levels were not significantly different between the worms fed OP50 or OP50 with 2x cholesterol, suggesting that *cebp-1* induction in *C. elegans* is not dependent on the dietary cholesterol level.

Working in concert with C/EBP to induce adipocyte differentiation in mammals is sterol response element binding protein (SREBP, *sbp-1* in worms) [38,39]. We have previously shown that *sbp-1* expression is increased 24 h after MeHg treatment in worms fed OP50. Here, we report that *sbp-1* levels in response to 10 or 20 μM MeHg remain elevated 72 h post-exposure in OP50-fed worms (Figure 5B). The expression of *sbp-1* followed a similar trend as *cebp-1* in the worms fed HT115 and HB101. These worms had significantly lower expression of *sbp-1* in response to MeHg than the OP50-fed worms. In mammalian systems, the SREBP transcription factor is regulated by dietary cholesterol levels; when there is low cholesterol, the transcription factor is active; however, when there are high levels of cholesterol in cells, SREBP is degraded and the transcription factor is inactivated [40]. Similar regulatory mechanisms are present in worms. Our data show that worms fed OP50 grown on 2x cholesterol plates had no induction of the *sbp-1* gene in response to MeHg.

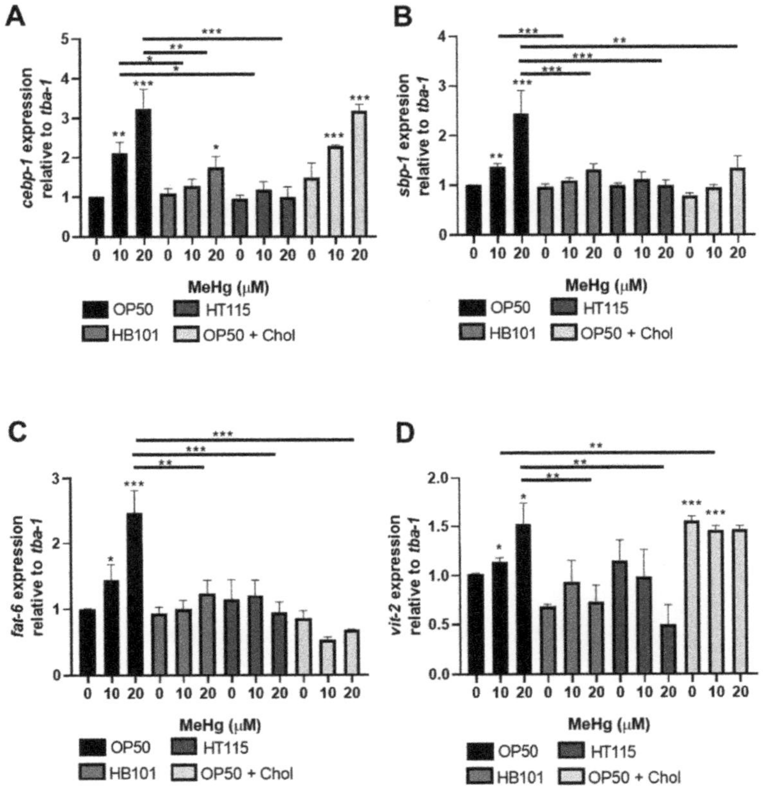

Figure 5. MeHg-induced pro-adipogenic gene expression is dependent on diet. Worms were treated with MeHg for 30 min and placed on NGM plates containing OP50, HB101, HT115, or OP50 supplemented with cholesterol. Then, 72 h after treatment, levels of (**A**) *cebp-1* (ortholog of human C/EBP), (**B**) *sbp-1* (ortholog of human SREBP-1), (**C**) *fat-6* (ortholog to glycerol-3-phosphate acyltransferase), (**D**) *vit-2* were measured by quantitative PCR and normalized to *tba-1* housekeeping gene. Data are expressed as mean relative expression ± SEM from five independent experiments. * $p < 0.05$, ** $p < 0.01$, *** $p < 0.001$ as compared to untreated OP50-fed control. Horizontal bars represent comparisons between Hg-treated worms fed different diets.

Concurrent with the upregulation of adipogenic transcription factor *sbp-1*, MeHg increased the expression of an *sbp-1*-responsive gene, *fat-6*, the worm ortholog of stearoyl-CoA desaturase 1 (SCD). SCD is the rate-limiting step in the formation of monounsaturated fatty acids and triglycerides. Due to its critical roles in in obesity and insulin resistance, SCD is emerging as a potential therapeutic target for these conditions [41]. The transcription of *fat-6* is controlled by multiple transcription factors shown to be affected by MeHg, including *nhr-49* and *sbp-1* [42,43]. As MeHg increases *sbp-1* expression, it is expected that *sbp-1* target genes, such as *fat-6* expression, would also change. Worms treated with MeHg and fed OP50 had increased expression of *sbp-1* (Figure 5B) and increased expression of *fat-6* (Figure 5C). Likewise, worms fed HB101 or HT115 after MeHg treatment had decreased *sbp-1* and *fat-6* expression as compared to OP50-fed worms treated with MeHg. Furthermore, worms fed OP50 raised on NGM plates containing 2x cholesterol did not display *sbp-1* expression or *fat-6* expression, further confirming the relationship of cholesterol levels and *sbp-1* activity in response to MeHg in worms.

Finally, we examined whether diet affected lipid transport proteins in response to MeHg in *C. elegans*. Vitellogenins (*vit-1*–*vit-6*) are yolk proteins with homology to human apolipoprotein B-100 [44] that deliver cholesterol to oocytes through a receptor-mediated endocytosis mechanism mediated by RME-2, a member of the LDL receptor superfamily. We have previously shown that the levels of *vit-2* were increased by MeHg treatment 24 h post-exposure in N2 worms fed OP50 [10]. In Figure 5D, *vit-2* is increased by MeHg 72 h post-exposure in OP50-fed worms. In worms fed HB101 or HT115, MeHg did not induce *vit-2* expression. Lastly, worms fed OP50 grown on 2x cholesterol NGM plates had increased expression of *vit-2* compared to MeHg-treated worms fed OP50 grown on standard NGM plates. Overall, these data suggest that the different diets affected lipid accumulation in the worms in response to MeHg, which were accommodated by a compensatory modulation of lipid binding and transport proteins.

3.5. Feeding Behavior in Response to MeHg Is Dependent on Diet

Feeding in *C. elegans* is linked to specific neuronal activity. We previously showed that MeHg, a known neurotoxin, increased feeding in worms as well as decreased locomotion and dopaminergic behavior [10]. As we noted differences in fat accumulation between worms fed the four test diets, we were interested in whether there were differences in feeding behaviors. Worms were treated with MeHg and fed for 72 h with either OP50, HT115, or HB101 spread on NGM plates or OP50 spread on 2x cholesterol NGM plates prior to behavioral analyses. Food consumption was measured by the pharynx pump assay. MeHg increased food consumption in worms fed OP50 seeded on standard NGM plates (Figure 6). MeHg did not increase the rate of feeding of worms fed the HB101 or HT115 diets (Figure 6A,B). There was no statistically significant effect of NGM plate cholesterol content on MeHg-induced food consumption; worms fed OP50 on either the standard or 2x cholesterol NGM plates had the same increase in pharynx pumps in response to MeHg (Figure 6C).

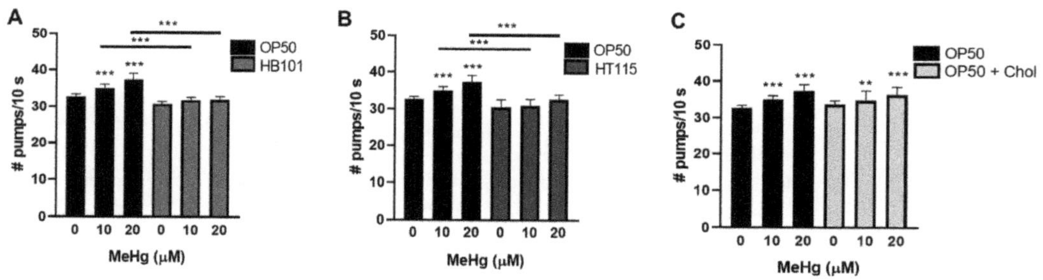

Figure 6. MeHg increases feeding behavior in worms fed a high-fat diet, but not a low-fat diet. Worms were treated with MeHg for 30 min and placed on NGM plates containing OP50 or (**A**) HB101, (**B**) HT115, or (**C**) OP50 supplemented with cholesterol. Then, 72 h after treatment, food consumption was analyzed by the pharynx pump assay. Data are expressed as means ± SEM from nine independent experiments. ** $p < 0.01$, *** $p < 0.001$ as compared with untreated, OP50-fed control. Horizontal bars represent comparisons between Hg-treated worms fed different diets.

Locomotion and dopaminergic-dependent behavior were also investigated in MeHg-treated worms 72 h following feeding on the four test diets. Previously, we showed that MeHg decreases locomotion rates by measuring forward-directed body-bends [10]. In comparing the four test diets, there was no significant difference in the rates of locomotion in response to MeHg treatment in worms fed either HB101, HT115, OP50, or OP50 on 2x cholesterol NGM plates (Figure 7A). We next measured the change in body-bends on bacteria vs. off bacteria (basal slowing rate, BSR). The BSR is a direct measure of dopaminergic function, as worms deficient in dopamine production (*cat-2* mutants) have no difference in the rates of locomotion on NGM plates with or without a bacterial food source [31]. Healthy N2 worms with functioning dopaminergic systems slow their locomotion on NGM plates spread with bacteria as compared to NGM plates unseeded with bacteria. MeHg is known to be toxic to dopaminergic neurons in mammals and in

C. elegans [28,29,45,46]; we were therefore curious as to whether changing the worm's diet might protect the dopaminergic neurons from MeHg-induced dysfunction. BSR was measured in worms treated with MeHg and fed one of the four test diets. MeHg decreased the BSR in worms fed either of the four test diets, and there was no statistical difference in the BSR between worms fed the HB101, HT115, or 2x cholesterol diets as compared to the standard OP50 diet (Figure 7B). This suggests that while MeHg damages the dopaminergic functioning in worms, dopaminergic behavior is not influenced by the bacterial diet strain or cholesterol level.

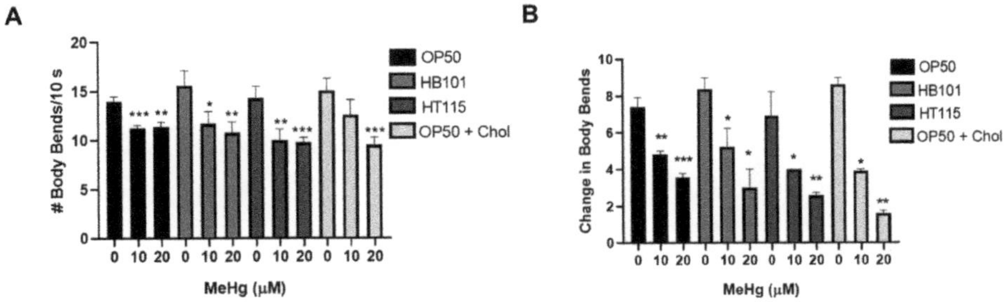

Figure 7. Locomotive and dopaminergic function in response to MeHg are not affected by diet. (**A**) Locomotion behavioral analysis was performed 72 h after MeHg treatment and feeding on the test diets. (**B**) Dopaminergic behavior was assessed by the basal slowing response (BSR) performed 72 h after MeHg treatment and feeding on the test diets. * $p < 0.05$, ** $p < 0.01$, *** $p < 0.001$ as compared with untreated, OP50-fed control.

3.6. Bacterial Diet Improves Measures of Oxidative Stress in Response to MeHg

Oxidative stress is a hallmark of MeHg exposure and can drive neurotoxicity and metabolic alterations. Dietary components can quench ROS [47–50]. We therefore investigated whether altering the bacterial diet fed to worms could prevent oxidative stress derived from MeHg treatment in *C. elegans*. Intracellular ROS were measured in worms treated with MeHg and fed either HB101, HT115, or OP50 on standard NGM plates or OP50 on 2x cholesterol NGM plates by DCFDA staining. MeHg increased intracellular ROS in worms fed OP50 on standard NGM plates (Figure 8A). Worms fed OP50 seeded on 2x cholesterol plates had exacerbated ROS levels in response to MeHg. Worms fed either the HB101 or HT115 diet had significantly less MeHg-induced ROS generation. These data suggest that the dietary components in the different *E. coli* strains produced differential oxidative stress in response to MeHg. ROS damages cellular biomacromolecules, leading to lipid peroxidation and subsequent carbonyl protein adduct formation on cysteine, lysine, and histidine amino acids through Michael addition chemistry. Levels of oxidized proteins were measured using a DNPH colorimetric assay that quantified carbonyl adducts on proteins in samples derived from lysates of N2 worms treated with 20 μM MeHg and fed one of the four test diets. MeHg significantly increased the protein carbonyl content in worms fed OP50 seeded on standard NGM plates, which was significantly decreased in worms fed either HB101 or HT115 (Figure 8B). In contrast, worms fed OP50 seeded on 2x cholesterol NGM plates had exacerbated protein carbonyl content as compared to worms fed OP50 seeded on standard NGM plates.

We next investigated the worms' ability to mount an antioxidant response to MeHg following feeding with the test diets. GSH is the main intracellular thiol that is responsible for maintaining the redox environment of the cell. MeHg readily binds free thiols, such as those present on GSH. As previously observed [28], treatment with 10 or 20 μM MeHg led to a 20% decrease in total GSH levels in worms fed OP50 seeded on standard NGM plates (Figure 8C). Worms fed HB101 had significantly higher basal levels of GSH than worms fed OP50. MeHg treatments decreased the levels of GSH in HB101-fed worms as compared to untreated worms fed HB101; however, the levels of GSH in MeHg-treated HB101 fed worms were significantly higher than those in OP50-fed MeHg-treated worms.

Worms fed HT115 showed a minimal decrease in GSH in response to MeHg; however, in comparing worms treated with MeHg, the HT115-fed worms contained more GSH than the OP50-fed worms. Similar to the oxidized protein and intracellular ROS data, worms fed OP50 seeded on 2x cholesterol plates had exacerbated GSH loss in comparison to worms fed OP50 seeded on standard NGM plates.

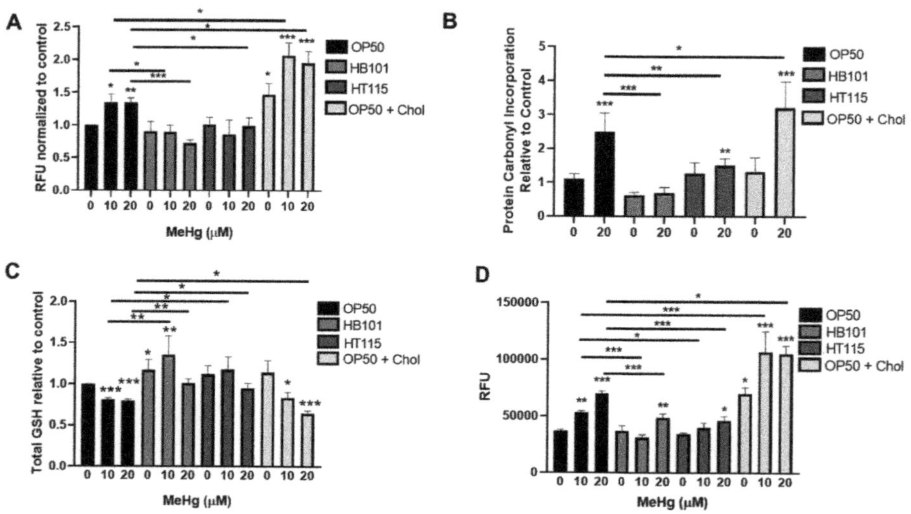

Figure 8. Diet affects MeHg-induced oxidative stress. Worms were treated with MeHg for 30 min and placed on agar plates spread with either OP50, HB101, Ht115, or OP50 supplemented with cholesterol. Measures of oxidative stress were assessed 24 h after MeHg treatment. (**A**) ROS levels were measured through DCFDA fluorescence. Data are expressed as mean fluorescence ± SEM for 6 independent experiments (**B**) Protein carbonyl levels were measured and normalized to protein content. Data are expressed as means ± SEM from 5 independent experiments. (**C**) Total GSH levels were measured and normalized to protein content. Data are expressed as mean ± SEM from five independent experiments. (**D**) Quantification of GFP fluorescence of VP596 transgenic worms expressing GFP under the *gst-4* promoter. Data are expressed as means fluorescence ± SEM from 5 independent experiments. * $p < 0.05$, ** $p < 0.01$, *** $p < 0.001$ as compared with untreated, OP50-fed control. Horizontal bars represent comparisons between Hg-treated worms fed different diets.

Finally, we examined the effect of the different diets on the ability of the worms to generate antioxidant defense proteins. Phase II metabolic genes and antioxidant defense enzymes are regulated by the activity of the Nrf2 (Nuclear factor erythroid 2-related factor 2) (SKN-1 in *C. elegans*) transcription factor. MeHg is known to induce Nrf2 in cell culture and in nematodes [51,52]. We used the VP596 strain, which expresses GFP under the control of the promoter for the SKN-1 target *gst-4*. MeHg treatment significantly increased the amount of GFP fluorescence in worms fed OP50 seeded on standard NGM plates (Figure 8D), indicating increased SKN-1 activity. Worms fed either HB101 or HT115 had significantly less GFP fluorescence in response to MeHg treatment than the OP50-fed worms. This suggested that there was less oxidative stress, and therefore less SKN-1 activity, in HB101- and HT115-fed worms than in the OP50-fed worms. In contrast, worms fed OP50 seeded on 2x cholesterol NGM plates had significantly more GFP fluorescence in response to MeHg than worms fed OP50 seeded on standard OP50, suggesting the presence of greater oxidative stress and SKN-1 activity.

4. Discussion

Herein, we demonstrate for the first time that the toxicity of MeHg in *C. elegans* is dependent on the strain of *E. coli* used as a food source. MeHg is a known neurotoxic agent that has long been understood to cause neurological changes and is emerging as a player in metabolic diseases. *C. elegans* is a useful model organism to study the effects of the metabolic changes induced by MeHg because of its evolutionarily conserved fat and sugar metabolic pathways [53]. We have previously shown that MeHg causes metabolic alterations in *C. elegans* that lead to decreased nicotinamide adenine dinucleotide (NAD^+) cofactor levels, mitochondrial dysfunction, and oxidative stress [28]. We also found that MeHg increases the transcription of *cepb-1* (ortholog to human C/EBP), *nhr-49* (ortholog to human peroxisome proliferator activated receptor gamma, PPARγ), and *sbp-1* (ortholog to human sterol response element binding protein-1, SREBP-1), pro-adipogenic transcription factors implicated in MS, as well as a number of other genes involved in lipid synthesis and transport [10]. All of these findings were under the context that the worms were consuming the standard OP50 *E. coli* diet. It was recently shown that feeding *C. elegans* dehydrated dead OP50 significantly decreased the susceptibility of worms to MeHg [54]. These data suggest that the diet fed to *C. elegans* is an important determinant of its susceptibility to the toxic effects of MeHg.

MeHg enters the human body primarily through fish consumption. For decades, studies of populations in the Seychelles, Faroe Islands, and Arctic Canada have followed cohorts of individuals exposed to MeHg in their diet and have produced, at times, conflicting data [55–57]. While there are multiple explanations for these discrepancies, from the types of fish eaten, to co-contaminants (such as polychlorinated biphenyls, PCBs) and/or genetic polymorphisms present in the populations, background diet composition has not been fully taken into consideration in these comparisons. Our data show in worms that the strain of *E. coli* can drastically alter the toxicity of MeHg independently of the amount of MeHg that accumulates in the worm. The strains HB101 and HT115, which are lower in triglycerides and free fatty acids than OP50 [36], conferred protection to worms from MeHg lethality, lower fat accumulation and adipogenic gene transcription, and decreased levels of oxidative stress than worms fed OP50. Conversely, worms fed OP50 supplemented with excess cholesterol were more susceptible to MeHg and had increased lipid accumulation and oxidative stress as compared to worms fed OP50 with a standard concentration of cholesterol in the NGM. Recent data have shown that *C. elegans* in the wild eat a varied diet comprising bacteria not only in the genus *Escherichia*, but also in the genera *Sphingomonas*, *Xanthomonas*, and *Methylobacterium* [58]. These diets confer differences in lifespan, development, reproduction, and gene expression [58]. It remains to be determined how feeding any of these non-*E. coli* species to worms might affect the toxicity of MeHg.

Our data examining lipid accumulation and MeHg toxicity in the context of higher- and lower-fat diets in *C. elegans* are in agreement with previous findings in rodents and humans. KK-Ay type 2 diabetic mice, which have higher body fat than C57BL/6J mice, had lower blood clearance of MeHg and increased neurological damage compared to C57BL/6J mice [59,60]. Likewise, high-fat diets and MeHg were shown to result in similar lipid and cholesterol accumulation and steatosis in the liver [61]. In a toxicokinetic study, Rowland et al. fed mice either a standard diet or a synthetic diet that was high in protein and low in fat. The mice fed the synthetic diet accumulated less Hg in their body than the standard-pellet-fed mice [62]. However, it was unclear whether the effect of this synthetic diet was due to the low fat, the high protein, or to both factors. Since Hg interacts with multiple amino acids, such as cysteine, or is antagonized by methionine, the protective effect that Rowland et al. observed may have been due to the amino acid content of the synthetic diet.

MeHg has been shown to increase cholesterol levels in humans and rodents. MeHg has been shown to inhibit paraoxonase activity, leading to increased low-density lipoproteins (LDL) in an Inuit population that regularly eats fish [63]. Additionally, chronic exposure to MeHg in wild-type and LDL receptor knockout mice causes hypercholesterolemia [64]. Our data support the notion that a diet rich in cholesterol potentiates the toxic effects of MeHg on lethality and lipid accumulation. While we did not specifically measure cholesterol levels in our worms following MeHg exposure and feeding on the 2x diet, our gene expression analysis showed that high cholesterol levels were indeed achieved in the worms, allowing for the inhibition of *sbp-1* transcription. SREBP1 (*sbp-1* in worms) is a transcription factor that upregulates steroid and cholesterol biosynthetic genes but is inactivated and degraded under high cholesterol levels [40].

The central nervous system plays an important role in sensing nutrients and integrating hormonal signals from the gastrointestinal tract and adipocytes to regulate caloric intake and energy expenditure [65]. In humans, the hypothalamic–pituitary axis (HPA) integrates hormonal signals including leptin, insulin, ghrelin, and adiponectin. The HPA is especially vulnerable to neurotoxicants, such as MeHg, as the blood–brain barrier is weak in certain areas, such as the arcuate of the hypothalamus, which produces neuropeptide Y [66]. In vitro studies of hypothalamic neuronal cell lines treated with MeHg show increased expression of neuropeptides pro-omiomelanocortin (Pomc) and Agouti-related peptide (Agrp), key regulators of homeostasis [67]. Just as the CNS controls feeding and energy expenditure, diet and nutrients can affect CNS function. Diet-induced obesity and high-fat diets reduce dopamine release and reuptake, leading to disruption of the satiety circuits between nucleus accumbens (NAc) dopamine terminals and projections to the hypothalamus [68,69]. Long-term feeding of high-fat diets in mice depletes dopamine in the NAc, which may contribute to the development of obesity [70]. Conversely, caloric restriction in rats has led to increased dopamine and serotonin levels in the striatum and increased leptin levels in the plasma [71].

The *C. elegans* nervous system shows simplified neuronal control of nutrient sensing. Both humans and nematodes use dopamine, glutamate, and serotonin to control foraging, locomotion, feeding, and nutrient sensing [72]. MeHg disrupts both dopamine and glutamate signaling, while little is known about MeHg's effects on serotonin. *C. elegans* use dopamine signaling to sense food, increase turn frequency when leaving food, and for defecation [31,73,74]. We have previously shown that MeHg decreases dopamine levels and behaviors in *C. elegans* fed OP50, leading to decreased locomotion and deficits in sensing the presence of food [28]. In our present study, we observed that MeHg decreased dopaminergic activity in the worms independently of which bacterial strain or cholesterol concentration was presented during the feeding. This was an unexpected result. Previously, we have shown that supplementing worms with excess nicotinamide adenine nucleotide can prevent dopaminergic damage in response to MeHg in *C. elegans* [28]. Since the low-fat HB101 and HT115 diets were more protective for MeHg lethality and metabolic dysfunction, we hypothesized that there would be less dopaminergic damage. However, MeHg is a well-characterized dopaminergic toxicant; the lipid level of the diet presented to the worm was irrelevant. MeHg caused significant dopaminergic damage that was not prevented.

The physical act of feeding in nematodes is measured by the rate of pharyngeal pumping. Serotonin regulates the pharynx muscles, allowing for food to be drawn through the mouth upon muscle contraction [75]. Serotonergic neurons coordinate the action of the cholinergic MC and glutamatergic M3 motor neurons that directly synapse on pharyngeal muscle cells [76]. Glutamate is released from the M3 neurons and activates glutamate-gated chloride channel AVR-15 expressed on pm4 and pm5 pharyngeal muscle cells, leading to the modulation of the duration and frequency of pharyngeal pumping [76,77]. Mutants deficient in glutamate signaling lose the ability to terminate action potentials on the pm4 and pm5 cells, resulting in a reduced pumping rate [78]. MeHg disrupts glutamate signaling, leading to an increased glutamate concentration in the synapses and neuronal excitotoxicity [79–81]. Therefore, disruption of glutamatergic signaling by MeHg can negatively affect the rate of pharynx pumping. Previously, we have reported increased feeding in response to MeHg in *C. elegans* fed OP50 [10]. In our present study, worms fed HB101 or HT115 showed no increase in feeding in response to MeHg. This suggests that there may have been less damage by MeHg to the glutamatergic or serotoniergic neurotransmitter systems in HB101- and HT115-fed worms than in worms fed OP50. Decreased feeding, as compared to OP50, may be one of the mechanisms by which the low-fat diets protected the worms from lipid dysregulation in response to MeHg. It is important to note that, basally, different bacterial species and bacterial strains can cause differential pharyngeal pumping in *C. elegans* [36,58]. Both HB101 and HT115 have been shown in L4 and young adult worms to lead to significantly lower rates of pharyngeal pumping as compared to worms fed OP50 [58,82,83]. Our data show no difference in basal pharynx pump rate between worms fed OP50, HB101, or HT115. This may be due to our use of older worms (adults ~72 h post L1 stage) than in previous reports. Our data also show that the cholesterol concentration in the diet did not affect the pharyngeal pumping rate basally or in response to MeHg. Worms fed OP50 grown on plates with 2x cholesterol concentration had a dose-dependent increase in feeding similar to worms fed OP50 grown on NGM plates with the standard cholesterol concentration.

While the nervous system is a key modulator of nutrient sensing, nutrients themselves actively signal to neurons to regulate feeding behaviors. Nematodes and mammals express neuropeptides that signal to neurons in response to the presence of nutrients. For example, FLP-20 regulates glutamatergic neurons to degrade fat and induce autophagy following starvation [84]. Nutrients send either feedforward or feedback modulation to the neurons that control the pharyngeal pump rate to either increase or decrease feeding [85]. It is unknown how these pathways are affected by MeHg exposure.

MeHg exerts many of its toxic and neurotoxic effects due to the induction of oxidative stress. ROS generation following MeHg exposure can damage membrane lipids, leading to lipid peroxidation, loss of membrane integrity, and dysfunction of neuronal signaling. Likewise, oxidative stress following MeHg exposure can alter gene expression or directly damage enzymes and transport proteins. Dietary factors can also influence ROS levels and oxidative stress. In our study, ROS levels were increased in worms fed OP50 following MeHg treatment and were exacerbated when worms were fed the 2x cholesterol diet. This is in agreement with previous studies demonstrating that high-fat diets are linked to increased oxidative stress, leading to mitochondrial dysfunction and additional ROS production [86,87]. Oxidative stress induced by high-fat diets has been shown to be blocked by dietary factors, such as yogurt, quercetin, geraniin (polyphenol derived from *Nephelium lappaceum* L. fruit rind), and *Terminalia arjuna* extract, to name a few [47–50]. Indeed, in our study, worms fed HB101 or HT115 following MeHg exposure had significantly lower ROS production than the worms fed OP50. ROS can damage lipids, leading to lipid peroxidation, and subsequently oxidizes proteins via a process known as protein carbonylation [88]. We observed that protein carbonylation in our study was consistent with our ROS production data. Worms fed OP50 following MeHg exposure had high levels of protein carbonylation, which was exacerbated by feeding the worms the 2x cholesterol diet. Conversely, worms fed the HB101 or HT115 diets following MeHg exposure had lower levels of protein carbonylation as compared to worms fed OP50. MeHg depletes cellular GSH, leading to a reduced capacity to buffer oxidative stress [89,90]. In *C. elegans*, GSH was significantly decreased in worms fed OP50 or the 2x cholesterol diet following MeHg treatment. Worms fed the HB101 or HT115 diet did not lose intracellular GSH content following MeHg exposure. While this may have been the result of less oxidative stress resulting from MeHg in the HB101- or HT115-fed worms, the bacterial strain fed to the worms cannot be ruled out. While total protein content is not significantly different between OP50, HT115, and HB101 [36], it remains to be determined whether the thiol levels between the three strains are similar. As a final measure of oxidative stress, we measured fluorescence from a GFP reporter strain that fluoresces upon activation of the antioxidant response element. Worms fed the OP50 or 2x cholesterol diet had increased induction of the GFP reporter in response to MeHg, suggesting high levels of antioxidant gene induction. Worms fed HB101 or HT115 following MeHg had significantly lower levels of GFP induction than worms fed OP50, corroborating that there were lower levels of oxidative stress in response to MeHg in worms fed either of these lower-fat bacterial strains.

5. Conclusions

Altogether, our study demonstrates that dietary lipid content and cholesterol content are major determinants in the response of *C. elegans* to MeHg. Worms fed diets low in lipids had reduced triglyceride and lipid accumulation in response to MeHg, ate less food, and experienced less oxidative stress than worms fed the standard OP50 diet that was higher in lipid content,. Conversely, worms fed a diet high in cholesterol had increased triglyceride and lipid accumulation in response to MeHg, and experienced more oxidative stress than worms fed the standard OP50 diet. Diet did not affect certain neurotoxicities in response to MeHg, such as dopaminergic dysfunction; however, diet did affect the rate of feeding. These data suggest that MeHg-induced lipid dysregulation and oxidative stress is influenced by dietary factors, such as triglycerides and cholesterol, leading to metabolic changes characteristic of obesity and metabolic syndrome.

Author Contributions: Conceptualization, S.W.C.; methodology, J.L.N.-C., J.B., T.S. and S.W.C.; formal analysis, J.L.N.-C., J.B. and S.W.C.; investigation, N.C., M.M., T.N., B.K., F.I., E.N., J.L.N.-C., J.B., T.S. and S.W.C.; resources, J.L.N.-C., T.S. and S.W.C.; writing—original draft preparation, J.L.N.-C., J.B. and S.W.C.; writing—review and editing, S.W.C.; supervision, J.L.N.-C., J.B. and S.W.C.; project administration, S.W.C.; Funding acquisition, J.B., T.S. and S.W.C. All authors have read and agreed to the published version of the manuscript.

Funding: This research was funded by the Husson University School of Pharmacy Research Grant, Husson University Research Fund Program, DFG Research Unit TraceAge (FOR 2558).

Institutional Review Board Statement: Not applicable.

Informed Consent Statement: Not applicable.

Data Availability Statement: Not applicable.

Conflicts of Interest: The authors declare no conflict of interest.

References

1. Chamorro-Garcia, R.; Blumberg, B. Current Research Approaches and Challenges in the Obesogen Field. *Front. Endocrinol.* **2019**, *10*, 167. [CrossRef] [PubMed]
2. Kassotis, C.D.; Stapleton, H.M. Endocrine-Mediated Mechanisms of Metabolic Disruption and New Approaches to Examine the Public Health Threat. *Front. Endocrinol.* **2019**, *10*, 39. [CrossRef]
3. Clarkson, T.W.; Magos, L. The Toxicology of Mercury and Its Chemical Compounds. *Crit. Rev. Toxicol.* **2006**, *36*, 609–662. [CrossRef] [PubMed]
4. Landrigan, P.J.; Sonawane, B.; Butler, R.N.; Trasande, L.; Callan, R.; Droller, D. Early Environmental Origins of Neurodegenerative Disease in Later Life. *Environ. Health Perspect.* **2005**, *113*, 1230–1233. [CrossRef] [PubMed]
5. Bulka, C.M.; Persky, V.W.; Daviglus, M.L.; Durazo-Arvizu, R.A.; Argos, M. Multiple metal exposures and metabolic syndrome: A cross-sectional analysis of the National Health and Nutrition Examination Survey 2011–2014. *Environ. Res.* **2019**, *168*, 397–405. [CrossRef] [PubMed]
6. Wang, X.; Mukherjee, B.; Park, S.K. Associations of cumulative exposure to heavy metal mixtures with obesity and its comorbidities among U.S. adults in NHANES 2003–2014. *Environ. Int.* **2018**, *121*, 683–694. [CrossRef]
7. Lee, K. Blood mercury concentration in relation to metabolic and weight phenotypes using the KNHANES 2011–2013 data. *Int. Arch. Occup. Environ. Health* **2018**, *91*, 185–193. [CrossRef] [PubMed]
8. Park, J.S.; Ha, K.H.; He, K.; Kim, D.J. Association between Blood Mercury Level and Visceral Adiposity in Adults. *Diabetes Metab. J.* **2017**, *41*, 113–120. [CrossRef] [PubMed]
9. Park, K.; Seo, E. Association between Toenail Mercury and Metabolic Syndrome is Modified by Selenium. *Nutrients* **2016**, *8*, 424. [CrossRef]
10. Caito, S.W.; Newell-Caito, J.; Martell, M.; Crawford, N.; Aschner, M. Methylmercury Induces Metabolic Alterations in *Caenorhabditis elegans*: Role for C/EBP Transcription Factor. *Toxicol. Sci.* **2020**, *174*, 112–123. [CrossRef] [PubMed]
11. Chen, D.Y.; Williams, V.J. Marine fish food in the United States and methylmercury risk. *Int. J. Environ. Health Res.* **2009**, *19*, 109–124. [CrossRef] [PubMed]
12. Simmons-Willis, T.A.; Koh, A.S.; Clarkson, T.W.; Ballatori, N. Transport of a neurotoxicant by molecular mimicry: The methylmercury–l-cysteine complex is a substrate for human L-type large neutral amino acid transporter (LAT) 1 and LAT2. *Biochem. J.* **2002**, *367*, 239–246. [CrossRef]
13. Caito, S.W.; Zhang, Y.; Aschner, M. Involvement of AAT transporters in methylmercury toxicity in *Caenorhabditis elegans*. *Biochem. Biophys. Res. Commun.* **2013**, *435*, 546–550. [CrossRef] [PubMed]
14. Adachi, T.; Yasutake, A.; Hirayama, K. Influence of dietary levels of protein and sulfur amino acids on the fate of methylmercury in mice. *Toxicology* **1994**, *93*, 225–234. [CrossRef]
15. Meydani, M.; Meydani, S.N.; Hathcock, J.N. Effects of dietary methionine, methylmercury, and atrazine on ex-vivo synthesis of prostaglandin E1 and thromboxane B2. *Prostaglandins Leukot. Med.* **1984**, *14*, 267–278. [CrossRef]

16. Cabañero, A.I.; Madrid, A.Y.; Cámara, C. Selenium Long-Term Administration and Its Effect on Mercury Toxicity. *J. Agric. Food Chem.* **2006**, *54*, 4461–4468. [CrossRef] [PubMed]
17. Folven, K.I.; Glover, C.N.; Malde, M.K.; Lundebye, A.-K. Does selenium modify neurobehavioural impacts of developmental methylmercury exposure in mice? *Environ. Toxicol. Pharmacol.* **2009**, *28*, 111–119. [CrossRef]
18. Wu, H.; Xu, L.; Ballantyne, C.M. Dietary and Pharmacological Fatty Acids and Cardiovascular Health. *J. Clin. Endocrinol. Metab.* **2020**, *105*, 1030–1045. [CrossRef]
19. Djuricic, I.; Calder, P. Beneficial Outcomes of Omega-6 and Omega-3 Polyunsaturated Fatty Acids on Human Health: An Update for 2021. *Nutrients* **2021**, *13*, 2421. [CrossRef] [PubMed]
20. Strain, J.J.; Davidson, P.W.; Thurston, S.W.; Harrington, D.; Mulhern, M.S.; McAfee, A.J.; van Wijngaarden, E.; Shamlaye, C.F.; Henderson, J.; Watson, G.; et al. Maternal PUFA Status but Not Prenatal Methylmercury Exposure Is Associated with Children's Language Functions at Age Five Years in the Seychelles. *J. Nutr.* **2012**, *142*, 1943–1949. [CrossRef]
21. Strain, J.J.; Yeates, A.J.; van Wijngaarden, E.; Thurston, S.W.; Mulhern, M.S.; McSorley, E.M.; Watson, G.E.; Love, T.M.; Smith, T.H.; Yost, K.; et al. Prenatal exposure to methyl mercury from fish consumption and polyunsaturated fatty acids: Associations with child development at 20 mo of age in an observational study in the Republic of Seychelles. *Am. J. Clin. Nutr.* **2015**, *101*, 530–537. [CrossRef]
22. Strain, J.J.; Love, T.M.; Yeates, A.J.; Weller, D.; Mulhern, M.S.; McSorley, E.M.; Thurston, S.W.; Watson, G.E.; Mruzek, D.; Broberg, K.; et al. Associations of prenatal methylmercury exposure and maternal polyunsaturated fatty acid status with neurodevelopmental outcomes at 7 years of age: Results from the Seychelles Child Development Study Nutrition Cohort 2. *Am. J. Clin. Nutr.* **2021**, *113*, 304–313. [CrossRef]
23. Nøstbakken, O.J.; Bredal, I.L.; Olsvik, P.A.; Huang, T.S.; Torstensen, B.E. Effect of Marine Omega 3 Fatty Acids on Methylmercury-Induced Toxicity in Fish and Mammalian CellsIn Vitro. *J. Biomed. Biotechnol.* **2012**, *2012*, 1–13. [CrossRef] [PubMed]
24. Guo, X.-F.; Li, X.; Shi, M.; Li, D. n-3 Polyunsaturated Fatty Acids and Metabolic Syndrome Risk: A Meta-Analysis. *Nutrients* **2017**, *9*, 703. [CrossRef]
25. Brenner, S. The genetics of *Caenorhabditis elegans*. *Genetics* **1974**, *77*, 71–94. [CrossRef] [PubMed]
26. Cheong, M.C.; Lee, H.-J.; Na, K.; Joo, H.-J.; Avery, L.; You, Y.; Paik, Y.-K. NSBP-1 mediates the effects of cholesterol on insulin/IGF-1 signaling in *Caenorhabditis elegans*. *Cell. Mol. Life Sci.* **2013**, *70*, 1623–1636. [CrossRef]
27. Stiernagle, T. Maintenance of *C. elegans*. In *C. Elegans: A Practical Approach*; Hope, I.A., Ed.; Oxford University Press: New York, NY, USA, 1999.
28. Caito, S.W.; Aschner, M. NAD+Supplementation Attenuates Methylmercury Dopaminergic and Mitochondrial Toxicity in *Caenorhabditis elegans*. *Toxicol. Sci.* **2016**, *151*, 139–149. [CrossRef]
29. Martinez-Finley, E.J.; Chakraborty, S.; Slaughter, J.C.; Aschner, M. Early-Life Exposure to Methylmercury in Wildtype and pdr-1/parkin Knockout, *C. elegans*. *Neurochem. Res.* **2013**, *38*, 1543–1552. [CrossRef] [PubMed]
30. Pino, E.C.; Webster, C.M.; Carr, C.E.; Soukas, A.A. Biochemical and High Throughput Microscopic Assessment of Fat Mass in *Caenorhabditis elegans*. *J. Vis. Exp.* **2013**, *73*, e50180. [CrossRef]
31. Sawin, E.R.; Ranganathan, R.; Horvitz, H. *C. elegans* Locomotory Rate Is Modulated by the Environment through a Dopaminergic Pathway and by Experience through a Serotonergic Pathway. *Neuron* **2000**, *26*, 619–631. [CrossRef]
32. Caito, S.W.; Aschner, M. Quantification of Glutathione in *Caenorhabditis elegans*. *Curr. Protoc. Toxicol.* **2015**, *64*, 6.18.1–6.18.6. [CrossRef]
33. Yoon, D.S.; Lee, M.H.; Cha, D.S. Measurement of Intracellular ROS in *Caenorhabditis elegans* Using 2′,7′-Dichlorodihydrofluorescein Diacetate. *Bio-Protocol* **2018**, *8*, e2774. [CrossRef] [PubMed]
34. Yasuda, K.; Adachi, H.; Fujiwara, Y.; Ishii, N. Protein carbonyl accumulation in aging dauer formation-defective (daf) mutants of *Caenorhabditis elegans*. *J. Gerontol. Ser. A Boil. Sci. Med. Sci.* **1999**, *54*, B47–B51. [CrossRef]
35. Stiernagle, T. Maintenance of *C. Elegans*. *WormBook* **2006**, *11*, 1–11. [CrossRef] [PubMed]
36. Brooks, K.K.; Liang, B.; Watts, J.L. The Influence of Bacterial Diet on Fat Storage in *C. elegans*. *PLoS ONE* **2009**, *4*, e7545. [CrossRef] [PubMed]
37. Ruszkiewicz, J.A.; de Macedo, G.T.; Miranda-Vizuete, A.; Bowman, A.B.; Bornhorst, J.; Schwerdtle, T.; Soares, F.A.A.; Aschner, M. Sex-Specific Response of *Caenorhabditis elegans* to Methylmercury Toxicity. *Neurotox. Res.* **2019**, *35*, 208–216. [CrossRef] [PubMed]
38. Lee, J.-E.; Schmidt, H.; Lai, B.; Ge, K. Transcriptional and Epigenomic Regulation of Adipogenesis. *Mol. Cell. Biol.* **2019**, *39*, e00601-18. [CrossRef]
39. Tang, Q.Q.; Otto, T.C.; Lane, M.D. CCAAT/enhancer-binding protein beta is required for mitotic clonal expansion during adipogenesis. *Proc. Natl. Acad. Sci. USA* **2003**, *100*, 850–855. [CrossRef]
40. Ye, J.; DeBose-Boyd, R.A. Regulation of Cholesterol and Fatty Acid Synthesis. *Cold Spring Harb. Perspect. Biol.* **2011**, *3*, a004754. [CrossRef] [PubMed]
41. Sampath, H.; Ntambi, J.M. The role of stearoyl-CoA desaturase in obesity, insulin resistance, and inflammation. *Ann. N. Y. Acad. Sci.* **2011**, *1243*, 47–53. [CrossRef] [PubMed]
42. Van Gilst, M.R.; Hadjivassiliou, H.; Jolly, A.; Yamamoto, K.R. Nuclear Hormone Receptor NHR-49 Controls Fat Consumption and Fatty Acid Composition in *C. elegans*. *PLoS Biol.* **2005**, *3*, e53. [CrossRef] [PubMed]
43. Nomura, T.; Horikawa, M.; Shimamura, S.; Hashimoto, T.; Sakamoto, K. Fat accumulation in *Caenorhabditis elegans* is mediated by SREBP homolog SBP-1. *Genes Nutr.* **2010**, *5*, 17–27. [CrossRef] [PubMed]

44. Baker, M. Is vitellogenin an ancestor of apolipoprotein B-100 of human low-density lipoprotein and human lipoprotein lipase? *Biochem. J.* **1988**, *255*, 1057–1060. [CrossRef] [PubMed]
45. Dreiem, A.; Shan, M.; Okoniewski, R.J.; Sanchez-Morrissey, S.; Seegal, R.F. Methylmercury inhibits dopaminergic function in rat pup synaptosomes in an age-dependent manner. *Neurotoxicology Teratol.* **2009**, *31*, 312–317. [CrossRef]
46. Shao, Y.; Chan, H.M. Effects of methylmercury on dopamine release in MN9D neuronal cells. *Toxicol. Mech. Methods* **2015**, *25*, 637–644. [CrossRef]
47. Kanthe, P.S.; Patil, B.S.; Das, K.K. Terminalia arjuna supplementation ameliorates high fat diet-induced oxidative stress in nephrotoxic rats. *J. Basic Clin. Physiol. Pharmacol.* **2021**. [CrossRef] [PubMed]
48. Chung, A.P.; Gurtu, S.; Chakravarthi, S.; Moorthy, M.; Palanisamy, U.D. Geraniin Protects High-Fat Diet-Induced Oxidative Stress in Sprague Dawley Rats. *Front. Nutr.* **2018**, *5*, 17. [CrossRef] [PubMed]
49. Lasker, S.; Rahman, M.; Parvez, F.; Zamila, M.; Miah, P.; Nahar, K.; Kabir, F.; Sharmin, S.B.; Subhan, N.; Ahsan, G.U.; et al. High-fat diet-induced metabolic syndrome and oxidative stress in obese rats are ameliorated by yogurt supplementation. *Sci. Rep.* **2019**, *9*, 20026. [CrossRef]
50. Pei, Y.; Otieno, D.; Gu, I.; Lee, S.-O.; Parks, J.S.; Schimmel, K.; Kang, H.W. Effect of quercetin on nonshivering thermogenesis of brown adipose tissue in high-fat diet-induced obese mice. *J. Nutr. Biochem.* **2021**, *88*, 108532. [CrossRef] [PubMed]
51. Ni, M.; Li, X.; Yin, Z.; Jiang, H.; Sidoryk-Wegrzynowicz, M.; Milatovic, D.; Cai, J.; Aschner, M. Methylmercury Induces Acute Oxidative Stress, Altering Nrf2 Protein Level in Primary Microglial Cells. *Toxicol. Sci.* **2010**, *116*, 590–603. [CrossRef] [PubMed]
52. Martinez-Finley, E.J.; Caito, S.; Slaughter, J.C.; Aschner, M. The Role of *skn-1* in Methylmercury-Induced Latent Dopaminergic Neurodegeneration. *Neurochem. Res.* **2013**, *38*, 2650–2660. [CrossRef] [PubMed]
53. Jones, K.T.; Ashrafi, K. *Caenorhabditis elegans* as an emerging model for studying the basic biology of obesity. *Dis. Model. Mech.* **2009**, *2*, 224–229. [CrossRef] [PubMed]
54. Ke, T.; Aschner, M. Bacteria affect *Caenorhabditis elegans* responses to MeHg toxicity. *NeuroToxicology* **2019**, *75*, 129–135. [CrossRef] [PubMed]
55. Davidson, P.; van Wijngaarden, E.; Shamlaye, C.; Strain, J.; Myers, G. Putting findings from the Seychelles Child Development Study into perspective: The importance of a historical special issue of the Seychelles Medical and Dental Journal. *NeuroToxicology* **2020**, *76*, 111–113. [CrossRef] [PubMed]
56. Rice, D.C. Identification of functional domains affected by developmental exposure to methylmercury: Faroe islands and related studies. *NeuroToxicology* **2000**, *21*, 1039–1044. [PubMed]
57. Bélanger, M.-C.; Dewailly, É.; Berthiaume, L.; Noël, M.; Bergeron, J.; Mirault, M.-É.; Julien, P. Dietary contaminants and oxidative stress in Inuit of Nunavik. *Metabolism* **2006**, *55*, 989–995. [CrossRef] [PubMed]
58. Stuhr, N.L.; Curran, S.P. Bacterial diets differentially alter lifespan and healthspan trajectories in *C. elegans*. *Commun. Biol.* **2020**, *3*, 653. [CrossRef] [PubMed]
59. Yamamoto, M.; Motomura, E.; Yanagisawa, R.; Hoang, V.A.T.; Mogi, M.; Mori, T.; Nakamura, M.; Takeya, M.; Eto, K. Evaluation of neurobehavioral impairment in methylmercury-treated KK-Ay mice by dynamic weight-bearing test. *J. Appl. Toxicol.* **2019**, *39*, 221–230. [CrossRef]
60. Yamamoto, M.; Yanagisawa, R.; Motomura, E.; Nakamura, M.; Sakamoto, M.; Takeya, M.; Eto, K. Increased methylmercury toxicity related to obesity in diabetic KK-Ay mice. *J. Appl. Toxicol.* **2014**, *34*, 914–923. [CrossRef]
61. Leocádio, P.C.L.; Dias, R.P.; Pinto, D.V.; Reis, J.M.; Nascimento, J.C.R.; Brito, G.A.D.C.; Valença, J.T.; Foureaux, G.; Ferreira, A.J.; Windmöller, C.C.; et al. Pollutants and nutrition: Are methylmercury effects on blood pressure and lipoprotein profile comparable to high-fat diet in mice? *Ecotoxicol. Environ. Saf.* **2020**, *204*, 111036. [CrossRef] [PubMed]
62. Rowland, I.R.; Mallett, A.K.; Wise, A. The Effect of Diet on the Mammalian Gut Flora and Its Metabolic Activities. *CRC Crit. Rev. Toxicol.* **1985**, *16*, 31–103. [CrossRef] [PubMed]
63. Ayotte, P.; Carrier, A.; Ouellet, N.; Boiteau, V.; Abdous, B.; Sidi, E.A.L.; Chateau-Degat, M.-L.; Dewailly, É. Relation between Methylmercury Exposure and Plasma Paraoxonase Activity in Inuit Adults from Nunavik. *Environ. Health Perspect.* **2011**, *119*, 1077–1083. [CrossRef] [PubMed]
64. Moreira, E.; de Oliveira, J.; Dutra, M.F.; Santos, D.B.; Gonçalves, C.A.; Goldfeder, E.M.; de Bem, A.F.; Prediger, R.; Aschner, M.; Farina, M. Does Methylmercury-Induced Hypercholesterolemia Play a Causal Role in Its Neurotoxicity and Cardiovascular Disease? *Toxicol. Sci.* **2012**, *130*, 373–382. [CrossRef] [PubMed]
65. Timper, K.; Brüning, J.C. Hypothalamic circuits regulating appetite and energy homeostasis: Pathways to obesity. *Dis. Model. Mech.* **2017**, *10*, 679–689. [CrossRef] [PubMed]
66. Huang, Y.; Lin, X.; Lin, S. Neuropeptide Y and Metabolism Syndrome: An Update on Perspectives of Clinical Therapeutic Intervention Strategies. *Front. Cell Dev. Biol.* **2021**, *9*, 695623. [CrossRef] [PubMed]
67. Ferrer, B.; Peres, T.V.; Dos Santos, A.A.; Bornhorst, J.; Morcillo, P.; Gonçalves, C.L.; Aschner, M.; Vieira, T.P. Methylmercury Affects the Expression of Hypothalamic Neuropeptides That Control Body Weight in C57BL/6J Mice. *Toxicol. Sci.* **2018**, *163*, 557–568. [CrossRef] [PubMed]
68. Wallace, C.; Fordahl, S. Obesity and dietary fat influence dopamine neurotransmission: Exploring the convergence of metabolic state, physiological stress, and inflammation on dopaminergic control of food intake. *Nutr. Res. Rev.* **2021**, 1–16. [CrossRef] [PubMed]

69. Geiger, B.; Haburcak, M.; Avena, N.; Moyer, M.; Hoebel, B.; Pothos, E. Deficits of mesolimbic dopamine neurotransmission in rat dietary obesity. *Neuroscience* **2009**, *159*, 1193–1199. [CrossRef]
70. Estes, M.K.; Bland, J.J.; Ector, K.K.; Puppa, M.J.; Powell, D.W.; Lester, D.B. A high fat western diet attenuates phasic dopamine release. *Neurosci. Lett.* **2021**, *756*, 135952. [CrossRef] [PubMed]
71. Rojic-Becker, D.; Portero-Tresserra, M.; Martí-Nicolovius, M.; Vale-Martínez, A.; Guillazo-Blanch, G. Effects of caloric restriction on monoaminergic neurotransmission, peripheral hormones, and olfactory memory in aged rats. *Behav. Brain Res.* **2021**, *409*, 113328. [CrossRef] [PubMed]
72. Dallière, N.; Holden-Dye, L.; Dillon, J.; O'Connor, V.; Walker, R.J. *Caenorhabditis elegans* Feeding Behaviors. In *Oxford Research Encyclopedia of Neuroscience*; Oxford University Press (OUP): Oxford, UK, 2017.
73. Hills, T.; Brockie, P.J.; Maricq, A.V. Dopamine and Glutamate Control Area-Restricted Search Behavior in *Caenorhabditis elegans*. *J. Neurosci.* **2004**, *24*, 1217–1225. [CrossRef]
74. Vidal-Gadea, A.; Topper, S.; Young, L.; Crisp, A.; Kressin, L.; Elbel, E.; Maples, T.; Brauner, M.; Erbguth, K.; Axelrod, A.; et al. *Caenorhabditis elegans* selects distinct crawling and swimming gaits via dopamine and serotonin. *Proc. Natl. Acad. Sci. USA* **2011**, *108*, 17504–17509. [CrossRef] [PubMed]
75. Horvitz, H.R.; Chalfie, M.; Trent, C.; Sulston, J.E.; Evans, P.D. Serotonin and Octopamine in the Nematode *Caenorhabditis elegans*. *Science* **1982**, *216*, 1012–1014. [CrossRef] [PubMed]
76. Niacaris, T.; Avery, L. Serotonin regulates repolarization of the *C. elegans* pharyngeal muscle. *J. Exp. Biol.* **2003**, *206*, 223–231. [CrossRef]
77. Dent, J.A.; Davis, M.; Avery, L. *avr-15* encodes a chloride channel subunit that mediates inhibitory glutamatergic neurotransmission and ivermectin sensitivity in *Caenorhabditis elegans*. *EMBO J.* **1997**, *16*, 5867–5879. [CrossRef] [PubMed]
78. Greer, E.R.; Pérez, C.L.; van Gilst, M.R.; Lee, B.H.; Ashrafi, K. Neural and Molecular Dissection of a *C. elegans* Sensory Circuit that Regulates Fat and Feeding. *Cell Metab.* **2008**, *8*, 118–131. [CrossRef] [PubMed]
79. Culbreth, M.; Aschner, M. Dysregulation of Glutamate Cycling Mediates Methylmercury-Induced Neurotoxicity. *Adv. Neurobiol.* **2016**, *13*, 295–305. [CrossRef] [PubMed]
80. Farina, M.; Dahm, K.C.S.; Schwalm, F.D.; Brusque, A.M.; Frizzo, M.E.; Zeni, G.; Souza, D.; da Rocha, J.B.T. Methylmercury Increases Glutamate Release from Brain Synaptosomes and Glutamate Uptake by Cortical Slices from Suckling Rat Pups: Modulatory Effect of Ebselen. *Toxicol. Sci.* **2003**, *73*, 135–140. [CrossRef] [PubMed]
81. Mutkus, L.; Aschner, J.L.; Fitsanakis, V.A.; Aschner, M. The In Vitro Uptake of Glutamate in GLAST and GLT-1 Transfected Mutant CHO-K1 Cells Is Inhibited by Manganese. *Biol. Trace Elem. Res.* **2005**, *107*, 221–230. [CrossRef]
82. Rodríguez-Palero, M.J.; López-Díaz, A.; Marsac, R.; Gomes, J.-E.; Olmedo, M.; Artal-Sanz, M. An automated method for the analysis of food intake behaviour in *Caenorhabditis elegans*. *Sci. Rep.* **2018**, *8*, 221–230. [CrossRef]
83. You, Y.J.; Kim, J.; Raizen, D.M.; Avery, L. Insulin, cGMP, and TGF-beta signals regulate food intake and quiescence in *C. elegans*: A model for satiety. *Cell Metab.* **2008**, *7*, 249–257. [CrossRef] [PubMed]
84. Kang, C.; Avery, L. The FMRFamide Neuropeptide FLP-20 Acts as a Systemic Signal for Starvation Responses in *Caenorhabditis elegans*. *Mol. Cells* **2021**, *44*, 529–537. [CrossRef]
85. Liu, H.; Qin, L.-W.; Li, R.; Zhang, C.; Al-Sheikh, U.; Wu, Z.-X. Reciprocal modulation of 5-HT and octopamine regulates pumping via feedforward and feedback circuits in *C. elegans*. *Proc. Natl. Acad. Sci. USA* **2019**, *116*, 7107–7112. [CrossRef] [PubMed]
86. Vial, G.; Dubouchaud, H.; Couturier, K.; Cottet-Rousselle, C.; Taleux, N.; Athias, A.; Galinier, A.; Casteilla, L.; Leverve, X.M. Effects of a high-fat diet on energy metabolism and ROS production in rat liver. *J. Hepatol.* **2011**, *54*, 348–356. [CrossRef] [PubMed]
87. Tan, B.L.; Norhaizan, M.E. Effect of High-Fat Diets on Oxidative Stress, Cellular Inflammatory Response and Cognitive Function. *Nutrients* **2019**, *11*, 2579. [CrossRef]
88. Fritz, K.S.; Petersen, D.R. Exploring the Biology of Lipid Peroxidation-Derived Protein Carbonylation. *Chem. Res. Toxicol.* **2011**, *24*, 1411–1419. [CrossRef]
89. Farina, M.; Aschner, M. Glutathione antioxidant system and methylmercury-induced neurotoxicity: An intriguing interplay. *Biochim. Biophys. Acta (BBA)-Gen. Subj.* **2019**, *1863*, 129285. [CrossRef] [PubMed]
90. Fujimura, M.; Usuki, F. Methylmercury-Mediated Oxidative Stress and Activation of the Cellular Protective System. *Antioxidants* **2020**, *9*, 1004. [CrossRef] [PubMed]

toxics

MDPI

Article

Cobalt Neurotoxicity: Transcriptional Effect of Elevated Cobalt Blood Levels in the Rodent Brain

Sara Gómez-Arnaiz [1], Rothwelle J. Tate [2] and Mary Helen Grant [1,*]

[1] Wolfson Centre, Biomedical Engineering Department, University of Strathclyde, Glasgow G4 0NW, UK; sara.gomez-arnaiz@strath.ac.uk
[2] Strathclyde Institute for Pharmacy & Biomedical Sciences, University of Strathclyde, 161 Cathedral Street, Glasgow G4 0RE, UK; r.j.tate@strath.ac.uk
* Correspondence: m.h.grant@strath.ac.uk

Abstract: Metal-on-metal (MoM) hip implants made of cobalt chromium (CoCr) alloy have shown early failure compared with other bearing materials. A consequence of the abnormal wear produced by these prostheses is elevated levels of cobalt in the blood of patients, which can lead to systemic conditions involving cardiac and neurological symptoms. In order to better understand the implications for patients with these implants, we carried out metal content and RNA-Seq analysis of excised tissue from rats treated intraperitonially for 28 days with low concentrations of cobalt. Cobalt blood levels in dosed rats were found to be similar to those seen in some patients with MoM implants (range: 4–38 μg/L Co in blood). Significant accumulation of cobalt was measured in a range of tissues including kidney, liver, and heart, but also in brain tissue. RNA-Seq analysis of neural tissue revealed that exposure to cobalt induces a transcriptional response in the prefrontal cortex (pref. cortex), cerebellum, and hippocampus. Many of the most up- and downregulated genes appear to correspond to choroid plexus transcripts. These results indicate that the choroid plexus could be the brain tissue most affected by cobalt. More specifically, the differentially expressed genes show a disruption of steroidogenesis and lipid metabolism. Several other transcripts also demonstrate that cobalt induces an immune response. In summary, cobalt exposure induces alterations in the brain transcriptome, more specifically, the choroid plexus, which is in direct contact with neurotoxicants at the blood–cerebrospinal fluid barrier.

Keywords: metal-on-metal (MoM) hip implants; cobalt; systemic cobaltism; neurotoxicity; RNA-Seq; RT-qPCR

Citation: Gómez-Arnaiz, S.; Tate, R.J.; Grant, M.H. Cobalt Neurotoxicity: Transcriptional Effect of Elevated Cobalt Blood Levels in the Rodent Brain. *Toxics* **2022**, *10*, 59. https://doi.org/10.3390/toxics10020059

Academic Editors: Richard Ortega and Asuncion Carmona

Received: 30 November 2021
Accepted: 13 January 2022
Published: 28 January 2022

1. Introduction

Hip arthroplasty procedures successfully lead to the reduction in pain and improved mobility in patients suffering from joint diseases such as osteoarthritis. However, a decade ago, consultants and regulatory bodies reported their concerns over the failures of certain models of metal-on-metal (MoM) hip implants, which resulted in the market withdrawal of the Articular Surface Replacement™ (ASR™) hip implant from DePuy Orthopaedics [1]. Recently, these concerns have been extended to all MoM implant models as it has been demonstrated that MoM implants induce adverse reaction to metal debris (ARMD), which includes metallosis, pseudotumours, aseptic lymphocytic vasculitis associated lesion (ALVAL), and even necrosis [2]. The metal debris is released from the bearing surface, and the taper junction in the case of the total hip replacements (THR), due to wear and corrosion of the metallic parts [3]. Metal ions eventually dissolve and are released into the bloodstream [4]. Co and Cr metal ions in blood are indicative of implant failure and tests for metal ion concentrations should be performed for all patients with MoM implants in the UK according to the MHRA updated guidelines [5]. The European Commission also recommends the monitoring of metal ion levels for MoM implant patients [6].

Neurological conditions thought to be caused by high levels of cobalt ions in blood have been demonstrated in several clinical reports in relation to patients with MoM implants [7–11]. These involve a range of symptoms such as cognitive decline, memory loss, and mood disturbances, in addition to other visual and auditory issues and peripheral neuropathy. There is little information about the actual effects of cobalt in the peripheral and the central nervous system, and the most appropriate response to diminish patients' symptoms to date is to remove the implant as the source of cobalt ions through revision surgery [8,12,13]. However, the recovery process from cobaltism is not well documented and there is no unified action considered appropriate to treat these patients. Our lack of knowledge on cobalt actions in the body extends further since we also fail to consider the implications for asymptomatic patients with high levels of cobalt in the blood [14].

The effects of cobalt have been extensively researched in vitro with models such as astrocytes [15,16], neurons [17–21], and glial cells [22,23], which are valuable tools to understand the modes of action of cobalt toxicity in the brain. Nevertheless, the nature of in vitro work requires highly controlled experimental conditions, and this can diminish an organism's complexity. Previous toxicogenomic analyses have demonstrated that in vitro systems fail to fully represent relevant processes occurring in rat liver tissues in vivo after exposure to toxic compounds [24]. Molecular and functional events occurring in vivo at the tissue and organ levels could be crucial mechanisms in the physiological response to cobalt. In this sense, transcriptomic applications have already been proven to be effective in toxicology, not only by understanding the mechanisms, but also by finding early end-points for the detection of toxicity and identification of biomarkers [25]. However, most in vivo cobalt studies have focused on the study of reactive oxygen species and the expression of hypoxia markers. Pregnant female rats dosed orally with 350 mg/L delivered pups with impaired levels of antioxidant proteins in the cerebrum and cerebellum [26]. Caltana et al. observed that the direct cortical injection of cobalt led to histological changes consistent with focal ischaemia involving neuronal and astrocyte morphological changes [27]. Another group applied cobalt dust directly into the dura mater [28] and discovered an elevated expression of proteins involved in thyroid transport and regulation of glycolysis. Nevertheless, it is difficult to establish the relevance of these studies for patients with MoM implants due to the high dosage of cobalt used in the animals, the different types of cobalt delivery methods, and the missing information on the resulting cobalt concentrations in blood or plasma. Our research mimicked the conditions that MoM implant patients endure to obtain a better representation of relevant cobalt toxic mechanisms for them.

The aim of this in vivo work was to identify how cobalt accumulated in tissues after long-term systemic circulation, and specifically answer whether cobalt concentration increased in the brain. For that, we carried out comprehensive time- and dose-response experiments to cobalt treatment. We also sought to understand the underlying molecular changes in the brain at the transcriptional level in laboratory rats after prolonged dose-response exposure. Research on MoM cobalt-induced systemic toxicity is scarce and to our knowledge, there have been few publications addressing neurological manifestations *in vivo*. We hope to generate and test hypotheses through RNA-Sequencing (RNA-Seq) to gain mechanistic insights into the modes of action of cobalt.

2. Materials and Methods

2.1. Experimental Animals and Research Design

Experiments were performed in adult male Sprague Dawley (SD) rats obtained from Charles River (UK). The body weight range at the start of the experiments was 210–280 g. Food and water were provided *ad libitum*. Their body weight and general aspects of health were monitored daily.

Freshly-made cobalt chloride hexahydrate ($CoCl_2.6H_2O$; Sigma-Aldrich, Dorset, UK) solutions dissolved in distilled water (dH_2O) and dH_2O were sterilised through a 0.22 μm syringe-driven filter (Merck Millipore, Watford, UK).

Two separate in vivo experiments were performed successively. The first experiment was a time-response experiment in which a 7-day cobalt treatment was compared against a 28-day treatment at a fixed cobalt dose. Animals were treated daily with either 1 mL/kg body weight dH_2O (controls) or 1 mg/kg body weight (BW) $CoCl_2$ doses injected i.p. (treated groups). The second in vivo experiment was a dose-response experiment. The animals were given 1 mL/kg dH_2O i.p. in the case of the control group, or a range of cobalt solutions- 0.1, 0.5, or 1 mg/kg BW i.p. injections for 28 days. Figure 1 depicts the sample size and design information diagrammatically.

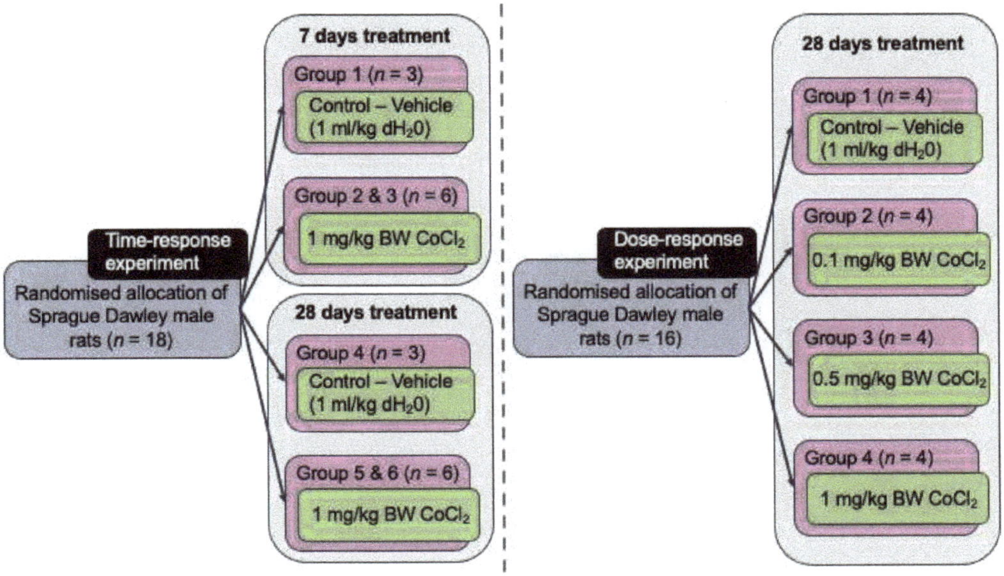

Figure 1. Design of in vivo time- and dose-response experiments showing group distribution and sample size. All injections, both control and Co-treated rats, were carried out intraperitonially.

2.2. Sacrifice of Animals and Tissue Harvest

At the end of the exposure time, animals were killed by carbon dioxide (CO_2) asphyxiation in a CO_2 chamber. Each organ or brain part was dissected separately, weighed, and sections of tissue were stored appropriately to preserve its metal content and molecular RNAs. Blood samples were collected through cardiac puncture immediately before death and mixed with 200 µL of heparin (1000 IU/mL diluted 1:10; Sigma-Aldrich, Dorset, UK) to avoid clotting.

2.3. Tissue Cobalt Content Measured by ICP-MS

Quantification of cobalt content in all organs was obtained via inductively coupled plasma mass spectrometry (ICP-MS) analysis. For this, 100 mg of tissue from each organ/area of interest was taken and stored at −80 °C until further sample digestion. To obtain a liquid solution of the samples suitable for metal detection, 0.5 mL concentrated nitric acid was used per sample (HNO_3; TraceSELECT^{TM} Ultra, Sigma-Aldrich (Fluka), Dorset, UK) followed by 0.25 mL 30% hydrogen peroxide (*w*/*w*) (H_2O_2; Sigma-Aldrich, Dorset, UK). Each reagent was to left act for 20 min at 100 °C in a hot block to ensure the decomposition of the matrix. A quantity of 0.25 mL was transferred together with 9.75 mL ultrapure water into acid-washed tubes to avoid trace metal contamination. Standard dilutions were prepared from Cobalt Standard for atomic absorption spectrometry stock (TraceCERT®, Sigma-Aldrich (Fluka), Dorset, UK). The 1:40 dilution samples were serially introduced to the Agilent 7700× octopole collision ICP-MS system (Agilent Technologies, Cheadle, UK). Scandium or indium was used as the internal standard and data were obtained from the maximum signal with the ICP-MS operating in helium mode.

2.4. RNA Extraction

2.4.1. Isolation of RNA

To reduce RNA degradation, 50 or 100 mg from specific segments of brain tissue were dissected and placed in RNase-free tubes with 0.5 mL of RNAlater (Ambion, Life Technologies, Paisley, UK). RNase-free aerosol-resistant filter tips were used for all pipetting steps. Tissues were further cut into thin pieces and completely submerged in the stabilisation solution for overnight storage at 4 °C. RNAlater reagent was removed the next day and the tissues were frozen at −80 °C until RNA extraction.

Tissue was then resuspended in 1 mL QIAzol lysis reagent (Qiagen, Crawley, UK) and a cone ball steel bead was placed inside the tube. The tissue was disrupted with a horizontal Retsch MM200 Mixer Mill (Retsch GmbH, Haan, Germany) set at 30 Hz for 1 min intervals (×3). The lysate was transferred to another tube with 4 mL of QIAzol lysis reagent and the contents vortexed. The pre-processed samples were further prepared according to the protocol of the RNeasy Plus Universal Midi Kit (Qiagen, Crawley, UK).

2.4.2. Quality Check of RNA Samples

RNA purity characterised by the ratio of the absorbances 260/280 nm and 260/230 nm as well as nucleic acid concentration were quantified with a Nanodrop-2000c spectrophotometer (Labtech International, Heathfield, UK). Sample integrity was assessed through microfluidic chip-based analysis (Experion RNA StdSensChip; Bio-Rad, Watford, UK) in the Experion Automated Electrophoresis System (Bio-Rad, Watford, UK) and evaluated with the RNA Quality Indicator (RQI) number [29]. The results of these analyses for pref. cortex, cerebellum, and hippocampus are displayed in Tables S1–S3, respectively, in the Supplementary Materials.

2.5. RNA Sequencing (RNA-Seq)

2.5.1. Sample Pooling for RNA-Sequencing

To reduce the RNA-Seq costs, four biological samples for each comparison group were pooled into a single sample. The necessary volume from each sample was adjusted to obtain an RNA quantity of 5–20 µg depending on the brain area yield. The individual sample volumes were pooled into the corresponding RNase-free tubes for each group and the final contents were briefly vortexed.

2.5.2. RNA-Sequencing Analysis by BGI

RNA-sequencing of the pooled samples was performed by BGI Tech Solutions (Hong Kong, China) Co. Ltd. using a BGISEQ-500 sequencing platform with depth of 20 million base pairs (Mb) clean reads per sample and a 50 single-end bases (50SE) read length. Filtering of clean reads and their mapping to the UCSC rn6 rat reference genome were carried out by BGI, in addition to the quantification of gene fold-change.

2.5.3. Gene Ontology (GO) and KEGG Pathway Enrichment Analysis

Further bioinformatic analyses such as hierarchical clustering or generation of Venn diagrams were performed in MATLAB software (The MathWorks Inc. Natick, MA, USA. Release R2021a Update 5, 9.10.0.1739362). The software tool Cytoscape and its plugin ClueGO were used to conduct the Gene Ontology analysis of the DEGs [30,31]. ClueGO allows for a comparison to different reference ontology sets such as Molecular Function (GO MF; 8 April 2016), Biological Process (GO BP; 8 April 2016), and Cellular Component (GO CC; 8 April 2016), which describe specific gene function and cellular location aspects of gene activity as well as the Kyoto Encyclopaedia of Genes and Genomes (KEGG; 14 June 2016) for the specific enrichment of known pathways. Additionally, other online software was used to create protein–protein interaction (PPI) networks of gene-protein products: STRING (https://string-db.org/) [32] (last accessed: 24 December 2020).

2.6. Quantitative Real-Time Polymerase Chain Reaction (RT-qPCR)

Real-time quantitative PCR (RT-qPCR) was used to obtain the expression of genes of interest in individual samples in order to validate the RNA-Seq results from the pooled samples. All RT-qPCR experiments were performed in accordance with the minimum information for the publication of quantitative real-time PCR experiments (MIQE) guidelines [33]. The checklist for RT-qPCR assays performed with samples from the in vivo experiments is fully displayed in Table S4 in the Supplementary Materials.

The LunaScript RT SuperMix Kit (New England Biolabs, Hitchin, UK) was utilised to generate cDNA for the dose-response experiment samples. To each RT+ or RT- reaction, enough volume of each sample was added to attain 1 μg RNA in each tube. The thermal cycler (Model 480, Perkin Elmer, Warrington, UK) was pre-heated at 25 °C, and primers were allowed to anneal for 2 min at this temperature, continuing at 55 °C for cDNA synthesis during 10 min, and the final RT inactivation at 95 °C for 1 min. After cooling down the samples, the cDNA samples were stored at −20 °C until further use for RT-qPCR reactions.

The primer sequences were designed to span an intron or exon–exon junction through the NCBI Primer-BLAST tool (https://blast.ncbi.nlm.nih.gov/Blast.cgi) to selectively amplify cDNA (last accessed: 3 March 2020). The primer design criteria followed is displayed in Table S5 in the Supplementary Materials. Oligos were synthesised and purchased from Integrated DNA Technologies (IDT, Leuven, Belgium). The genes with their accession number and primer sequence, amplicon length, and melting temperature are presented in Table S6.

PCR experiments were performed with triplicate RT+ per sample, one no-reverse transcriptase control (RT-) per sample, and one no-template control (NTC) per gene and with the corresponding reference gene controls. The SYBR Green detection method was used for the detection of amplification with the kit PowerUp™ SYBR™ Green Master Mix (Applied Biosystems, Thermo Fisher Scientific, Paisley, UK). The forward and reverse primers (10 pM/μL) and molecular-grade water were mixed with Master Mix (2×). Samples consisting of 1 μL cDNA were pipetted into MicroAmp Fast Optical 96-well reaction plates (Applied Biosystems, Paisley, UK) with the previous mix to a final volume of 20 μL per well. The specific thermal cycling parameters (Table S7) were set according to the optimised PowerUp™ SYBR™ Green Master Mix for fast cycling mode in the StepOnePlus Real-Time PCR system (Applied Biosystems, Paisley, UK). At the end of the cycling process, a melt curve was produced and inspected for the occurrence of primer dimers, misprimes, and possible contamination of genomic DNA.

The method used to calculate the relative fold-change was the comparative C_T method [34], with C_T being the threshold cycle detected over the 40 run cycles. For gene normalisation, the expression of typical brain reference genes *Ywhaz*, *Tbp*, and *Pes1* was studied. Primer sequences are shown in Table S8. After quantification, the RefFinder web-based tool: https://www.heartcure.com.au/reffinder/ [35] was used to define the most stable reference gene for each tissue (last accessed: 2 March 2020). Full results of RefFinder analyses for each tissue are shown in Figures S1 and S2 in the Supplementary Materials.

2.7. Statistics

Shapiro–Wilk and Levene's tests were used as preliminary methods to evaluate data normality and homogeneity of variances, respectively. Independent samples *t*-test was performed to establish statistical comparisons between the two groups (i.e., control and treatment groups). For statistical comparisons between more groups, the tests selected were one-way analysis of variance (ANOVA) together with the Dunnett's post-hoc test. Statistical significance was declared when p-value < 0.05. The statistical software used was IBM SPSS Statistics 25 (IBM Corp. Armonk, NY, USA. Released 2017. IBM SPSS Statistics for Macintosh, version 25.0).

3. Results

3.1. Cobalt Accumulates in Organs and in Different Brain Structures

Figure 2 displays cobalt accumulation in all tissues tested and blood after 7- and 28-day treatments. ICP-MS results reveal that kidney, liver, and heart incorporated the highest Co content among the rats' organs in that order. After 28 days, pref. cortex, cerebellum, and hippocampus assimilated significant amounts of cobalt ($p < 0.01$, compared with the control rats). This was not the case for the 7-day treatment. Co levels detected in blood from the treated rats by ICP-MS were within the range found in MoM patients. After 28 days, Co detected in rat blood was 27.14 ± 2.70 μg/L compared with the average levels of 1–2 μg/L reported in MoM hip resurfacing patients [36]. The maximum cobalt concentration found in a patient with a THR prosthesis was 6521 μg/L [37].

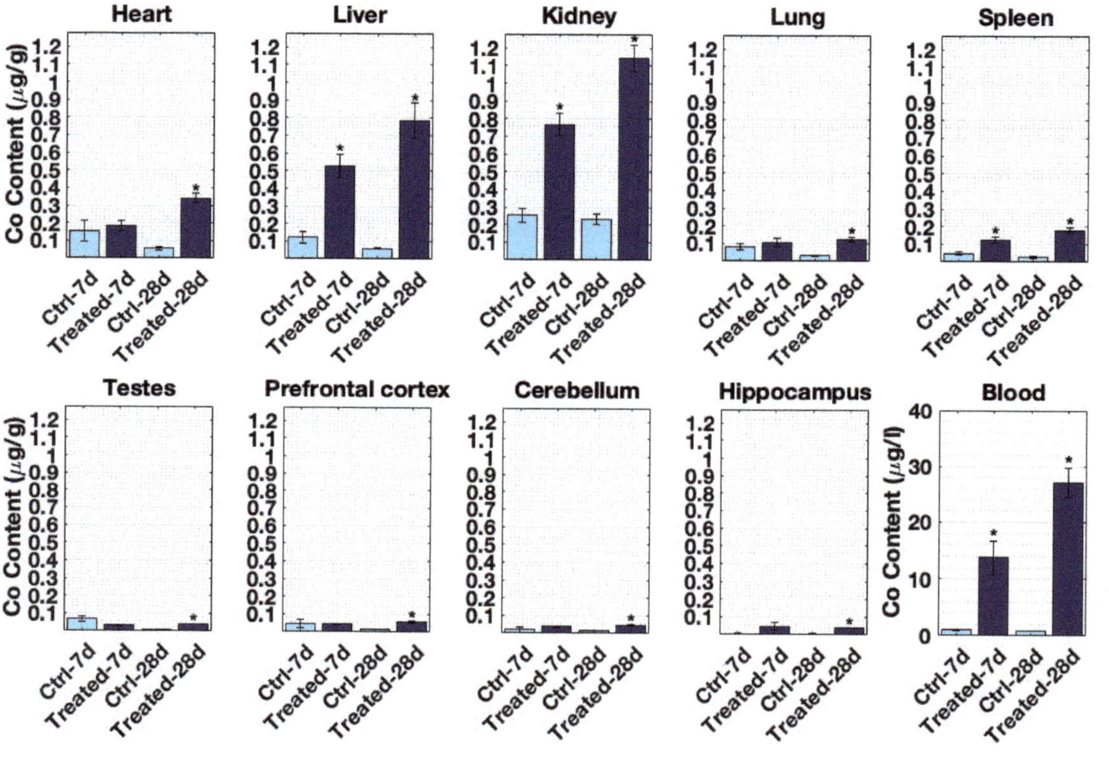

Figure 2. Cobalt content in SD male rats' tissues (ng/g) and blood (μg/L) at 7- and 28-days of daily i.p. $CoCl_2$ injection treatment as assessed by ICP-MS analysis. Control groups were instead injected with distilled water following the same procedures. Figure presents mean ± SEM calculated from $n = 3$ samples in control groups (dH_2O) and $n = 6$ in treatment groups (1 mg/kg B.W. $CoCl_2$). * significantly different between control group and treatment group at a given time-point as assessed by two sample t-test ($p < 0.05$).

Several tissues in Figure 2 presented increased cobalt accumulation from 7- to 28-days, indicating possible cobalt time-dependent accumulation in the heart, liver, kidney, spleen, pref. cortex, and blood. However, many of the control samples in the 7-day treatment indicated high levels of cobalt (heart, liver, kidney, lung, spleen, testes, pref. cortex, and cerebellum). This augmentation is likely an artefact of ICP-MS measurements. For values closer to the Co detection limit, it could be that cobalt content is pushed to higher values.

Figure 3 shows the cobalt content detected by ICP-MS in all organs after 28 days of cobalt dose-treatment: 0.1, 0.5, and 1 mg/kg B.W. $CoCl_2$. This metal content analysis revealed an incremental accumulation of cobalt concentrations in tissues with increased doses. Thus, there is a dose-response accumulation of cobalt, which was significant after 0.5 mg/kg B.W. $CoCl_2$ in most tissues. Kidney, liver, and heart in that order accumulated most cobalt in line with the time-response experiment trend (Figure 2). The pref. cortex and hippocampus also accumulated significant levels of cobalt ($p < 0.01$). The same issue was observed with regard to the offset in the control samples, which were closer to the detection limit of the ICP-MS due to their low cobalt content. This effect became obvious in the case of the cerebellum control group.

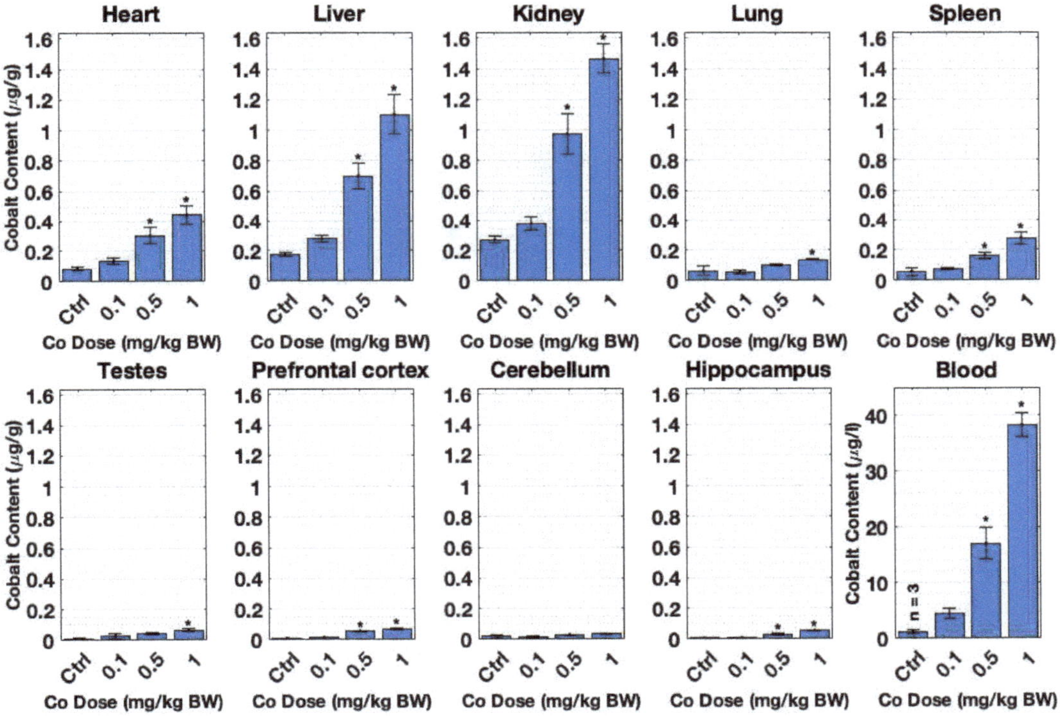

Figure 3. Organ cobalt content (μg/g, tissue; μg/L, blood) obtained by ICP-MS after tissue and blood collection. SD male rats were treated with dH_2O (control group) or different doses of $CoCl_2$: 0.1, 0.5, and 1 mg/kg B.W. Animals were dosed daily with i.p. injections for 28 days. Each group presents mean ± SEM from $n = 4$ rats, and * significant differences in control and treatment means as tested by one-way ANOVA ($p < 0.05$).

3.2. *The Transcriptional Response to the Cobalt Doses Selected Is Non-Proportional*

The cobalt tissue content measured through ICP-MS determined that the pref. cortex and hippocampus of 0.5 and 1 mg/kg BW $CoCl_2$-treated groups had significantly greater accumulated cobalt compared to their control groups (Figure 3). However, to evaluate whether the incremental dose cobalt treatment resulted in a progressive transcriptomic response, the number of genes was plotted for all brain areas and the three cobalt treatments: 0.1, 0.5, and 1 mg/kg B.W. $CoCl_2$ in Figure 4 (threshold = $|$fold change$|$ > 2). Although the number of DEGs progressively increased in the hippocampus, there was no clear dose-response in terms of the number of DEGs in the pref. cortex and cerebellum.

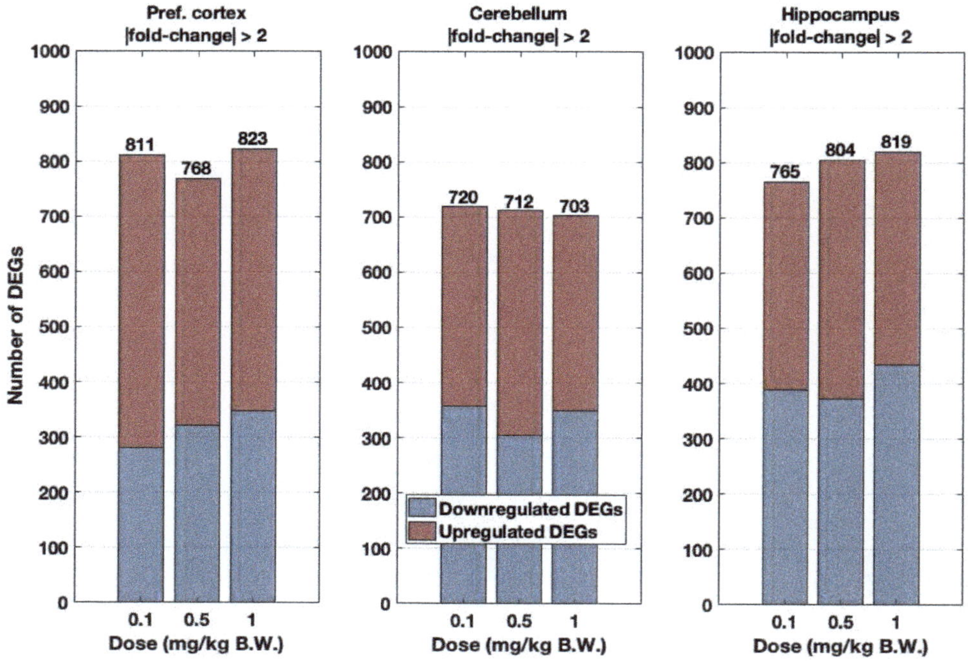

Figure 4. Number of upregulated (red) and downregulated (blue) DEGs (cutoff |fold change| > 2 only) in the pref. cortex, cerebellum, and hippocampus according to cobalt dose treatment: 0.1, 0.5, and 1 mg/kg B.W. $CoCl_2$. Animals were dosed i.p. daily for 28 days with those doses or dH_2O. Data were extracted from RNA-Seq experiments in which $n = 4$ samples were pooled to obtain $n' = 1$, except in the case of the hippocampus treatment group 0.5 mg/kg B.W. $CoCl_2$, where $n' = n = 3$ as well as for 1 mg/kg B.W. $CoCl_2$ where $n' = n = 1$.

Figure 5 shows the Venn diagram of DEGs found in the pref. cortex, cerebellum, and hippocampus. The diagram intersections display common DEGs between groups, with the pref. cortex and hippocampus demonstrating a higher number of common genes than the cerebellum for every dose. Although the number of overlapping genes between brain areas indicate that the hippocampus, cerebellum, and pref. cortex are all part of the same tissue (i.e., the brain), the different DEGs also point towards regionalised areas with specific functions. It can also be appreciated that there were few differences in the number of common genes between $CoCl_2$ dose regimes. Moreover, it was not possible to identify sets of DEGs that followed a dose-response fold change after hierarchical clustering of the common genes (Figure S3). Thus, we can conclude that the number and range of the doses used does not prompt a dose-response. This does not mean that the transcriptional response elicited is not relevant.

3.3. Global Transcriptional Response in the Pref. Cortex and Hippocampus

To further explore the effects on gene expression according to the dosage, this study focused on common DEGs for tissues with significant metal content accumulation. According to the results from the ICP-MS analyses, these tissues are the pref. cortex and hippocampus of rats dosed with 0.5 and 1 mg/kg B.W. $CoCl_2$ (Figure 3). The result of this hierarchical clustering is displayed in Figure 6. The gene enrichment analysis of these genes generated with Cytoscape is displayed in Table 1. Several GO terms of importance involved in immunity and hormone activity could be observed. In addition, the protein–protein interaction (PPI) of DEGs–protein products was created to observe the possible links between the overlapped DEGs (Figure 7). The immune axis centred around interleukin-6

(IL6) was clearly separated from other clusters related to growth factors and hormone activity as well as some UDP-glucuronosyltransferases, specifically UGT enzymes that are involved in glucuronidation. Another cluster of interest reflected on the PPI is the glycosylphosphatidylinositol (GPI) anchor biosynthesis.

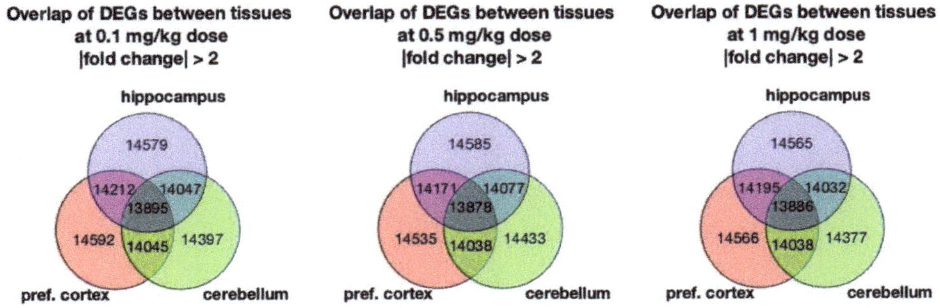

Figure 5. Venn diagrams showing the number of overlapping DEGs between the pref. cortex, cerebellum, and hippocampus at the different cobalt treatment doses: 0.1, 0.5, and 1 mg/kg B.W. $CoCl_2$. Rats were treated by daily i.p. injection for 28 days. DEGs were obtained through RNA-Seq by comparing the brain parts' mRNA abundance of the treatment groups against the controls (dH_2O-treated).

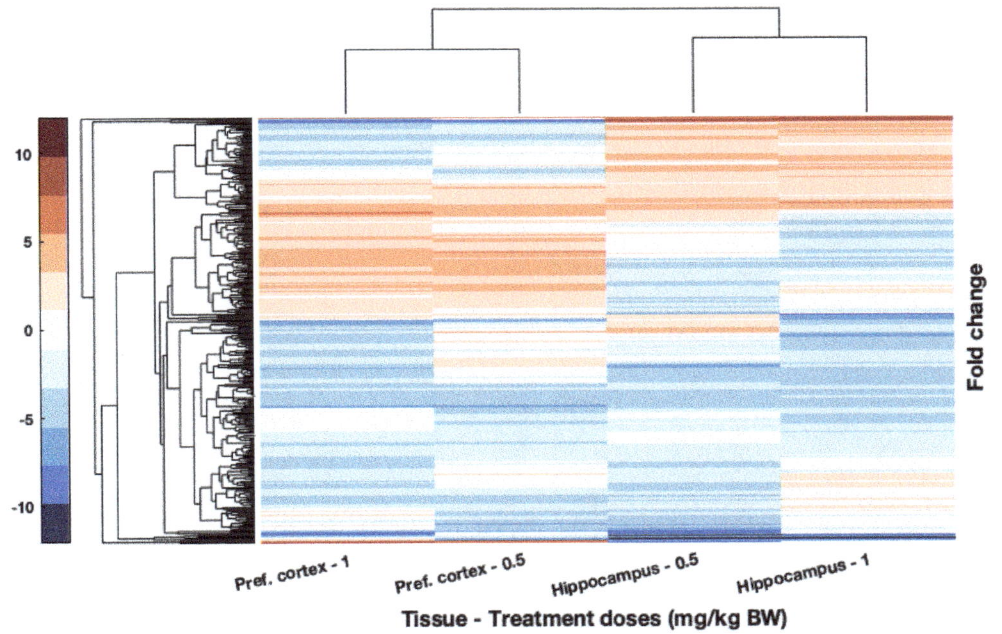

Figure 6. Hierarchical clustering of DEGs from brain tissues with significant accumulation of cobalt: the pref. cortex and hippocampus from rats treated with 0.5 and 1 mg/kg B.W. $CoCl_2$. DEGs were obtained from RNA-Seq comparing the RNA isolated from those tissues with those of controls treated with dH_2O. Condition applied is for fold change to be over 2. Upregulated genes are shown in red while downregulated are displayed in blue. Hierarchical clustering and resulting dendrogram were generated with Euclidian distance. Samples analysed through RNA-Seq were pooled ($n' = 1$) from $n = 4$ pref. cortex samples, $n = 3$ in hippocampus from 0.5 mg/kg B.W. $CoCl_2$ treatment group, and $n = 1$ from 1 mg/kg B.W. $CoCl_2$ treatment group.

Table 1. Enriched GO terms obtained from DEGs of the pref. cortex and hippocampus in response to cobalt treatment with 0.5 and 1 mg/kg B.W. $CoCl_2$ compared to control animals (dH_2O). Rats were dosed for 28 days with i.p. injections. GO terms annotated were significantly enriched with $p < 0.05$. GO terms enriched belong to the Molecular Function (MF; 8 April 2016), Biological Process (BP; 8 April 2016), Cellular Component (CC; 8 April 2016), and KEGG (14 June 2016) GO databases.

Gene Ontologies (GO) and GO Terms	Number of Genes with Annotations in the Ontology (%)	Number of Genes Represented in GO Terms (%)
Biological Process (BP) GO terms (8 April 2016)	194 (72.12%)	93 (34.57%)
Regulation of hormone levels	*Afp, Bco1, Bik, Cga, Cyp11a1, Cyp17a1, Fam3b, Ffar2, Fga, Fgf23, Gh1, Ghrh, Ghrhr, Il6, Nr0b2, Nt5c1b, Pax8, Slc30a8, Slco1a5, Sult1e1*	
Response to interleukin-6	*Crp, Fga, Fgf23, Ghrh, Hamp, Il6, Pck1*	
T-helper 17 cell lineage commitment	*Batf, Il6, Ly9*	
Response to vitamin	*Cyp11a1, Fgf23, Folr1, Hamp, Orm1, Otc, Sult2a1, Tshb*	
Response to pH	*Acer1, Gh1, Gja3, Kcnk18, Pck1*	
Response to interleukin-1	*Ccl21, Ccl25, Cyp11a1, Il6, Mmp3, Pck1, Slc30a8*	
Cell chemotaxis	*Ccl21, Ccl25, Ccr6, Cxcl13, Cxcl9, Cxcr2, Ffar2, Hrg, Stap1*	
Organ formation	*Folr1, Foxh1, Gdnf, Ntf4, Pax8*	
Cell fate commitment	*Batf, Elf5, Gata5, Gsx1, Gsx2, Il6, Ly9, Myl2, Ntf4, Olig3, Sostdc1*	
Molecular Function (MF) GO terms (8 April 2016)	185 (68.77%)	60 (22.3%)
Cytokine receptor binding	*Bmp10, Ccl21, Ccl25, Cxcl13, Cxcl9, Gh1, Il6, Inhbc, Ntf4, Stap1*	
Heparin binding	*Ang2, Comp, Cxcl13, Hrg, Mcpt4, Serpind1, Wisp3*	
Growth factor activity	*Areg, Bmp10, Fgf23, Gdnf, Il6, Inhbc, Ntf4*	
Steroid binding	*Comp, Crp, Cyp11a1, Fabp1, Sult1e1, Ugt1a1*	
Hormone activity	*Bmp10, Cga, Gh1, Ghrh, Gpha2, Hamp, Inhbc, Pyy, Rln3, Tshb*	
KEGG terms (14 June 2016)	97 (36.06%)	29 (10.78%)
Steroid hormone biosynthesis	*Cyp11a1, Cyp17a1, Cyp2c7, Cyp3a23/3a1, Sult1e1, Ugt1a1, Ugt2b17*	
Hematopoietic cell lineage	*Cd19, Cd3g, Cd8a, Fcer2, Il1r2, Il6*	
Cellular Component terms (8 April 2016)	216 (80.3%)	18 (6.69%)
External side of plasma membrane	*Cd19, Cd8a, Cxcl9, Fcer2, Fga, Folr1, Hyal5, Il6, Itgad, Trpm8*	

To better observe genes possibly involved in cobalt toxic mechanisms, the DEGs expressed over 2-fold change from the pref. cortex and hippocampus were plotted, as shown in Figures 8 and 9, respectively. For display purposes only, DEGs with a significance threshold <0.05 obtained from the Poisson distribution and provided in the original RNA-Seq data are shown. These are displayed as arranged by hierarchical clustering analyses of the DEGs (dendrogram not shown). Figure 8 displays the common genes of the pref. cortex across the three doses, and presents several genes whose proteins have a role in inflammation and immunity such as *Crp*, *Tnf*, and *Cxcl13*. Figure 9 shows significant DEGs of the hippocampus over 2-fold change. Surprisingly, there are several markers attributed to the choroid plexus such as *Clic6*, *Ttr*, *Kl*, *Col8a1*, and others [38–40].

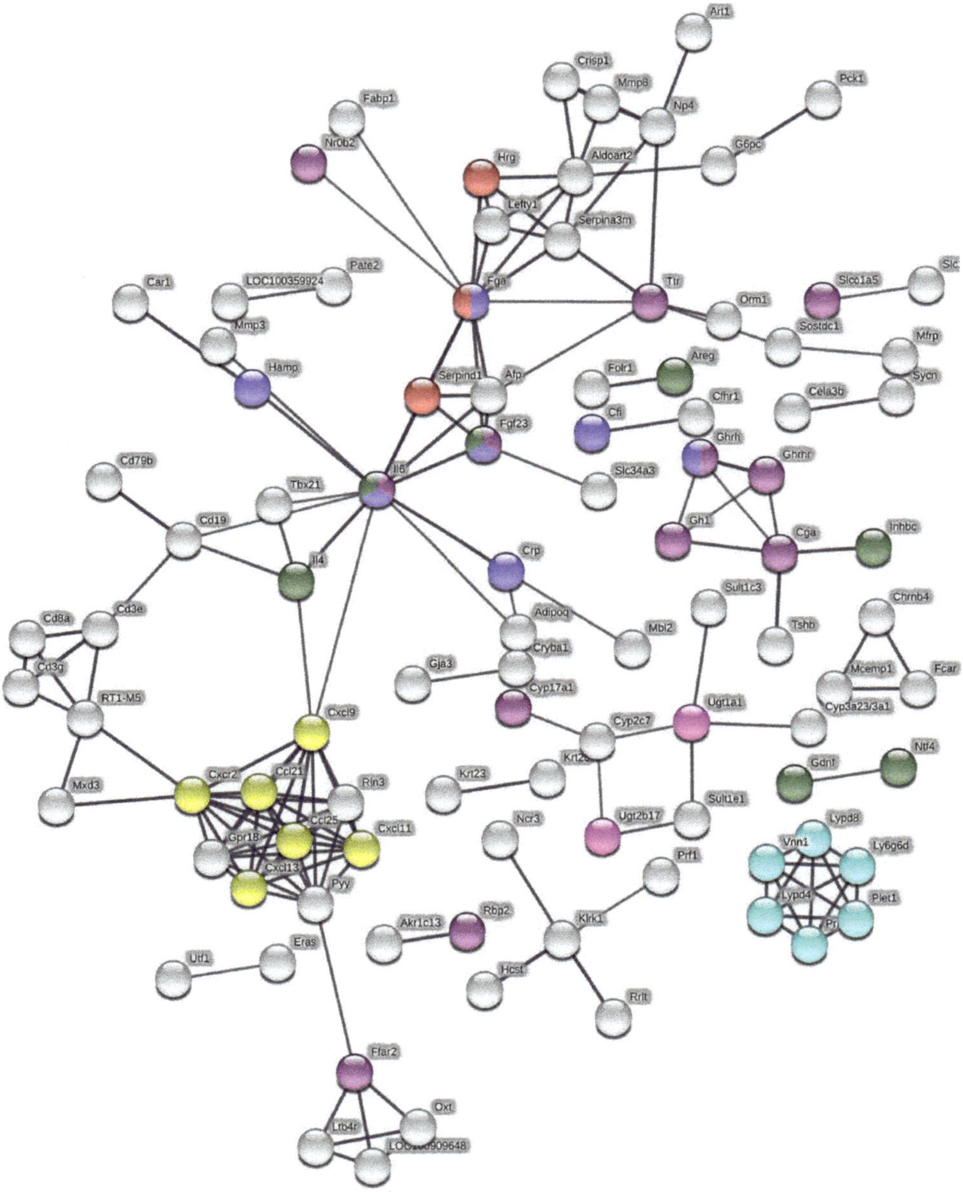

Figure 7. Protein–protein interaction (PPI) network obtained from the STRING web tool by analysing DEGs as their protein products. DEGs were obtained from RNA-Seq analyses of the pref. cortex and hippocampus from rats treated with 0.5 and 1 mg/kg B.W. $CoCl_2$ against controls treated with dH_2O for 28 days of i.p. injections. The following terms/keywords have been highlighted: cellular response to interleukin-6 (blue), regulation of hormone levels (purple), chemokine receptors bind chemokines (yellow), blood coagulation (red), UDP-glucuronosyltransferase activity (pink), growth factor activity (green), and post-translational modification: synthesis of GPI-anchored protein (cyan). The thickness of links between nodes represent the confidence in the interaction, only nodes connected with high confidence (0.7) are displayed.

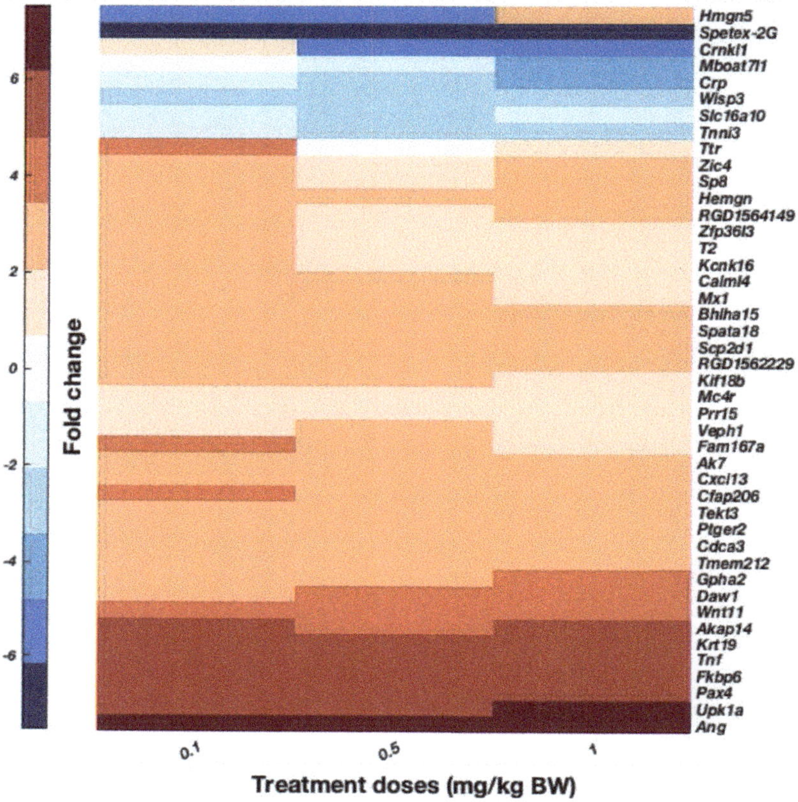

Figure 8. DEGs obtained from the comparison of pref. cortex from rats dosed with 0.1, 0.5, and 1 mg/kg B.W. $CoCl_2$ against the control group (dH_2O). Animals were treated for 28 days with daily i.p. injections. DEGs displayed were obtained from the RNA-Seq analysis of pooled samples ($n' = 1$ from $n = 4$ samples per group). Fold-change gene expression is indicated by colour, as described by the bar in the right side, upregulated genes are displayed in red while downregulated are blue. Genes are displayed as determined by the hierarchical clustering of DEGs over 2-fold-change ($p < 0.05$ from RNA-Seq Poisson distribution), dendrogram not shown (Euclidian distance).

A number of genes were evaluated via RT-qPCR according to their high expression level in RNA-Seq data as well as their function. The genes selected were *Tnf*, *Spata18*, *Ttr*, and *Akap14* in the pref. cortex and *Kl* in the hippocampus (see Figures S4 and S5 in the Supplementary Materials). In general, the RT-qPCR gene expression results agreed with the RNA-Seq data, and the fold expression of evaluated DEGs from RNA-Seq and RT-qPCR in this study did correlate. This is consistent with previous high-throughput comparisons between the two technologies [41] as well as research with pooled samples [42]. Thus, it was considered that the RT-qPCR results validate the RNA-Seq results.

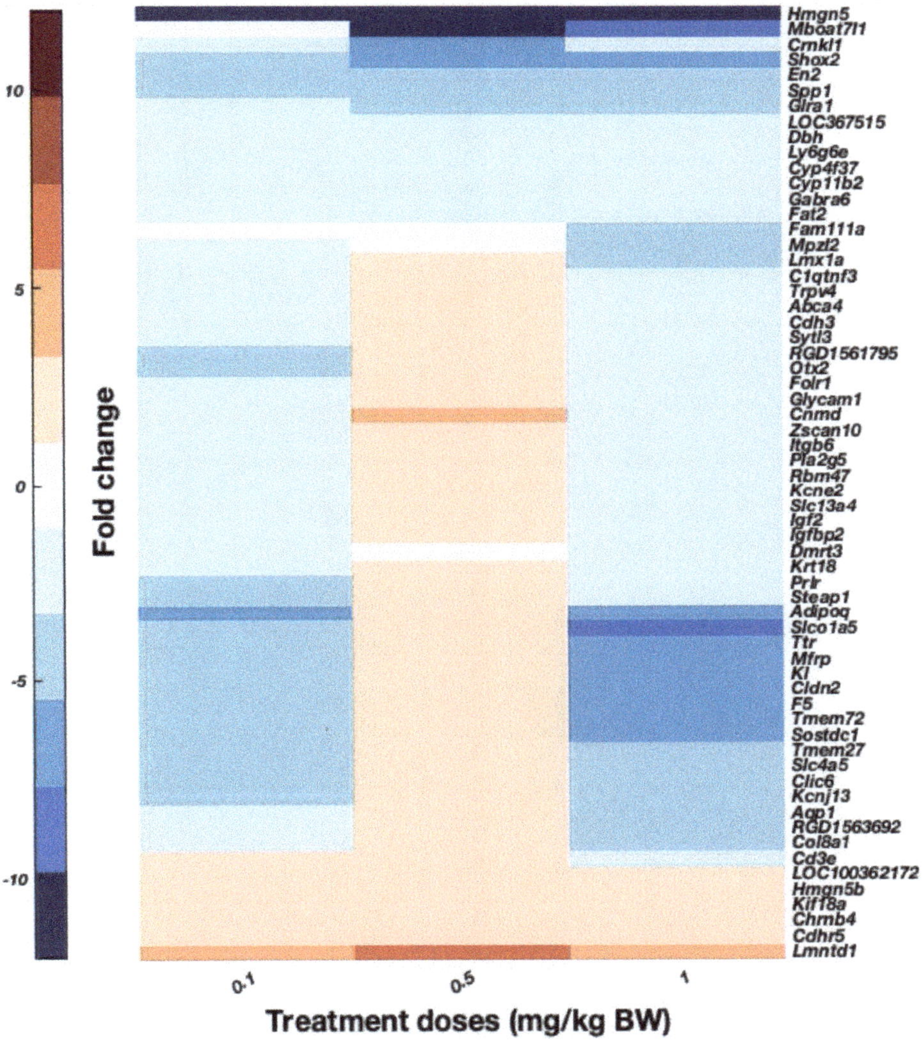

Figure 9. DEGs expressed in hippocampus of rats treated via i.p. with daily injections of 0.1, 0.5, and 1 mg/kg B.W. $CoCl_2$ or dH_2O (control groups) for 28 days. Pooled samples (n' = 1 from n = 4 samples in group 0.1 mg/kg B.W. $CoCl_2$, n = 3 from 0.5 mg/kg B.W. $CoCl_2$, and n = 1 from 1 mg/kg B.W. $CoCl_2$ group) were analysed through RNA-Seq and data are presented as the result of hierarchical clustering. DEGs in the graph are only those with fold-change >2 and $p < 0.05$ from RNA-Seq Poisson distribution, dendrogram is not shown. Colour bar presents fold-change: upregulated genes in red and downregulated genes in blue.

4. Discussion

4.1. Metal Distribution Pattern Is Consistent with Previous Research and Cobalt Accumulates Significantly in the Brain

Figure 2 shows that there is a time-dependent cobalt accumulation trend in most tissues. This time-dependent trend might indicate that the longer cobalt levels remain elevated in blood, with a higher deposition of cobalt in nearly all organs tested, particularly in the heart, liver, and kidney. The accumulation of cobalt in most organs also appears to be proportional to the dose used, as displayed in Figure 3. In addition, cobalt levels were significantly increased in the pref. cortex and hippocampus of rats dosed with 0.5 and 1 mg/kg B.W. $CoCl_2$ at 28 days.

The liver, kidney, and heart accumulated more metal than other organs (Figures 2 and 3), which correlates with cobalt organ concentration in humans following cobalt radioisotope distribution [13]. This cobalt distribution follows the same pattern as previous experiments conducted by this research group [43,44] and other teams investigating cobaltism [45]. Most cobalt is excreted through urine [13,46], and given that both kidney and liver are involved in toxin and waste detoxification, it is not surprising that cobalt is predominantly concentrated in these tissues.

With regard to the cobalt in blood after the 28-day treatment with 1 mg/kg B.W., the level was 27.15 ± 2.70 μg/L in the time-response experiment while it was 38.24 ± 2.14 μg/L in the dose-response experiment. Given that the blood cobalt levels were under 100 μg/L, only subtle or moderate neurotoxicity was expected [1].

The cobalt content of two hearts from MoM patients severely affected by systemic cobalt toxicity has been reported in the literature as 4.75 [47] and 8.32 μg/g [48] (reference values for heart cobalt content: 0.06 μg/g) [49]. In addition, other post-mortem analyses on patients with metal-on-polyethylene hip prostheses also revealed significantly elevated cobalt content averaging 0.12 μg/g (range: 0.006–6.299 μg/g) [49]. Table 2 shows a compact presentation of these figures and results for comparison. The higher values of 0.34 ± 0.03 μg/g and 0.44 ± 0.06 μg/g in the rats' hearts at 28 days with 1 mg/kg B.W. $CoCl_2$ treatment were also close to the average content in asymptomatic metal-on-polyethylene (MoP) patients (0.12 μg/g) [49]. Hence, the gradual dosing model used by this study and in previous studies [43] is comparable to the long-term systemic exposure of cobalt through circulating blood in MoM patients.

Table 2. Cobalt concentrations of whole blood (WB), brain and heart tissues in the dose and time-response experiments and other studies that mimic gradual cobalt release through daily treatment [45], cobalt tissue analyses in the cardiac tissue, and serum of MoM patients with cobaltism [47,48] as well as post-mortem heart tissue from metal-on-polyethylene (MoP) patients [49]. Cobalt content values in unexposed human brain (<0.025 μg/g), heart (0.060 μg/g) and blood (<1 μg/L) were obtained from [36,49,50] in that order. The abbreviations are: i.p., intraperitoneal injections; WB, whole blood; avg., average; C, control; T, treatment; THA, total hip arthroplasty. * significantly different control and treatment groups as assessed by one-way ANOVA with Dunnett's multiple comparison.

Studies		**Time-Response Study**	**Dose-Response Study**	**[45]**	**[47]**	**[48]**	**[49]**
Study outline		$CoCl_2$ SD rats 28 days 1 mg/kg BW i.p.	$CoCl_2$ SD rats 28 days 1 mg/kg BW i.p.	$CoCl_2$ rabbits 18 days 1354 μg/mL Intravenous infusion	MoM patient	MoM patient	MoP Patients
Tissues		*n* = 3 (C) and 6 (T)	*n* = 4 (C) and 4 (T)	*n* = 2 (C) and 4 (T)	case report	case report	*n* = 73 (C), 75 (THA)
WB or serum (μg/L)	Cobalt exposed	27.15 ± 2.70 * (WB)	38.24 ± 2.14 * (WB)	420.9 ± 154.5 (WB)	192 (serum)	287.6 (serum)	Unknown
	Controls/unexposed	0.87 ± 0.00	1.17 ± 0.32	11.7 ± 2.7		<1	
Brain (μg/g)	Cobalt exposed	0.06 ± 0.00 * (pref. cortex)	0.07 ± 0.01 * (pref. cortex)	0.2 ± 0.2		Unknown	
	Controls	0.02 ± 0.00	0.01 ± 0.00	0.06 ± 0.04		<0.025	
Heart (μg/g)	Cobalt exposed	0.34 ± 0.03 *	0.44 ± 0.06 *	0.7 ± 0.5	4.75	8.32	0.12 (avg.); range: 0.006–6.299
	Controls/unexposed	0.06 ± 0.01	0.09 ± 0.01	0.07 ± 0.1		0.06	

To our knowledge, there is no information on the cobalt concentrations of MoM patients in the brain, probably due to the difficulty of obtaining samples from patients. In this study, the pref. cortex, and hippocampus had significant cobalt accumulated at 28 days (Figures 2 and 3). A study by Apostoli et al. with rabbits dosed with cobalt for

18 days intravenously reported elevated brain levels, 0.2 ± 0.2 μg/g, from the average control levels, 0.06 ± 0.04 μg/g dry weight, and cobalt whole blood, 420.9 ± 154.5 μg/L (see Table 2 for reference [45]). The histology in several organs only reported damage to the eyes and the auditory systems. The model used for this study had much lower cobalt blood concentration and animals remained in good health, while Apostoli et al. described balance disturbance in rabbits due to vestibular damage. Given the literature and observations of this study, cobalt appeared to induce only subtle or moderate neurotoxicity in this study even when cobalt had significantly accumulated in the rats' brains.

4.2. The Low Range Cobalt Dosage Used Does Not Lead to a Dose-Response

No dose-response was demonstrated either in the number or in the average fold change of DEGs elicited by cobalt treatment (Figure 4 and Figure S3). Previous studies suggest that the fraction of ionic cobalt remains constant throughout a wide range of cobalt concentrations in the blood due to its albumin binding capacity [13]. If cobalt is being sequestered by albumin, it is likely that the response to any administered cobalt treatment will be dampened. However, it is also possible for cobalt toxic effects to follow a dose response, but the few concentrations used here did not cover such a range of toxicity.

4.3. Overall Transcriptional Effects of Cobalt and the Choroid Plexus as a Target of Cobalt Toxicity

This study found some GO terms of interest that could be associated with cobalt toxicity in Table 1. This is the case of 'steroid hormone biosynthesis', which is comprised of a few genes of the Cytochrome P450 (*Cyp* prefix) family as well as by a couple UDP-glucuronosyltransferases (*Ugt* prefix) and a sulfotransferase (*Sult* prefix). The protein products of these gene families form part of the drug metabolism and detoxification pathways. CYPs are also involved in the biosynthesis of serotonin, dopamine [51], and steroid hormones [52]. The findings suggest that CYPs could be a target of cobalt since they normally bind to the heme metal substrate as some metal ions have been observed to inactivate members of the CYP family [53]. Nevertheless, there are other large GO terms linked with hormone homeostasis such as 'regulation of hormone levels', 'steroid binding', and 'hormone activity' in Table 1. Some of these are not directly related to steroid hormones e.g., essential thyroid related genes (*Tshb*), whose deletion leads to hypothyroidism, and other GO terms such as 'response to retinoic acid'. Thus, nuclear receptors could possibly be regulated by cobalt. Nuclear receptors may bind steroids, retinoic acid, or thyroid hormones, and this binding depends on the nuclear receptor zinc finger domain. The CYP family synthesises some of the ligands that the nuclear receptors bind to, and further experiments could be performed to ascertain whether cobalt binds to CYPs or to nuclear receptors [54].

The PPI analysis (Figure 7) displayed a regulation of hormone levels consistently with the GO enrichment analysis as well as a network of drug-metabolising enzymes consisting of a few *Cyp*, *Ugt*, and *Sult* genes. The immune and haematopoietic axis centred around *Il6* with important chemokine presence can also be observed. This immune response was also reflected in the GO enrichment from Table 1. Some of the immune-related GO terms are 'response to interleukin-6', 'T-helper 17 cell lineage commitment', 'response to interleukin-1', 'cell chemotaxis', and 'cytokine receptor binding', which suggest immune cell differentiation, activity, and migration. Other factors related to blood coagulation such as *Hrg*, *Serpind1*, and *Fga* are displayed in Table 1. Activation of the immune system and dysregulation of haematopoietic transcriptional programmes could lead to several autoimmune and blood disease syndromes. Cobalt was a historical treatment of anaemia, and polycythaemia and skin rashes have been documented as a sporadic result of cobalt treatment [13]. Finally, the PPI also reported the presence of 'GPI-anchored protein' related transcripts. This is a post-transcriptional modification that mainly occurs in the endoplasmic reticulum, where most glycosylphosphatidylinositol (GPI) synthesis proteins function [55]. GPI-anchored protein synthesis depends on phospholipids, while the synthesis of steroid hormones is determined by cholesterol, hence, it is suggested here that cobalt modulates lipid metabolism.

It is possible that cobalt could modulate lipid metabolism directly as cobalt has been seen to affect the rigidity of lipid membranes such as liposomes [56]. Lipid droplets were present in the spleen of a patient with a CoCr prosthesis [57]. Moreover, intracytoplasmic lipid and lipofuscin accumulation were found in a recent heart biopsy of a patient with arthroprosthetic cobaltism [48].

This study also found genetic markers in the pref. cortex (Figure 8) and hippocampus (Figure 9) almost exclusively attributable to the choroid plexus (e.g., *Clic6*, *Klotho* (*Kl*), transthyretin (*Ttr*), *Veph1*, some cilia markers, and *Scl* transporters). The molecular characterisation of the choroid plexus has only been achieved recently [38], and unfortunately the GO ontologies have not been adequately updated to report its presence. The choroid plexus is anatomically attached to the hippocampus and its joint dissection can go unnoticed when doing a fast isolation, as reported by specialists in the choroid plexus [39,58]. In fact, several studies investigating the effect of drugs or other interventions in the hippocampus have knowingly or unwittingly reported choroid plexus markers [58–60]. There are also markers of the choroid plexus such as transthyretin (*Ttr*) in the pref. cortex, and it is possible that part of the choroid plexus has also been included in brain samples other than the hippocampus, since the choroid plexus is distributed through all brain ventricles [58]. Different studies have revealed that heavy metals preferentially accumulate early on in the choroid plexus, which appears to retain them, thus protecting the brain [61–66]. Harrison-Brown et al. discovered that the penetration of cobalt in the cerebrospinal fluid (CSF) of MoM patients was limited to 15% of the cobalt in plasma [67]. They also found a nonlinear trend with a ceiling effect in the CSF cobalt accumulation in relation to Co plasma levels in blood. Thus, the choroid plexus could function as an absorptive barrier, and early cobalt accumulation and damage in the brain might occur in the choroid plexus.

The choroid plexus is also a place for steroid hormone biosynthesis [68] and it hosts metabolising enzymes to deal with and metabolise xenobiotics [69]. Many immune cells are resident in the choroid plexus, which works as the site for immune trafficking with the brain [70]. B and T lymphocytes can infiltrate the choroid plexus and affect its function under certain challenges, thus leading to inflammation [71–73]. In particular, the 'cell chemotaxis' GO term includes *Cxcl13*, a chemokine that recruits B lymphocytes, and it has been involved in lymphoid infiltration in the choroid plexus in a mouse model of neuropsychiatric lupus [73]. Moreover, very recently, modulation of *Otx2* expression in the choroid plexus has been seen to regulate anxiogenic behaviour in mice [74], and in two studies, the choroid plexus transcriptome quickly responded to stress tests in mice [39,58]. Given that the choroid plexus has been implicated in depression disorders [75] and that some patients with elevated cobalt levels in their blood showed signs of neuropsychiatric symptoms such as depression [8,10], one might speculate that cobalt toxicity in the choroid plexus could impair its function, and contribute towards mood dysregulation.

5. Conclusions

In summary, although the rat pref. cortex and hippocampus accumulated lower amounts of cobalt than other tissues, these accumulations were still significant at similar elevated circulating cobalt levels to those found in some patients with MoM implants. We found that the common transcriptional response to cobalt in the brain areas analysed involved hormone and drug-metabolising activity, in addition to also describing a powerful immune response, perhaps mediated by inteleukin-6 (IL-6). An underlying dysfunction in lipid metabolism is also likely. We suggest that these mechanisms could be instigated as a consequence of cobalt ion binding and substitution of native metal ions of CYPs or nuclear receptors. In the future, researchers should consider evaluating the markers of inflammation and lymphoid cell activation as well as the steroidogenic activity in the choroid plexus in response to cobalt. Thus, we have generated a mechanistic hypothesis for cobalt neurotoxicity that could be explored further and have relevant implications for patients with MoM implants who develop neurological health issues.

Supplementary Materials: The following are available online at https://www.mdpi.com/article/10.3390/toxics10020059/s1, Table S1. Quality check of RNA samples from the pref. cortex tissue obtained from the dose-response in vivo experiments, Table S2. Quality check of RNA samples from the cerebellum tissues obtained from the in vivo dose-response experiments, Table S3. Quality check of RNA samples from the hippocampus tissues obtained from the in vivo dose-response experiments, Table S4. MIQE checklist from the MIQE guidelines for reproducibility and assessment of experimental RT-qPCR conditions, Table S5. Criteria for the design of the primers selected through NCBI Primer-BLAST tool, Table S6. Primer sequences of targeted genes designed for the in vivo dose-response experiment, Table S7. Fast PCR thermal cycling steps based on PowerUp™ SYBR™ Green Master Mix instructions for StepOnePlus Real-Time PCR system, Table S8: Primer sequences of control genes, Figure S1. RefFinder ranking of three reference genes Ct values in pref. cortex samples from the in vivo dose-response experiment, Figure S2. RefFinder ranking of three reference genes Ct values in hippocampus samples from the in vivo dose-response experiment, Figure S3. Hierarchical clustering of DEGs over 2-fold-change from RNA-Seq data obtained from the comparison of the pref. cortex, cerebellum, and hippocampus of rats treated with three concentrations of cobalt against those of controls treated with dH_2O, Figure S4: Fold change mRNA gene expression levels obtained from RNA-Seq and RT-qPCR, Figure S5: Quantification of *Tnf*, *Spata18*, *Ttr*, and *Akap14* delta (CT) values (CT target gene–CT internal control) in the pref. cortex and *Kl* in the hippocampus through RT-qPCR.

Author Contributions: Conceptualization, S.G.-A., R.J.T. and M.H.G.; Methodology, S.G.-A., R.J.T. and M.H.G.; Software, S.G.-A.; Validation, S.G.-A.; Formal analysis, S.G.-A.; Investigation, S.G.-A.; Resources, R.J.T. and M.H.G.; Data curation, S.G.-A.; Writing—original draft preparation, S.G.-A.; Writing—review and editing, S.G.-A., R.J.T. and M.H.G.; Visualization, S.G.-A.; Supervision, R.J.T. and M.H.G.; Project administration, M.H.G.; Funding acquisition, M.H.G. All authors have read and agreed to the published version of the manuscript.

Funding: This research was funded by an Engineering the Future (ETF) studentship awarded by the University of Strathclyde (internal funding, Faculty of Engineering, University of Strathclyde).

Institutional Review Board Statement: All animal procedures were conducted under the UK Home Office project licenses, 60/4341 and PDE5626B67. Ethic Committee Name: UK Home Office and AWERB. Approval Code: 60/4341 and PDE5626B67 (UK Home Office licence numbers). Approval Date: 25th April 2017.

Informed Consent Statement: Not applicable.

Data Availability Statement: All data underpinning this publication are openly available from the University of Strathclyde KnowledgeBase at https://doi.org/10.15129/cbbe122f-4ee1-4bfb-9ec7-46c664bf10f2 (accessed on 29 November 2021).

Acknowledgments: Katie Henderson, Sarunya Laovitthayanggoon, Laura Beattie, and Ibrahim Alanazi provided invaluable technical and human support during the experiments. We are also grateful for the assistance and technical support of the Biological Procedures Unit (BPU) from the University of Strathclyde.

Conflicts of Interest: The authors declare no conflict of interest. The funders had no role in the design of the study; in the collection, analyses, or interpretation of data; in the writing of the manuscript, or in the decision to publish the results.

References

1. Kovochich, M.; Finley, B.L.; Novick, R.; Monnot, A.D.; Donovan, E.; Unice, K.M.; Fung, E.S.; Fung, D.; Paustenbach, D.J. Understanding outcomes and toxicological aspects of second generation metal-on-metal hip implants: A state-of-the-art review. *Crit. Rev. Toxicol.* **2018**, *48*, 853–901. [CrossRef] [PubMed]
2. Langton, D.J.; Jameson, S.S.; Joyce, T.J.; Hallab, N.J.; Natu, S.; Nargol, A.V.F. Early failure of metal-on-metal bearings in hip resurfacing and large-diameter total hip replacement. *J. Bone Jt. Surg. Br.* **2010**, *92*, 38–46. [CrossRef] [PubMed]
3. Sidaginamale, R.P.; Joyce, T.J.; Bowsher, J.G.; Lord, J.K.; Avery, P.J.; Natu, S.; Nargol, A.V.F.; Langton, D.J. The clinical implications of metal debris release from the taper junctions and bearing surfaces of metal-on-metal hip arthroplasty. *Bone Jt. J.* **2016**, *98*, 925–933. [CrossRef] [PubMed]
4. Goode, A.E.; Perkins, J.M.; Sandison, A.; Karunakaran, C.; Cheng, H.; Wall, D.; Skinner, J.A.; Hart, A.J.; Porter, A.E.; McComb, D.W.; et al. Chemical speciation of nanoparticles surrounding metal-on-metal hips. *Chem. Commun.* **2012**, *48*, 8335. [CrossRef] [PubMed]

5. MDA/2017/018. Medical Device Alert. All Metal-On-Metal (MoM) Hip Replacements: Updated Advice For Follow-Up of Patients. Available online: http://www.mhra.gov.uk/ (accessed on 29 November 2021).
6. Epstein, M.; Emri, I.; Hartemann, P.; Hoet, P.; Leitgeb, N.; Martínez Martínez, L.; Proykova, A.; Rizzo, L.; Rodriguez-Farré, E.; Rushton, L.; et al. Final Opinion on the Safety of Metal-on-Metal Joint Replacements with a Particular Focus on Hip Implants. Sci. Comm. Emerg. New. Identified Heal. Risks, no. September, 2014. Available online: https://ec.europa.eu/health/sites/default/files/scientific_committees/emerging/docs/scenihr_o_042.pdf (accessed on 29 November 2021).
7. Rizzetti, M.C.; Liberini, P.; Zarattini, G.; Catalani, S.; Pazzaglia, U.; Apostoli, P.; Padovani, A. Loss of sight and sound. Could it be the hip? *Lancet* **2009**, *373*, 1052. [CrossRef]
8. Tower, S.S. Arthroprosthetic Cobaltism: Neurological and Cardiac Manifestations in Two Patients with Metal-on-Metal Arthroplasty: A Case Report. *J. Bone Jt. Surg. Am.* **2010**, *92*, 2847–2851. [CrossRef]
9. Mao, X.; Wong, A.A.; Crawford, R.W. Cobalt toxicity–an emerging clinical problem in patients with metal-on-metal hip prostheses? *Med. J. Aust.* **2011**, *194*, 649–651. [CrossRef]
10. Green, B.; Griffiths, E.; Almond, S. Neuropsychiatric symptoms following metal-on-metal implant failure with cobalt and chromium toxicity. *BMC Psychiatry* **2017**, *17*, 33. [CrossRef]
11. Catalani, S.; Rizzetti, M.C.; Padovani, A.; Apostoli, P. Neurotoxicity of cobalt. *Hum. Exp. Toxicol.* **2012**, *31*, 421–437. [CrossRef]
12. Machado, C.; Appelbe, A.; Wood, R. Arthroprosthetic Cobaltism and Cardiomyopathy. *Hear. Lung. Circ.* **2012**, *21*, 759–760. [CrossRef]
13. Paustenbach, D.J.; Tvermoes, B.E.; Unice, K.M.; Finley, B.L.; Kerger, B.D. A review of the health hazards posed by cobalt. *Crit. Rev. Toxicol.* **2013**, *43*, 316–362. [CrossRef] [PubMed]
14. Ho, J.H.; Leikin, J.B.; Dargan, P.I.; Archer, J.R.; Wood, D.M.; Brent, J. Metal-on-Metal Hip Joint Prostheses: A Retrospective Case Series Investigating the Association of Systemic Toxicity with Serum Cobalt and Chromium Concentrations. *J. Med. Toxicol.* **2017**, *13*, 321–328. [CrossRef] [PubMed]
15. Karovic, O.; Tonazzini, I.; Rebola, N.; Edström, E.; Lövdahl, C.; Fredholm, B.B.; Dare, E. Toxic effects of cobalt in primary cultures of mouse astrocytes. Similarities with hypoxia and role of HIF-1α. *Biochem. Pharmacol.* **2007**, *73*, 694–708. [CrossRef] [PubMed]
16. Wang, P.; Li, L.; Zhang, Z.; Kan, Q.; Chen, S.; Gao, F. Time-dependent homeostasis between glucose uptake and consumption in astrocytes exposed to CoCl2 treatment. *Mol. Med. Rep.* **2016**, *13*, 2909–2917. [CrossRef] [PubMed]
17. Gómez-Arnaiz, S.; Tate, R.J.; Grant, M.H. Cytotoxicity of cobalt chloride in brain cell lines-a comparison between astrocytoma and neuroblastoma cells. *Toxicol. Vitr.* **2020**, *68*, 104958. [CrossRef] [PubMed]
18. Li, P.; Ding, D.; Salvi, R.; Roth, J.A. Cobalt-Induced Ototoxicity in Rat Postnatal Cochlear Organotypic Cultures. *Neurotox. Res.* **2015**, *28*, 209–221. [CrossRef]
19. Kikuchi, S.; Ninomiya, T.; Kohno, T.; Kojima, T.; Tatsumi, H. Cobalt inhibits motility of axonal mitochondria and induces axonal degeneration in cultured dorsal root ganglion cells of rat. *Cell Biol. Toxicol.* **2018**, *34*, 93–107. [CrossRef]
20. Chimeh, U.; Zimmerman, M.A.; Gilyazova, N.; Li, P.A. B355252, a novel small molecule, confers neuroprotection against cobalt chloride toxicity in mouse hippocampal cells through altering mitochondrial dynamics and limiting autophagy induction. *Int. J. Med. Sci.* **2018**, *15*, 1384–1396. [CrossRef]
21. Naves, T.; Jawhari, S.; Jauberteau, M.O.; Ratinaud, M.H.; Verdier, M. Autophagy takes place in mutated p53 neuroblastoma cells in response to hypoxia mimetic $CoCl_2$. *Biochem. Pharmacol.* **2013**, *85*, 1153–1161. [CrossRef]
22. Fung, F.K.; Law, B.Y.; Lo, A.C. Lutein Attenuates Both Apoptosis and Autophagy upon Cobalt (II) Chloride-Induced Hypoxia in Rat Műller Cells. *PLoS ONE* **2016**, *11*, e0167828. [CrossRef]
23. Mou, Y.H.; Yang, J.Y.; Cui, N.; Wang, J.M.; Hou, Y.; Song, S.; Wu, C.F. Effects of cobalt chloride on nitric oxide and cytokines/chemokines production in microglia. *Int. Immunopharmacol.* **2012**, *13*, 120–125. [CrossRef] [PubMed]
24. McMullen, P.D.; Pendse, S.N.; Black, M.B.; Mansouri, K.; Haider, S.; Andersen, M.E.; Clewell, R.A. Addressing systematic inconsistencies between in vitro and in vivo transcriptomic mode of action signatures. *Toxicol. Vitr.* **2019**, *58*, 1–12. [CrossRef] [PubMed]
25. Joseph, P. Transcriptomics in toxicology. *Food Chem. Toxicol.* **2017**, *109*, 650–662. [CrossRef] [PubMed]
26. Garoui, E.; Amara, I.B.; Driss, D.; Elwej, A.; Chaabouni, S.E.; Boudawara, T.; Zeghal, N. Effects of Cobalt on Membrane ATPases, Oxidant, and Antioxidant Values in the Cerebrum and Cerebellum of Suckling Rats. *Biol. Trace Elem. Res.* **2013**, *154*, 387–395. [CrossRef]
27. Caltana, L.; Merelli, A.; Lazarowski, A.; Brusco, A. Neuronal and Glial Alterations Due to Focal Cortical Hypoxia Induced by Direct Cobalt Chloride (CoCl2) Brain Injection. *Neurotox. Res.* **2009**, *15*, 348–358. [CrossRef]
28. Kajiwara, K.; Sunaga, K.; Tsuda, T.; Sugaya, A.; Sugaya, E.; Kimura, M. Peony root extract upregulates transthyretin and phosphoglycerate mutase in mouse cobalt focus seizure. *Biochem. Biophys. Res. Commun.* **2008**, *371*, 375–379. [CrossRef]
29. Denisov, V.; Strong, W.; Walder, M.; Gingrich, J.; Wintz, H. Development and Validation of RQI: An RNA Quality Indicator for the Experion Automated Electrophoresis System. *Bio-Rad Bull* **2008**, 5761. Available online: http://www.gene-quantification.org/Bio-Rad-bulletin-5761.pdf (accessed on 29 November 2021).
30. Shannon, P.; Markiel, A.; Ozier, O.; Baliga, N.S.; Wang, J.T.; Ramage, D.; Amin, N.; Schwikowski, B.; Ideker, T. Cytoscape: A software environment for integrated models of biomolecular interaction networks. *Genome Res.* **2003**, *13*, 2498–2504. [CrossRef]

31. Bindea, G.; Mlecnik, B.; Hackl, H.; Charoentong, P.; Tosolini, M.; Kirilovsky, A.; Fridman, W.-H.; Pagès, F.; Trajanoski, Z.; Galon, J. ClueGO: A Cytoscape plug-in to decipher functionally grouped gene ontology and pathway annotation networks. *Bioinformatics* **2009**, *25*, 1091–1093. [CrossRef]
32. Szklarczyk, D.; Gable, A.L.; Lyon, D.; Junge, A.; Wyder, S.; Huerta-Cepas, J.; Simonovic, M.; Doncheva, N.T.; Morris, J.H.; Bork, P.; et al. STRING v11: Protein-protein association networks with increased coverage, supporting functional discovery in genome-wide experimental datasets. *Nucleic Acids Res.* **2019**, *47*, D607–D613. [CrossRef]
33. Bustin, S.A.; Benes, V.; Garson, J.A.; Hellemans, J.; Huggett, J.; Kubista, M.; Mueller, R.; Nolan, T.; Pfaffl, M.W.; Shipley, G.L. The MIQE Guidelines: Minimum Information for Publication of Quantitative Real-Time PCR Experiments. *Clin. Chem.* **2009**, *55*, 611–622. [CrossRef]
34. Schmittgen, T.D.; Livak, K.J. Analyzing real-time PCR data by the comparative CT method. *Nat. Protoc.* **2008**, *3*, 1101–1108. [CrossRef] [PubMed]
35. Xie, F.; Xiao, P.; Chen, D.; Xu, L.; Zhang, B. miRDeepFinder: A miRNA analysis tool for deep sequencing of plant small RNAs. *Plant Mol. Biol.* **2012**, *80*, 75–84. [CrossRef] [PubMed]
36. Sidaginamale, R.P.; Joyce, T.J.; Lord, J.K.; Jefferson, R.; Blain, P.G.; Nargol, A.V.F.; Langton, D.J. Blood metal ion testing is an effective screening tool to identify poorly performing metal-on-metal bearing surfaces. *Bone Jt. Res.* **2013**, *2*, 84–95. [CrossRef] [PubMed]
37. Zywiel, M.G.; Brandt, J.M.; Overgaard, C.B.; Cheung, A.C.; Turgeon, T.R.; Syed, K.A. Fatal cardiomyopathy after revision total hip replacement for fracture of a ceramic liner. *Bone Jt. J.* **2013**, *95*, 31–37. [CrossRef] [PubMed]
38. Lun, M.P.; Johnson, M.B.; Broadbelt, K.G.; Watanabe, M.; Kang, Y.J.; Chau, K.F.; Springel, M.W.; Malesz, A.; Sousa, A.M.M.; Pletikos, M.; et al. Spatially Heterogeneous Choroid Plexus Transcriptomes Encode Positional Identity and Contribute to Regional CSF Production. *J. Neurosci.* **2015**, *35*, 4903–4916. [CrossRef] [PubMed]
39. Mathew, R.S.; Mullan, H.; Blusztajn, J.K.; Lehtinen, M.K. Comment on 'Multiple repressive mechanisms in the hippocampus during memory formation. ' *Science* **2016**, *353*, 453. [CrossRef] [PubMed]
40. Sathyanesan, M.; Girgenti, M.J.; Banasr, M.; Stone, K.; Bruce, C.; Guilchicek, E.; Wilczak-Havill, K.; Nairn, A.; Williams, K.; Sass, S.; et al. A molecular characterization of the choroid plexus and stress-induced gene regulation. *Transl. Psychiatry* **2012**, *2*, e139. [CrossRef]
41. Everaert, C.; Luypaert, M.; Maag, J.L.; Cheng, Q.X.; Dinger, M.E.; Hellemans, J.; Mestdagh, P. Benchmarking of RNA-sequencing analysis workflows using whole-transcriptome RT-qPCR expression data. *Sci. Rep.* **2017**, *7*, 1559. [CrossRef]
42. Assefa, A.T.; Vandesompele, J.; Thas, O. On the utility of RNA sample pooling to optimize cost and statistical power in RNA sequencing experiments. *BMC Genom.* **2020**, *21*, 312. [CrossRef]
43. Laovitthayanggoon, S.; Henderson, C.J.; McCluskey, C.; MacDonald, M.; Tate, R.J.; Grant, M.H.; Currie, S. Cobalt Administration Causes Reduced Contractility with Parallel Increases in TRPC6 and TRPM7 Transporter Protein Expression in Adult Rat Hearts. *Cardiovasc. Toxicol.* **2019**, *19*, 276–286. [CrossRef]
44. Afolaranmi, G.A.; Akbar, M.; Brewer, J.; Grant, M.H. Distribution of metal released from cobalt–chromium alloy orthopaedic wear particles implanted into air pouches in mice. *J. Biomed. Mater. Res. A* **2012**, *100*, 1529–1538. [CrossRef] [PubMed]
45. Apostoli, P.; Catalani, S.; Zaghini, A.; Mariotti, A.; Poliani, P.L.; Vielmi, V.; Semeraro, F.; Duse, S.; Porzionato, A.; Macchi, V.; et al. High doses of cobalt induce optic and auditory neuropathy. *Exp. Toxicol. Pathol.* **2013**, *65*, 719–727. [CrossRef] [PubMed]
46. Daniel, J.; Ziaee, H.; Pradhan, C.; Pynsent, P.B.; McMinn, D.J. Renal Clearance of Cobalt in Relation to the Use of Metal-on-Metal Bearings in Hip Arthroplasty. *J. Bone Jt. Surg. Am.* **2010**, *92*, 840–845. [CrossRef] [PubMed]
47. Martin, J.R.; Spencer-Gardner, L.; Camp, C.L.; Stulak, J.M.; Sierra, R.J. Cardiac cobaltism: A rare complication after bilateral metal-on-metal total hip arthroplasty. *Arthroplast. Today* **2015**, *1*, 99–102. [CrossRef] [PubMed]
48. Allen, L.A.; Ambardekar, A.V.; Devaraj, K.M.; Maleszewski, J.J.; Wolfel, E.E. Missing elements of the history. *N. Engl. J. Med.* **2014**, *370*, 559–566. [CrossRef]
49. Wyles, C.C.; Wright, T.C.; Bois, M.C.; Amin, S.; Fayyaz, A.; Jenkins, S.M.; Wyles, S.P.; Day, P.L.; Murray, D.L.; Trousdale, R.T.; et al. Myocardial cobalt levels are elevated in the setting of total hip arthroplasty. *J. Bone Jt. Surg. Am.* **2017**, *99*, e118. [CrossRef]
50. Garcia, F.; Ortega, A.; Domingo, J.L.; Corbella, J. Accumulation of metals in autopsy tissues of subjects living in Tarragona County, Spain. *J. Environ. Sci. Heal. Part. A* **2001**, *36*, 1767–1786. [CrossRef]
51. Ferguson, C.S.; Tyndale, R.F. Cytochrome P450 enzymes in the brain: Emerging evidence of biological significance. *Trends Pharmacol. Sci.* **2011**, *32*, 708–714. [CrossRef]
52. Pikuleva, I.A.; Waterman, M.R. Cytochromes P450: Roles in Diseases. *J. Biol. Chem.* **2013**, *288*, 17091–17098. [CrossRef]
53. Dixit, V.A.; Warwicker, J.; De Visser, S.P. How Do Metal Ions Modulate the Rate-Determining Electron-Transfer Step in Cytochrome P450 Reactions? *Chem. A Eur. J.* **2020**, *26*, 15270–15281. [CrossRef]
54. Honkakoski, P.; Negishi, M. Regulation of cytochrome P450 (CYP) genes by nuclear receptors. *Biochem. J.* **2000**, *347*, 321–337. [CrossRef] [PubMed]
55. Kinoshita, T. Biosynthesis and biology of mammalian GPI-anchored proteins. *Open Biol.* **2020**, *10*, 190290. [CrossRef] [PubMed]
56. Umbsaar, J.; Kerek, E.; Prenner, E.J. Cobalt and nickel affect the fluidity of negatively-charged biomimetic membranes. *Chem. Phys. Lipids* **2018**, *210*, 28–37. [CrossRef] [PubMed]
57. Urban, R.M.; Tomlinson, M.J.; Hall, D.J.; Jacobs, J.J. Accumulation in liver and spleen of metal particles generated at nonbearing surfaces in hip arthroplasty. *J. Arthroplast.* **2004**, *19*, 94–101. [CrossRef] [PubMed]

58. Stankiewicz, A.M.; Goscik, J.; Majewska, A.; Swiergiel, A.H.; Juszczak, G.R. The Effect of Acute and Chronic Social Stress on the Hippocampal Transcriptome in Mice. *PLoS ONE* **2015**, *10*, e0142195. [CrossRef]
59. Cho, J.; Yu, N.K.; Choi, J.H.; Sim, S.E.; Kang, S.J.; Kwak, C.; Lee, S.-W.; Kim, J.-I.; Choi, D.I.; Kim, V.N.; et al. Multiple repressive mechanisms in the hippocampus during memory formation. *Science* **2015**, *350*, 82–87. [CrossRef]
60. Schneider, J.S.; Anderson, D.W.; Sonnenahalli, H.; Vadigepalli, R. Sex-based differences in gene expression in hippocampus following postnatal lead exposure. *Toxicol. Appl. Pharmacol.* **2011**, *256*, 179–190. [CrossRef]
61. Watanabe, T.; Natt, O.; Boretius, S.; Frahm, J.; Michaelis, T. In vivo 3D MRI staining of mouse brain after subcutaneous application of $MnCl_2$. *Magn. Reson. Med.* **2002**, *48*, 852–859. [CrossRef]
62. Sudarshana, D.M.; Nair, G.; Dwyer, J.T.; Dewey, B.; Steele, S.U.; Suto, D.J.; Wu, T.; Berkowitz, B.A.; Koretsky, A.P.; Cortese, I.C.M.; et al. Manganese-enhanced MRI of the brain in healthy volunteers. *Am. J. Neuroradiol.* **2019**, *40*, 1309–1316. [CrossRef]
63. Steuerwald, A.J.; Blaisdell, F.S.; Geraghty, C.M.; Parsons, P.J. Regional Distribution and Accumulation of Lead in Caprine Brain Tissues Following a Long-Term Oral Dosing Regimen. *J. Toxicol. Environ. Heal. Part. A* **2014**, *77*, 663–678. [CrossRef]
64. Manton, W.I.; Kirkpatrick, J.B.; Cook, J.D. Does the choroid plexus really protect the brain from lead? *Lancet* **1984**, *324*, 351. [CrossRef]
65. Takeda, A.; Takefuta, S.; Ijiro, H.; Okada, S.; Oku, N. ^{109}Cd transport in rat brain. *Brain Res. Bull.* **1999**, *49*, 453–457. [CrossRef]
66. Nakamura, M.; Yasutake, A.; Fujimura, M.; Hachiya, N.; Marumoto, M. Effect of methylmercury administration on choroid plexus function in rats. *Arch. Toxicol.* **2011**, *85*, 911–918. [CrossRef] [PubMed]
67. Harrison-Brown, M.; Scholes, C.; Field, C.; McQuilty, R.; Farah, S.B.; Nizam, I.; Kerr, D.; Kohan, L. Limited penetration of cobalt and chromium ions into the cerebrospinal fluid following metal on metal arthroplasty: A cross-sectional analysis. *Clin. Toxicol.* **2020**, *58*, 233–240. [CrossRef]
68. Quintela, T.; Goncalves, I.; Carreto, L.C.; Santos, M.A.S.; Marcelino, H.; Patriarca, F.M.; Santos, C.R.A. Analysis of the Effects of Sex Hormone Background on the Rat Choroid Plexus Transcriptome by cDNA Microarrays. *PLoS ONE* **2013**, *8*, e60199. [CrossRef]
69. Gradinaru, D.; Minn, A.L.; Artur, Y.; Minn, A.; Heydel, J.M. Drug metabolizing enzyme expression in rat choroid plexus: Effects of in vivo xenobiotics treatment. *Arch. Toxicol.* **2009**, *83*, 581–586. [CrossRef]
70. Ghersi-Egea, J.F.; Strazielle, N.; Catala, M.; Silva-Vargas, V.; Doetsch, F.; Engelhardt, B. Molecular anatomy and functions of the choroidal blood-cerebrospinal fluid barrier in health and disease. *Acta Neuropathol.* **2018**, *135*, 337–361. [CrossRef]
71. Zhu, L.; Stein, L.R.; Kim, D.; Ho, K.; Yu, G.Q.; Zhan, L.; Larsson, T.E.; Mucke, L. Klotho controls the brain–immune system interface in the choroid plexus. *Proc. Natl. Acad. Sci. USA* **2018**, *115*, E11388–E11396. [CrossRef]
72. Baruch, K.; Ron-Harel, N.; Gal, H.; Deczkowska, A.; Shifrut, E.; Ndifon, W.; Mirlas-Neisberg, N.; Cardon, M.; Vaknin, I.; Cahalon, L.; et al. CNS-specific immunity at the choroid plexus shifts toward destructive Th2 inflammation in brain aging. *Proc. Natl. Acad. Sci. USA* **2013**, *110*, 2264–2269. [CrossRef]
73. Stock, A.D.; Der, E.; Gelb, S.; Huang, M.; Weidenheim, K.; Ben-Zvi, A.; Putterman, C. Tertiary lymphoid structures in the choroid plexus in neuropsychiatric lupus. *JCI Insight* **2019**, *4*, e124203. [CrossRef]
74. Vincent, C.; Gilabert-Juan, J.; Gibel-Russo, R.; Alvarez-Fischer, D.; Krebs, M.O.; Le Pen, G.; Prochiantz, A.; Di Nardo, A.A. Non-cell-autonomous OTX2 transcription factor regulates anxiety-related behavior in the mouse. *Mol. Psychiatry* **2021**, *26*, 1–12. [CrossRef]
75. Turner, C.A.; Thompson, R.C.; Bunney, W.E.; Schatzberg, A.F.; Barchas, J.D.; Myers, R.M.; Akil, H.; Watson, S.J. Altered choroid plexus gene expression in major depressive disorder. *Front. Hum. Neurosci. 2014. 8*, 238. [CrossRef]

toxics

MDPI

Article

Manganese-Induced Neurotoxicity through Impairment of Cross-Talk Pathways in Human Neuroblastoma Cell Line *SH-SY5Y* Differentiated with Retinoic Acid

Raúl Bonne Hernández [1,2,*], Nadja C. de Souza-Pinto [3], Jos Kleinjans [4], Marcel van Herwijnen [4], Jolanda Piepers [4], Houman Moteshareie [2], Daniel Burnside [2] and Ashkan Golshani [2]

1 Laboratory of Bioinorganic and Environmental Toxicology—LABITA, Department of Chemistry, Federal University of São Paulo, Rua Prof. Artur Riedel, 275, Diadema 09972-270, SP, Brazil
2 Department of Biology, Carleton University, 209 Nesbitt Biology Building, 1125 Colonel by Drive, Ottawa, ON K1S 5B6, Canada; houman_mot@yahoo.com (H.M.); daniel.j.burnside@gmail.com (D.B.); ashkan_golshani@carleton.ca (A.G.)
3 Departamento de Bioquímica, Instituto de Química, Universidade de São Paulo (USP), Av. Prof. Lineu Prestes, 748, Butantã, São Paulo 05508-900, SP, Brazil; nadja@iq.usp.br
4 Department of Toxicogenomics, Maastricht University, Universiteitssingel 50, Room 4.112 UNS 50, 6229 ER Maastricht, The Netherlands; j.kleinjans@maastrichtuniversity.nl (J.K.); m.vanherwijnen@maastrichtuniversity.nl (M.v.H.); j.piepers@maastrichtuniversity.nl (J.P.)
* Correspondence: rbhernandez@unifesp.br; Tel.: +55-11-3385-4137 (ext. 3522)

Citation: Hernández, R.B.; de Souza-Pinto, N.C.; Kleinjans, J.; van Herwijnen, M.; Piepers, J.; Moteshareie, H.; Burnside, D.; Golshani, A. Manganese-Induced Neurotoxicity through Impairment of Cross-Talk Pathways in Human Neuroblastoma Cell Line *SH-SY5Y* Differentiated with Retinoic Acid. *Toxics* **2021**, *9*, 348. https://doi.org/10.3390/toxics9120348

Academic Editors: Richard Ortega and Asuncion Carmona

Received: 4 October 2021
Accepted: 30 November 2021
Published: 9 December 2021

Abstract: Manganese (Mn) is an important element; yet acute and/or chronic exposure to this metal has been linked to neurotoxicity and neurodegenerative illnesses such as Parkinson's disease and others via an unknown mechanism. To better understand it, we exposed a human neuroblastoma cell model (*SH-SY5Y*) to two Mn chemical species, $MnCl_2$ and Citrate of Mn(II) (0–2000 μM), followed by a cell viability assay, transcriptomics, and bioinformatics. Even though these cells have been chemically and genetically modified, which may limit the significance of our findings, we discovered that by using RA-differentiated cells instead of undifferentiated *SH-SY5Y* cell line, both chemical species induce a similar toxicity, potentially governed by disruption of protein metabolism, with some differences. The $MnCl_2$ altered amino acid metabolism, which affects RNA metabolism and protein synthesis. Citrate of Mn(II), however, inhibited the E3 ubiquitin ligases–target protein degradation pathway, which can lead to the buildup of damaged/unfolded proteins, consistent with histone modification. Finally, we discovered that Mn(II)-induced cytotoxicity in RA-*SH-SY5Y* cells shared 84 percent of the pathways involved in neurodegenerative diseases.

Keywords: manganese speciation; *SH-SY5Y*; neurotoxicity; neurodegeneration; protein metabolism

1. Introduction

Manganese (Mn) is the twelfth most abundant element in the crust. Environmentally, it ranges from 1–200 g/L in fresh water to 410–6700 mg/kg (dry weight) in sediments [1]. Mn is a trace mineral that is found in low concentrations in legumes, pineapples, beans, nuts, tea, and cereals [2,3]. Mn is also a key cofactor for enzymes such as glutamine synthetase, pyruvate decarboxylase, serine/threonine protein phosphatase I, Mn-superoxide dismutase, and arginase [4]. Consequently, it is an essential element to maintain normal physiological development including the metabolism of lipid, protein, and carbohydrate; blood sugar regulation; bone formation; immunological response; reproduction; neurotransmitter synthesis and metabolism, as well as neuronal and glial function [3].

Local levels of Mn in the environment can be dramatically increased due to natural and human causes [5–10]. Additionally, it can be found as Mn(II), Mn(III), and Mn(IV) in aquatic systems owing to oxi-reductive processes [1,11–15]; these chemical species are environmentally and toxicologically relevant [10,16,17]. Occupational exposure [17] and consumption of contaminated well water [7] represent the most relevant means by which humans are exposed to Mn, with high risks for health [7,17–20], consistent with increasing evidence of developmental neurotoxicity due to oral parenteral nutrition [7,21].

Mn neurotoxicity, which is characterized by motor and sensory problems, known as manganism, as well as neuropsychiatric and cognitive impairments [3], is the most serious adverse consequence of this metal. Hypertonia with cogwheel stiffness, bradykinesia, "cock-gait", fast postural tremor, and a tendency to stumble when walking backwards are all symptoms of parkinsonism [3]. These symptoms have been linked to an excess of Mn in the basal ganglia, especially the globus pallidus, subthalamic nucleus, substantia nigra, and striatum, which are involved in motor control and nonmotor functions [3,4]. However, additional brain areas, such as the cerebellum, red nucleus, pons, cortex, thalamus, and anterior horn of the spinal cord, may be altered by Mn exposure [14]. These cellular features, including Mn-induced mitochondrial dysfunction, inflammation, autophagy, overexpression of α-synuclein (αSyn) in vitro, and their aggregation in vivo in neurons and glial cells, have been linked to Parkinson's disease (PD) [4]. These traits distinguish parkinsonism, which is characterized by the lack of Lewy bodies (another hallmark of PD). Surprisingly, dopaminergic neurons of the substantia nigra pars compacta are particularly destroyed following chronic or acute Mn exposure [3].

Furthermore, Mn toxicity has been connected to Huntington's disease (HD), since cultured striatal cells surprisingly reduced the vulnerability of mutant expressing cells ST*Hdh*Q111/Q111 [22]. Furthermore, pre-manifest YAC128 transgenic mice, another model of HD, exposed to $MnCl_2$ had a decreased response to transcriptional and protein alterations, whereas manifest YAC128 animals had a suppressed metabolic response, despite equivalent elevations in whole striatal Mn [23]. Mn has also been linked to Amyotrophic Lateral Sclerosis (ALS), since certain ALS patients have T1-weighted hyperintensity during MRI, a neuroradiological signal associated with Mn overload, as well as an increase in MnSOD levels in motor neurons and genetic variations of two melastatins, TRPM2 and 7. Whereas early study in macaques has suggested that chronic Mn treatment induces upregulation of amyloidlike protein 1 and diffuse amyloid-β plaques in the frontal cortex, perhaps implying a relationship between advanced-stage manganism and Alzheimer's disease (AD) [4]. In addition, Mn exposure in dogs enhanced the expression of nuclear neuronal NF-B and iNOS, as well as changed blood–brain barrier (BBB) function, diffuse Aβ plaques, neurofibrillary tangles, and alteration of Mn-dependent antioxidant enzyme, as reported in nonhuman primates and humans [24]. All these alterations can be influenced by chemical fractionation and speciation, developmental stage [14,16,25], and cell type [14].

The cytotoxicity of chemical compounds of Mn and their mechanisms has been supported by several works in silico, in vitro, and in vivo. In this manner, both undifferentiated and differentiated human neuroblastoma cell models (*SH-SY5Y*) were employed to investigate the role of Mn in neurotoxicity [26–28]. However, some issues must be considered. For example, undifferentiated *SH-SY5Y* cells can have fluctuations in the cell cycle and are considered immature catecholaminergic neurons [29]. While retinoic acid (RA) synchronizes the cell cycle and generates a modest rate of proliferation, RA differentiates cell morphologically close to primary neurons and increases electrical excitability of the plasma membrane [29], which could leave axons more susceptible for chemical injury [30]. Furthermore, NoRA *SH-SY5Y* cells are disabled in ATP production [31] while RA induces survival of *SH-SY5Y* cells. Consequently, RA-differentiated cells are more resilient to toxins [29]. Altogether, the in vitro cell model should mimic the phenotypes and be sensitive to cellular alterations commonly verified in vivo and specifically in humans [30,31]. It this respect, it has been shown that RA differentiates human neuroblastoma *SH-SY5Y* cells, which generate a largely mature dopaminergic (DAergic)-like neurotransmitter phenotype found in vivo as well as other neurotransmitters in lower expression, such as noradrenaline, acetylcholine, glutamate, serotonin, and histamine [31]. This allows the study of PD, ALS, AD, and HD [31–33] together with the relationship between these illnesses and Mn-induced neurotoxicity [26–28], potentially governed by disturbance of protein synthesis [25,34,35].

Protein synthesis is an energy-intensive process that is highly controlled and tightly linked to other cellular activities such as the cell cycle and metabolic pathways [36,37]. Previous research with NoRA *SH-SY5Y* cells revealed that $MnCl_2$ causes endoplasmic reticulum (ER) stress [38,39], accompanied by autophagy [38], and accumulation of parkin protein and its redistribution to aggregated Golgi complex [39]. Two other independent transcriptomics studies verified that $MnCl_2$ induces cytotoxicity in *SH-SY5Y* cells by promoting mitophagy through BNIP3-mediated oxidative stress [40] and/or upregulation of apoptotic pathways, neuronal differentiation, and synaptic transmission [26]. However, they have not identified alteration in the ER–Golgi system, involved in protein metabolism before energy–mitochondrial dysfunction and cell death [28].

However, the research described above has not contemplated the significance of chemical speciation. It is well known that the $MnCl_2$ (aqua-complex of Mn or Mn-free) can cross the brain–blood barrier easier than Citrate of Mn(II) or other species of Mn(II) or Mn(III) [41]. In cerebellar granule neurons, the $MnCl_2$ was more bioconcentrated than Citrate of Mn(II), although both species displayed similar cytotoxicity, associated with energy–mitochondrial impairment [14]. However, this energetic dysfunction reduced the influx of the two species of Mn in rat brain [42]. In zebrafish embryos, after exposure to species of Mn(II) or Mn(III), the Mn appeared in fluids mainly as Mn-free, followed by Citrate of Mn(II), a species that induced more bioaccumulation of Mn and gene overexpression than $MnCl_2$, but both species perturbed the calcium homeostasis and protein metabolism [16,35].

Finally, considering the issues stated above, we hypothesized that impairment of pathways linked to protein biosynthesis drives Mn-induced neurotoxicity and potentially neurodegeneration, which can be affected by chemical speciation. Thus, we aimed to develop a toxicogenomics study in the RA-differentiated *SH-SY5Y*-DAergic cell model, exposed to $MnCl_2$ and Citrate of Mn(II). Using system biology approaches, we provided additional evidence that connects Mn-induced impairment of protein metabolism to Mn-neurotoxicity to neurodegenerative disorders (AD, ALS, HD, and PD).

2. Materials and Methods

2.1. Preparation of Manganese Species

Compounds of manganese ($MnCl_2$ and Citrate of Mn(II) or Mn(II)Cit, 10 mM each) were prepared and characterized according to previous works of our group [14,16] and stored at 4 °C. For these experiments we used manganese(II) chloride tetrahydrate, $MnCl_2{\cdot}4H_2O$ (99.99% trace metals basis, Merck, São Paulo, SP, Brazil) and sodium citrate tribasic dihydrate (Cit; $HOC(COONa)(CH_2COONa)_2{\cdot}2H_2O$) (99.99% trace metals basis, Merck, São Paulo, SP, Brazil). The work solution was prepared the same day of cell exposure by diluting an aliquot to reach the desired final concentration, according to each experiment.

2.2. Human Neuroblastoma SH-SY5Y Cell Line Experimental Setup

Human neuroblastoma *SH-SY5Y* cells, a third generation subclone of *SK-SN-SH* cells, are a brain-derived catecholaminergic neuroblastoma cell line. *SH-SY5Y* cells differentiate into neuronlike cells and cease proliferating [43]. *SH-SY5Y* cells were grown in Dulbecco's Modified Eagle's Medium/Nutrient Mixture F-12 Ham (DMEM/F12 nutritional mixture (1:1), Merck, São Paulo, SP, Brazil) supplemented with 10% of Fetal Bovine Serum (FBS, Merck, São Paulo, SP, Brazil) and penicillin/streptomycin (50 IU/mL, Merck, São Paulo, SP, Brazil) in a humidified environment of 5% CO_2 and 95% air at 37 °C. The medium was changed every 4–5 days. Serum was lowered to 2% FBS for all experimental conditions. Cells were differentiated for 7 days in the presence of 10 nM of trans-Retinoic Acid (trans-RA, Merck, São Paulo, SP, Brazil). Cultures were inspected under a microscope prior to treatment to determine differentiation. When 80 percent or more of the cells in a culture showed neurite outgrowth extensions >2–3 times longer than the cell's body diameter, the culture was declared differentiated. In the presence or absence of the Mn species, the cells were seeded at a density of 5×10^4 cells per 100 mL in a 96-well plate format for cell viability assay and 6-well plates for array experiment.

2.3. Cell Viability Assay

Cell viability was determined in quintuplicate from three separate cell cultures using the MTT test, which measures the reduction of 3-[4,5-dimethythiazol-2-il]-2,5-diphenyl-tetrazolium bromide (MTT, Merck, São Paulo, SP, Brazil) by live cells. We conducted the assay according to Hernández et al. [14] after 1 day of treatment with $MnCl_2$ or Citrate of Mn(II) from 0 to 2000 µM.

2.4. Transcriptomics

2.4.1. RNA Extraction and Purification

The RA-*SH-SY5Y* cells were treated for 24 h with 500 µM of $MnCl_2$ or Mn(II)Cit. Then the cells were collected in RNA*later* (Qiagen, Toronto, ON, Canada), and total RNA was extracted and purified from three different pools of cells using the miRNeasy mini kit (Qiagen, Toronto, ON, Canada), followed by a DNase I (Qiagen, Toronto, ON, Canada) treatment, as directed by the manufacturer. The amount of RNA was quantified using a spectrophotometer, and the quality was assessed using BioAnalyzer equipment (Agilent Technologies, Palo Alto, CA, USA). All RNA samples showed clear 18S and 28S rRNA peaks and demonstrated an RNA integrity number (RIN) level higher than 8.

2.4.2. Microarray Assays

Using an accessible and commercial human gene expression microarray kit, studies on differential gene expression were done in triplicates from three independent biological groups to discover the mode of action of each chemical species of manganese in RA-differentiated *SH-SY5Y* cells. The experiment was carried out in accordance with the manufacturer's procedures for one-color microarray-based gene expression analysis (Agilent), which are accessible in: http://www.agilent.com/cs/library/usermanuals/Public/G4140-90040_GeneExpression_OneColor_6.9.pdf, accessed on 29 November 2021).

2.4.3. Validation of Toxicogenomics Results through Real-Time Reverse Transcription-PCR (qRT-PCR)

The qRT-PCR assay has been used for identification of gene alterations in previous studies [44,45]. Thus, we used this approach to verify our results about Microarray (item 2.2.2.2). Table 1 shows the primer sequences that were employed. To complete the qRT-PCR experiment, we extracted total RNA from each sample using the Qiagen RNA extraction kit, followed by cDNA production using the iScript cDNA synthesis kit with SYBR green supermix (Bio-Rad, Hercules, CA, USA), as directed by the manufacturer. qRT-PCR was used to quantify mRNA using Rotor-Gene RG-300 from Corbett research [46]. All investigations were carried out in triplicate by three different biological groups.

2.5. Bioinformatics and Data Analysis

The results were presented as the mean ± SEM of at least three separate trials. Fitting sigmoidal curves (Hill slope) to concentration–response data of individual replicates and computing the mean of those replicates yielded the LC50. The D'Agostino and Pearson omnibus normality test verified that our data had a normal or statistical distribution. ANOVA (analysis of variance) and Bonferroni's tests were employed to discover statistically significant differences. GraphPad Prism was used for fitting and statistical analysis (GraphPad 4.0 Software Inc., San Diego, CA, USA). The arrays were examined using Babelomics' Gene Expression Pattern Analysis Suite (https://babelomics.bioinfo.cipf.es/, accessed on 13 May 2016) [47], which is an integrated web-based pipeline designed for the analysis of data generated in microarray studies. Normalization, grouping, differential gene expression, class prediction, and functional annotation are all included in the suite.

Table 1. Genes selected for qRT-PCR analysis. Primer sequences and associated parameters are included.

Gene Symbol	Gene Name	PL		Sequence (5′→3′)	TS	Length	Start	Stop	Tm	GC%	Self C.	Self 3′ C.
RPS29	RPS29 ribosomal protein S29	135	FP	ACACTGGCGGCACATATTGA	Plus	20	49,585,986	49,586,005	60.04	50	4	2
			RP	GGTAGTAGCCGTCTGAGTGC	Minus	20	49,586,120	49,586,101	59.9	60	4	3
MT-CO1	mitochondrially encoded cytochrome c oxidase I	107	FP	CCCCGATGCATACACCACAT	Plus	20	7232	7251	60.18	55	6	2
			RP	TCGAAGCGAAGGCTTCTCAA	Minus	20	7338	7319	59.68	50	7	3
COX4I2	cytochrome c oxidase subunit 4I2	118	FP	GATGAACCGTCGCTCCAATG	Plus	20	30,135,127	30,135,146	59.35	55	5	3
			RP	GATGAGGTGTTGCCACTCAC	Minus	20	30,135,244	30,135,225	58.84	55	4	3
MT-CYB	mitochondrially encoded cytochrome b	134	FP	ACCCCCTAGGAATCACCTCC	Plus	20	15,366	15,385	60.03	60	6	1
			RP	GCCTAGGAGGTCTGGTGAGA	Minus	20	15,499	15,480	60.03	60	6	2
MT-ND4	mitochondrially encoded NADH dehydrogenase 4	296	FP	CCCCATCGCTGGGTCAATAG	Plus	20	11,428	11,447	60.25	60	8	2
			RP	TAAGCCCGTGGGCGATTATG	Minus	20	11,723	11,704	60.25	55	6	1
Gapdh	glyceraldehyde-3-phosphate dehydrogenase	117	FP	AAAGGGCCCTGACAACTCTTT	Plus	21	6,538,069	6,538,089	59.78	47.62	8	3
			RP	GGTGGTCCAGGGGTCTTACT	Minus	20	6,538,185	6,538,166	60.55	60	5	1

Product Length (PL), Forward primer (FP), Reverse primer (RP), Template Strand (TS), Self Complementarity (Self C.), Self 3′ Complementarity (Self 3′ C.).

2.6. Prediction of Protein–Protein Interaction (PPI) and Gene Ontology (GO) Analysis

A PPI map is a heterogeneous network of proteins connected by interactions as edges. PPI and GO enrichment analyses were performed utilizing data from the current study, the String database (http://string-db.org, accessed on 30 November 2019) [48], and the Comparative Toxicogenomic Database—CTD (http://ctdbase.org/, accessed on 30 November 2019) [49]. Particularly, String discovers cellular pathways that are enriched in a target list of genes, proteins, or metabolites that are not approximated by the original omics data (using hypergeometric testing against either the entire genome or a user-supplied background gene list), allowing for the extraction of strong mechanistic information [48,50].

3. Results

Multiple genetic and environmental variables contribute to the onset and progression of Parkinson's disease, which is related with the degradation of DAergic neurons in the substantia nigra pars compacta [31]. Indeed, epidemiological studies have connected the development of neurodegeneration to Mn toxicity [22–24,51,52]. This can be mediated by impairment of protein metabolism, according to findings in nonhuman models [25,34,35]. To test the validity of this hypothesis, we conducted an unbiased toxicogenomics study on Mn-induced acute neurotoxicity in the human model, RA-differentiated *SH-SY5Y*-DAergic cells.

3.1. Manganese-Induced Toxicity and Differential Gene Expression in the RA-Differentiated SH-SY5Y-DAergic Cell Model

Through MTT assay we verified that both Mn species ($MnCl_2$ and Mn(II)Cit) induced similar cytotoxicity, in a dose dependent manner, where approximately 900 and 500 μM induces 50% and 10% of cell death, respectively (Figure 1).

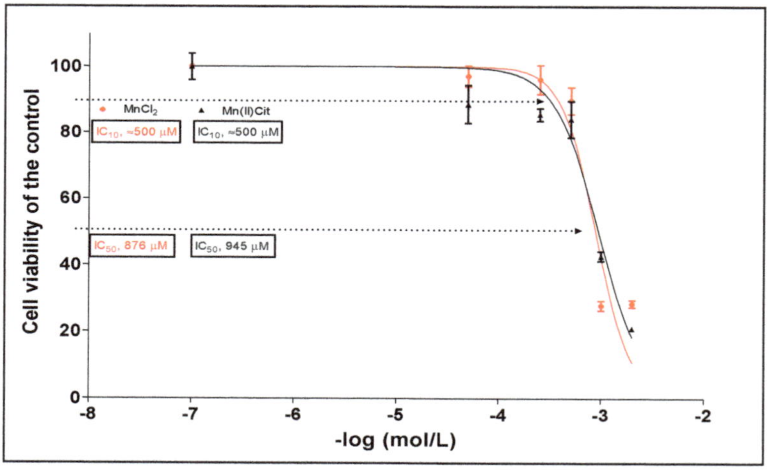

Figure 1. Cytotoxicity induced by $MnCl_2$ and Mn(II)Cit in RA-differentiated *SH-SY5Y* cells, after exposure for 24 h. Cell viability was quantified by MTT assay. The plot is representative of three biological replicates (mean ± sem); each experiment was made with five analytical replicates.

To find the mode of action of manganese, we designed a gene expression analysis in *SH-SY5Y* cells treated for 24 h to 500 μM of $MnCl_2$ or Mn(II)Cit, which might reflect an environmental state of acute Mn exposure owing to water intake polluted with this metal [15,53,54]. In this way, we verified a significant ($p < 0.05$) impairment of 406 genes (Tables S1 and S2). There, we discovered 117 genes that are affected by both chemical species of manganese, which suppress rather than stimulate gene expression. In this case,

$MnCl_2$ reduced the expression of 169 genes (80%), while Mn(II)Cit decreased the expression of 102 genes (70%), Figure 2.

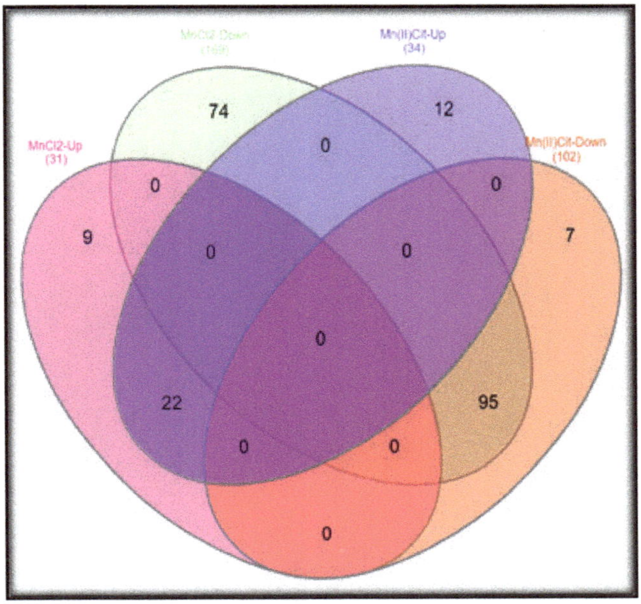

Figure 2. Three microarray assays in *SH-SY5Y* after exposure to $MnCl_2$ or Mn(II)Cit (0–500 μM) for 24 h yielded overlapping results of 117 genes differentially expressed and affected by both chemical species of Mn. Significant differences were estimated through two-way ANOVA, followed by false discovery rate and Bonferroni ($p < 0.05$).

3.2. Manganese-Induced Cross-Impairment of Several Pathways Associated with Neurotoxicity and Neurodegeneration

Additionally, based on all genes significantly affected under Mn stress, and using the String database [48], we used systems biology to create an enriched protein–protein interaction (PPI) map for each Mn species and an enriched gene ontology analysis. The String database studies a defined group of target proteins using physical interactions and functional relationships between proteins, and then expands the set by incorporating linked proteins to investigate toxicological pathways [48]. The network of functional Mn interactors was expanded to around 700 edges (proteins) (p-value 1.0×10^{-16}), with approximately 85 percent of the new interactions experimentally validated [48]. A schematic representation of these interactors is shown in Figure 3A,C, including several processes/pathways potentially impaired by Mn (Tables 2 and 3).

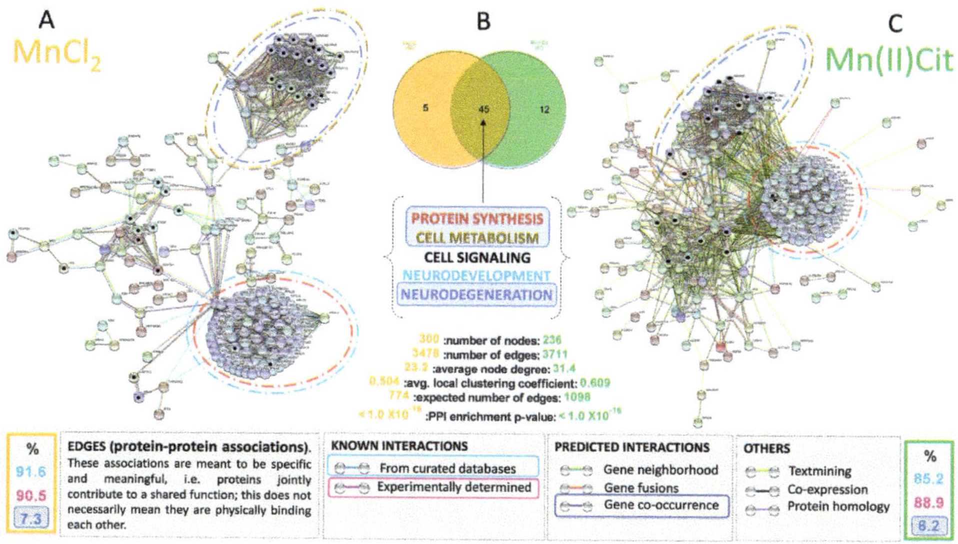

Figure 3. (**A**,**C**) represent an inferred enriched protein–protein interactions (PPI) network for cells stressed with $MnCl_2$ and Mn(II)Cit, respectively; *p*-value $< 1.0 \times 10^{-16}$. More than 85% of these interactions are derived experimentally. The predicted PPIs were calculated with high confidence score (0.9) and proteins/genes with no connections were removed. (**B**) Overlapping representation of 45 pathways that were affected by $MnCl_2$ and Mn(II)Cit. These interactions can be categorized in Protein Synthesis, Cell Metabolism, Cell Signaling, Neurodevelopment, and Neurodegeneration groups.

Table 2. Enriched cellular pathways influenced by $MnCl_2$ in RA-differentiated *SH-SY5Y* cells, after gene ontology analysis using the String database.

Cell Pathways	Term Description	Observed Gene Count	Background Gene Count	FDR
Protein Synthesis	Peptide chain elongation	75	86	4.37×10^{-92}
	Viral mRNA Translation	75	86	4.37×10^{-92}
	SRP-dependent cotranslational protein targeting to membrane	78	109	1.15×10^{-91}
	Eukaryotic Translation Termination	75	90	1.22×10^{-91}
	Nonsense Mediated Decay (NMD) independent of the Exon Junction Complex (EJC)	75	92	2.25×10^{-91}
	Formation of a pool of free 40S subunits	75	98	5.96×10^{-90}
	L13a-mediated translational silencing of Ceruloplasmin expression	75	107	5.93×10^{-88}
	GTP hydrolysis and joining of the 60S ribosomal subunit	75	108	9.04×10^{-88}
	Ribosome	76	130	3.15×10^{-84}
	Nonsense Mediated Decay (NMD) enhanced by the Exon Junction Complex (EJC)	75	112	6.29×10^{-87}
	Major pathway of rRNA processing in the nucleolus and cytosol	76	179	1.10×10^{-76}
	Metabolism of amino acids and derivatives	77	354	2.98×10^{-59}
	Metabolism of RNA	78	652	7.72×10^{-43}
	Formation of the ternary complex and subsequently the 43S complex	31	49	1.27×10^{-34}
	Translation initiation complex formation	31	55	1.85×10^{-33}
	Ribosomal scanning and start codon recognition	31	55	1.85×10^{-33}
	Metabolism of proteins	102	1948	1.12×10^{-27}
	Biosynthesis of amino acids	9	72	5.98×10^{-05}

Table 2. *Cont.*

Cell Pathways	Term Description	Observed Gene Count	Background Gene Count	FDR
Cell Metabolism	Selenocysteine synthesis	75	90	1.22×10^{-91}
	Selenoamino acid metabolism	76	112	2.37×10^{-88}
	Metabolism	137	2032	6.11×10^{-52}
	Oxidative phosphorylation	26	131	9.48×10^{-18}
	The citric acid (TCA) cycle and respiratory electron transport	28	173	7.83×10^{-18}
	Thermogenesis	29	228	1.68×10^{-15}
	Complex I biogenesis	17	55	8.42×10^{-15}
	Metabolic pathways	50	1250	2.11×10^{-08}
	Glycolysis/Gluconeogenesis	12	68	5.27×10^{-08}
	Fructose and mannose metabolism	9	33	1.85×10^{-07}
	Metabolism of carbohydrates	20	266	2.71×10^{-07}
	Carbon metabolism	10	116	2.90×10^{-04}
Cell Signaling	Retrograde endocannabinoid signaling	18	148	2.25×10^{-09}
	HIF-1 signaling pathway	14	98	3.22×10^{-08}
	Negative regulation of MAPK pathway	6	40	1.00×10^{-03}
Neurodevelopment	Axon guidance	79	541	3.00×10^{-49}
Neurodegeneration	Parkinson's disease	26	142	3.69×10^{-17}
	Alzheimer's disease	20	168	3.22×10^{-10}
	Huntington's disease	19	193	1.57×10^{-08}

False Discovery Rate—FDR.

Table 3. Enriched cellular pathways influenced by Mn(II)Cit in RA-differentiated *SH-SY5Y* cells, after gene ontology analysis using String-db.

Cell Pathways	Term Description	Observed Gene Count	Background Gene Count	FDR
Protein Synthesis	Peptide chain elongation	76	86	7.69×10^{-103}
	Viral mRNA Translation	76	86	7.69×10^{-103}
	Eukaryotic Translation Termination	76	90	2.96×10^{-102}
	Nonsense Mediated Decay (NMD) independent of the Exon Junction Complex (EJC)	76	92	4.89×10^{-102}
	Formation of a pool of free 40S subunits	76	98	1.34×10^{-100}
	L13a-mediated translational silencing of Ceruloplasmin expression	76	107	1.58×10^{-98}
	GTP hydrolysis and joining of the 60S ribosomal subunit	76	108	2.37×10^{-98}
	SRP-dependent cotranslational protein targeting to membrane	76	109	3.59×10^{-98}
	Nonsense Mediated Decay (NMD) enhanced by the Exon Junction Complex (EJC)	76	112	1.52×10^{-97}
	Ribosome	77	130	6.14×10^{-95}
	Formation of the ternary complex and subsequently the 43S complex	31	49	5.38×10^{-38}
	Translation initiation complex formation	31	55	7.97×10^{-37}
	Ribosomal scanning and start codon recognition	31	55	7.97×10^{-37}
	Metabolism of proteins	88	1948	2.38×10^{-27}
	Biosynthesis of amino acids	8	72	6.26×10^{-05}
	Senescence-Associated Secretory Phenotype (SASP)	6	78	6.50×10^{-03}
	Amyloid fiber formation	6	78	6.50×10^{-03}
	E3 ubiquitin ligases ubiquitinate target proteins	5	53	7.80×10^{-03}

Table 3. *Cont.*

Cell Pathways	Term Description	Observed Gene Count	Background Gene Count	FDR
Cell Metabolism	Selenocysteine synthesis	76	90	2.96×10^{-102}
	Oxidative phosphorylation	26	131	2.08×10^{-20}
	Thermogenesis	28	228	2.41×10^{-17}
	Metabolic pathways	45	1250	1.71×10^{-09}
	Glycolysis/Gluconeogenesis	12	68	3.66×10^{-09}
	Carbon metabolism	10	116	3.67×10^{-05}
	Pentose phosphate pathway	5	30	6.60×10^{-04}
	Starch and sucrose metabolism	5	33	9.30×10^{-04}
	Galactose metabolism	4	31	8.10×10^{-03}
	Endocrine resistance	6	95	1.26×10^{-02}
Cell Signaling	Retrograde endocannabinoid signaling	16	148	2.85×10^{-09}
	HIF-1 signaling pathway	12	98	1.42×10^{-07}
	Negative regulation of MAPK pathway	5	40	2.50×10^{-03}
Neurodevelopment	Axon guidance	78	541	5.39×10^{-57}
Neurodegeneration	Parkinson's disease	26	142	8.36×10^{-20}
	Alzheimer's disease	20	168	3.71×10^{-12}
	Huntington's disease	19	193	2.88×10^{-10}

False Discovery Rate—FDR.

We identified that almost 50% of the pathways affected by Mn are directly linked to protein biosynthesis, including ribosomes, translation initiation, and termination. Additionally, we verified that $MnCl_2$-induced impairment of protein metabolism involves alteration of the metabolism of amino acids. The Mn(II)Cit appears to affect the E3 ubiquitin ligases–target protein degradation pathway, which can lead to damaged/unfolded protein accumulation. This is followed by pathways associated with cell metabolism, especially energy metabolism (~30%) and cell signaling pathways (6%), Tables 2 and 3. Although the chemical speciation has influenced the Mn-induced toxicity, we found that both species, $MnCl_2$ and Mn(II)Cit shared impaired pathways (Figure 3B), which are similar to molecular changes linked with neurodegenerative illnesses such as AD, HD, and PD (Figure 3A,C and Tables 2 and 3).

Furthermore, through an additional analysis using CTD [49], among the genes that were directly affected by Mn species (Tables S1 and S2) and the genes/proteins added by String (Figure 3A,C), we identified 34 curated genes (Figure 4A), previously affected by chemical species of Mn, including $MnCl_2$. Again, these genes are involved in the pathways inferred in this study (Tables 2 and 3). Moreover, we confirmed using qRT-PCR analysis that both $MnCl_2$ and Mn(II)Cit disrupted the expression of the genes RPS29, MT-CO1, MT-ND4, MT-CYB, and COX4I2, which are associated with protein synthesis, energy metabolism, and neurodegeneration [49] (Figure 4B).

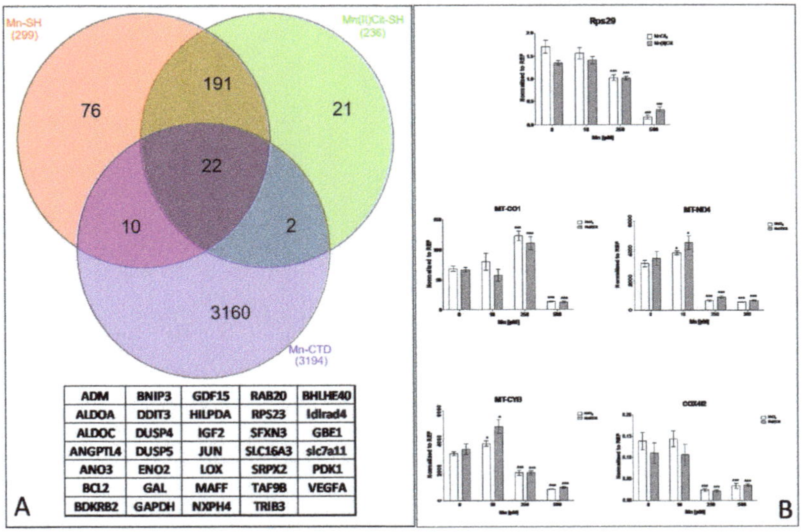

ADM	BNIP3	GDF15	RAB20	BHLHE40
ALDOA	DDIT3	HILPDA	RPS23	ldlrad4
ALDOC	DUSP4	IGF2	SFXN3	GBE1
ANGPTL4	DUSP5	JUN	SLC16A3	slc7a11
ANO3	ENO2	LOX	SRPX2	PDK1
BCL2	GAL	MAFF	TAF9B	VEGFA
BDKRB2	GAPDH	NXPH4	TRIB3	

Figure 4. (**A**) The enriched set of proteins identified using PPI data, including genes identified by transcriptomics analysis and added by String were analyzed using the comparative toxicogenomic data. A set of 34 curated genes were identified that previously were affected by $MnCl_2$ or other Mn compounds. (**B**) Representative genes associated with protein synthesis, cell metabolism, and neurodegeneration were analyzed by qRT-PCR analysis (mean ± sem, $n = 3$). Statistical differences were verified by two-way ANOVA, followed by Bonferroni post test, (*, $p < 0.05$; ***, $p < 0.001$).

Lastly, our findings suggest that chemical species of Mn induce cytotoxicity in the RA-differentiated *SH-SY5Y*-DAergic cell model (Figures 1 and 5A) through alterations of the expression in a notable number of genes (Tables S1 and S2 and Figure 5B), because of which numerous pathways are harmed (Figures 3 and 5C). There, disruption of protein metabolism, especially protein synthesis, appears to be a key event for Mn-induced cell perturbation (Figure 5C) responsible for Mn-induced neurotoxicity, which may result in neurodegeneration (Figure 5D).

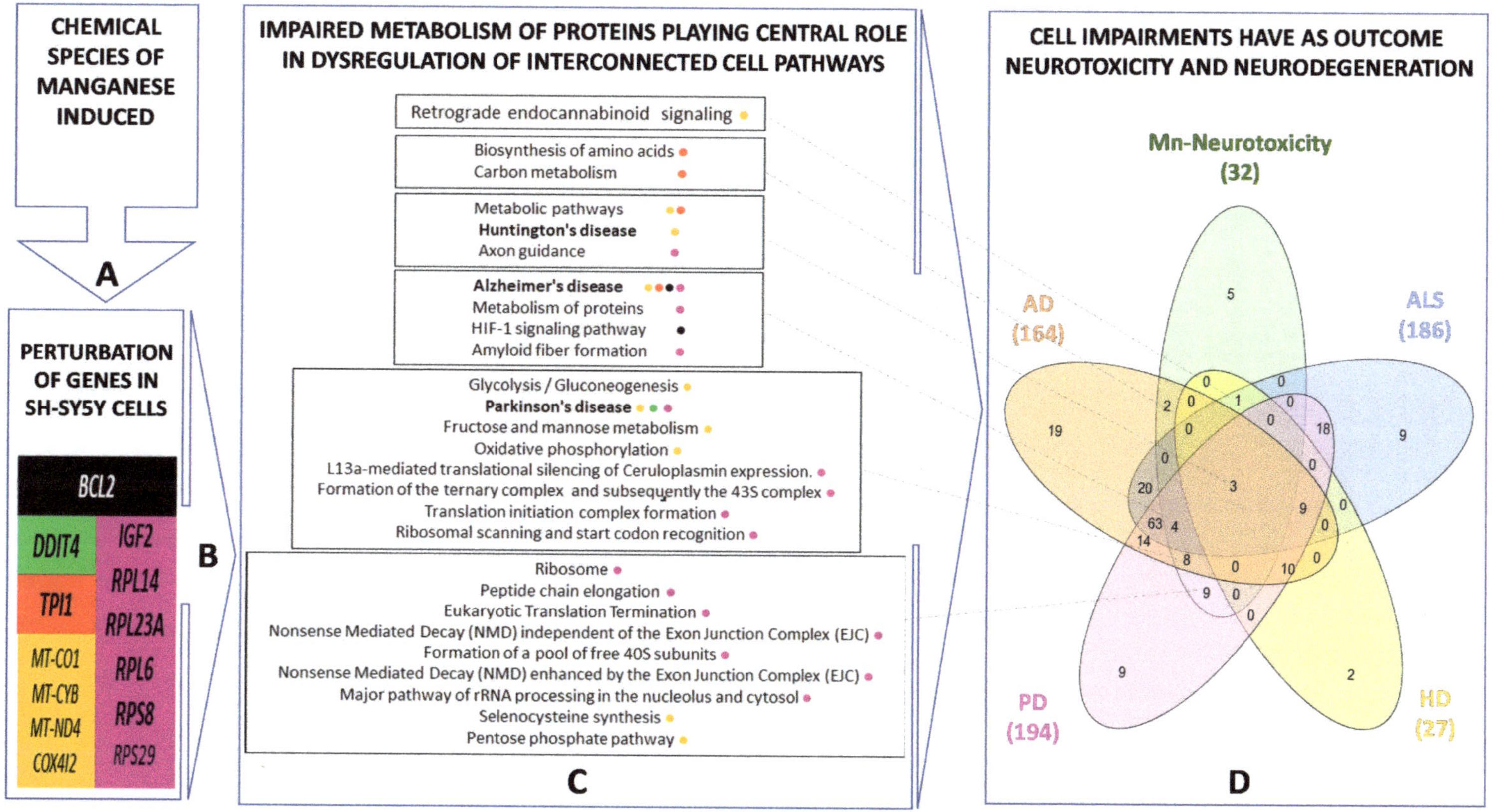

Figure 5. Comparative toxicogenomics analysis for Mn-induced cytotoxicity in *SH-SY5Y*. Exposure of *SH-SYSY* cells to different chemical species of Mn (**A**) results in disruption of genes shown in (**B**). This can lead to impairment of protein metabolism that plays a central role in the dysregulation of several pathways such as cell signaling; cell metabolism, including energy pathway; Alzheimer's Disease (AD); Huntington's Disease (HD) and Parkinson's Disease (PD) (**C**). Overlapping pathways among Mn-induced neurotoxicity and Amyotrophic Lateral Sclerosis (ALS), AD, HD, and PD is shown in (**D**).

4. Discussion

In this work, we found that the 876 μM of $MnCl_2$ or 945 μM of Mn(II)Cit decreased by 50% the viability of RA-differentiated *SH-SY5Y*-DAergic cells. These concentrations are higher than found in studies with NoRA-*SH-SY5Y* cells, where 800 μM [55] and approximately 600 μM [40] of $MnCl_2$, respectively, induced 50% of cytotoxicity after exposure for 24 h. It is known that RA-differentiated *SH-SY5Y* cells show activated survival pathways, including Nrf2 (nuclear factor erythroid 2-related factor 2) or Akt (serine-threonine protein kinase) signaling [29], while the NoRA-differentiated cells are ATP-deficient compared to the RA-differentiated cells [31]. This could justify why NoRa *SH-SY5Y* cells are more susceptible for Mn treatment, as observed by Kemsheh and Oblitey [56]. On the other hand, the RA-differentiated *SH-SY5Y*-DAergic cell model shows increased dopamine and other neurotransmitters, which allowed the screening of Mn-induced neurotoxicity and their association with neurodegeneration [31–33,57], after exposure for 500 μM of both Mn-species, a concentration that induced approximately 10% of cell death and was used in previous studies with NoRA-differentiated cells [58]. In this toxicologic context, the CTD have systematized approximately 140 Mn compounds associated with 379 pathways [49], where the top fifty are dominated by eleven cellular processes, including cell signaling, diseases, immune system, metabolism, hemostasis, metabolism of proteins, apoptosis, endocrine resistance, cellular responses to stress, and neurodevelopment. These processes are potentially connected to 1522 diseases, of which 21 are associated with the nervous system [49]. Other functional and omics studies [25–28,34,35] have identified these potential mechanisms of Mn-induced neurotoxicity, reviewed by Tinkov et al. [52]. In line with this, using microarray analysis and systems biology approaches, we verified a connection between Mn-induced acute cytotoxicity in the RA-differentiated *SH-SY5Y*-DAergic cell model and impairment of several pathways, including protein synthesis (18 processes), cell metabolism (10 processes), cell signaling (3 processes), neurodevelopment (1 process), and neurodegeneration (3 processes) (Tables 2 and 3).

Presynaptic local protein synthesis, the process by which mRNAs are translated in axons and terminals, is well shown to be mediated by retrograde endocannabinoid (eCB) transmission [59]. The eCBs are lipids that are mobilized by postsynaptic action and travel backward across the synapse to bind presynaptic Gi/o-coupled type 1 cannabinoid (CB1) receptors, inhibiting neurotransmitter release. CB1 activation, in turn, boosts protein synthesis via the mTOR pathway [59]. Indeed, we identified impairment of retrograde endocannabinoid signaling, which can contribute to the disruption of axon guidance, after exposure to Mn (Tables 2 and 3). Furthermore, it has been proposed that chemical stresses can either upregulate or downregulate the endocannabinoid system. This can disrupt synaptic transmission and brain circuit functions [60], leading to neurodegenerative illnesses such as AD, ALS, HD, and PD [61,62], as well as activate other signaling systems [60] such as hypoxia-inducible factor 1 (HIF-1), whose activation contributes to recovery of synaptic functions [63]. However, depending on the chemical speciation and the biological model investigated, HIF-1 can be inactivated due to downregulation of the BCL2 gene in response to Mn stress. Mn deficiency reduces the production of BCL2 mRNA and protein [64]. Manganese (III)-tetrakis(4-benzoic acid)porphyrin has been shown in studies with isolated human endothelial cells to inhibit the decrease of protein BCL2 expression [65]. Exposure to maneb and mancozeb, both Mn-based pesticides, results in increased expression of *BCL2* mRNA and BCL2 protein abundance [66,67].

In agreement with our findings (Figures 3 and 5, Tables 2 and 3), researchers discovered altered genes related to neurogenesis, neurodevelopment, synaptic transmission, and apoptosis after exposing NoRa-differentiated SH-SY5Y cells to 100 μM $MnCl_2$ for 30 days, which may be linked to neurodegeneration [26]. However, this work did not identify that these alterations can be preceded by impaired protein metabolism. Indeed, before changes in the mitochondrial–energy pathway and neurotransmission system, Fernandes et al. discovered increased abundance of the genes BET1 (Golgi vesicular membrane-trafficking protein), ADAM10 (ADAM metallopeptidase domain 10), and ARFGAP3 (ADP-ribosylation factor GTPase-activating protein 3) [28]. Accordingly, we confirmed that 500 μM of Mn(II)Cit or $MnCl_2$ caused neurotoxicity in RA-differentiated SH-SY5Y-DAergic cells by disrupting approximately 50% of protein metabolism-related pathways, followed by energy dysfunction and other related molecular cell changes (Tables 2 and 3). This is consistent with previous observations that $MnCl_2$-induced neurotoxicity in NoRA SH-SY5Y was caused by changes in the ER–Golgi complex [38,39]. Hernández's group's research with a variety of models ranging from yeast to mammals, different Mn-species, and different exposure times has also suggested that impaired protein metabolism is a key event in Mn-induced neurotoxicity [25,34,35].

Although RA-differentiated *SH-SY5Y* can have altered axons [29,30], certainly Mn induces disturbance of axonal functions potentially linked to protein synthesis impairment [68]. Studies with primary cerebellar granule neurons exposed for $MnCl_2$ identified potential axonal alteration due to hypokalemia and overexpression of STX1A, implicated in the production of presynaptic local proteins that regulate neurotransmitter release [25]. Here, in RA-differentiated SH-SY5Y exposed for Mn, we identified significant alterations of inferred and curated genes such as *IGF2, RPL14, RPL23A, RPL6, RPS8,* and *RPS29* (Figures 2B and 4D), transcripts hit revealed using the String database [48] and CTD [49]. These genes are involved in an enriched set of pathways linked to translation, post-translation modifications, and protein degradation (Figure 3A,C, Tables 2 and 3), that can be perturbed by disruption of energy pathways [36,37] as well. Here, we found that the species of Mn can increase or arrest the expression of genes such as *MT-CO1, MT-CYB, MT-ND4,* and *COX4I2* (Figure 4B), which are associated with glycolysis and glycogenesis, and oxidative phosphorylation, along with neurodegenerative diseases such as AD, HD, and PD [49]. Concomitantly, we identified impairment of *TPI1* gene, which is associated with one-carbon metabolism pathway, biosynthesis of amino acid, and AD [49]. Furthermore, disruption of amino acid biosynthesis has been linked to the alteration of translation efficiency [69], similar to that observed in yeast exposed to $MnCl_2$ for 24 h, which showed decreased β-galactosidase reporter expression, heavy polysomes fractions, and the expression of *NOP1* and *NSR1* genes, associated with ribosome biogenesis [34]. Altogether, the *RPL14* gene impaired by Mn(II) is a marker of PD, whose alteration has been detected already, after domestic exposure to maneb (manganese(2+);*N*-[2-(sulfidocarbothioylamino)ethyl]carbamodithioate) [70]. We have discovered maneb-impaired protein metabolism in cerebellar granule neurons [25]. Another study identified Mn-induced ER-stress and increased phosphorylation of translation initiation factor *eIF2α* in *SH-SY5Y* cells [70]. Collectively, these findings confirm that the network among protein metabolism, energy, cell development, and metabolic pathways [36,37] may be disrupted by chemical species of Mn [34,35,71–75].

Contrary to the transcriptomics studies conducted by Gandhi et al. [26] and Fernandes et al. [28], which considered only $MnCl_2$, we verify that chemical speciation is important for Mn-induced impairment of gene expression and protein metabolism on RA-differentiated *SH-SY5Y* cells. For example, the $MnCl_2$-induced alteration of the metabolism of amino (AA) acids, which influences RNA metabolism and protein synthesis [76], similar to that identified in yeast exposed for $MnCl_2$ [34]. It has been demonstrated that amino acids are a regulator of late endosomes/lysosomes anterograde transport, which operate as mRNA translation platforms to produce new proteins necessary to support mitochondria function in axons growth, which is sensible to energy stress [77]. Indeed, we verified cross-talk alterations of the metabolism of AAs, proteins, cell energy, and axon guidance in response to cytotoxic concentrations of Mn. A metabolomics study with NoRA *SH-SY5Y* cells exposed to noncytotoxic concentrations of $MnCl_2$ revealed a positive association between Mn exposure and glutamate and *N*-acetylglutamate semialdehyde, and a negative relationship with other AAs such as leucine/isoleucine, 4-imidazoleacetate, histidine, arginine, and valine. As a result, neurotransmitter-related metabolites such as GABA, adrenochrome, N4-acetylaminobutanal, N-methyl salsolinol, and dopamine sulfate were dramatically altered [27]. This corroborates the importance of the AAs for local protein synthesis, during neurodevelopment, which can be impaired by Mn.

Although RA can be anti-proteasome inhibitor in SH-SY5Y cells [33], we deduced that Mn(II)Cit induces neurotoxicity associated with impairment of E3 ubiquitin ligases–target protein degradation pathway, which is involved in the proteasome's identification of substrates and proteins for degradation [78]. This can lead to damaged/unfolded protein accumulation, consistent with altered expression of the genes *HIST1H2BB*, *HIST1H2BH*, and *HIST1H2BO* (Figure 3A,C). Histone modification is linked to the creation of amyloid fibers (Table 3) and, as a result, the development of neurodegenerative diseases [79]. Other research has shown that Mn-induced neurotoxicity can target the ubiquitin system. For example, in cultured astrocytes treated with $MnCl_2$, SNAT3 protein degradation and Gln homeostasis disruption occur via the ubiquitin-mediated proteolytic mechanism [80]. Additionally, in cerebellar granule cells stressed with $MnCl_2$ or maneb, the Mn induces impairment of protein metabolism, involving dysregulation of the ubiquitin system as well [25]. However, the ubiquitin system was not altered in dopaminergic cells (SH-SY5Y and CATH.a), with impaired ER–Golgi complex, under the effect of Mn [39].

5. Conclusions

Lastly, we identified some advantages to using the RA-differentiated cells instead of the undifferentiated *SH-SY5Y* cell line to study Mn-induced neurotoxicity in humans. These cells are chemically and genetically modified; consequently, it cannot be considered normal [31], which might restrict the experiment's outcomes in a variety of ways. Our findings, however, are equivalent to those obtained with primary culture of mouse cerebellar granule neurons. At the same time, we revealed that RA-differentiated *SH-SY5Y* cells respond differently to distinct chemical species of Mn, which was not considered in previous studies with NoRA-differentiated *SH-SY5Y* cells. Indeed, Mn causes disruption of cross-talk networks of pathways in the RA-differentiated *SH-SY5Y*-DAergic cells, which may be mediated by protein metabolism disturbance, evidently influenced by chemical speciation, for example, the $MnCl_2$-altered amino acid metabolism, which affects RNA metabolism and protein synthesis. Mn(II)Cit altered the E3 ubiquitin ligases–target protein degradation pathway, potentially leading to the accumulation of damaged/unfolded proteins, which is consistent with histone modification. These findings support the relevance of chemical speciation in understanding the process behind Mn-induced neurotoxicity and neurodegeneration, which appears to be conserved from yeast [34], to zebrafish [35], to mammals [25,28]. A functional analysis of cross-species translation in the presence of Mn could either confirm or refute our findings.

Supplementary Materials: The following are available online at https://www.mdpi.com/article/10.3390/toxics9120348/s1, Table S1: Genes differentially expressed in RA-differentiated *SH-SY5Y* cells after exposure to $MnCl_2$ are listed below. Table S2: Genes differentially expressed in RA-differentiated *SH-SY5Y* cells after exposure to $MnCl_2$ are listed below.

Author Contributions: R.B.H. was responsible for the principal idea of this study, coexecutor of all designed experiments, and coordinators of the grants that supported this work. J.K. together with M.v.H. and J.P. participated of the design, execution, and data analysis of microarrays assays. N.C.d.S.-P. participated in data analysis. H.M. participated directly in the experiments on gene expression and data analysis. A.G. and D.B. participated actively in the execution of bioinformatic approaches and data analysis. All authors wrote and reviewed the manuscript. All authors have read and agreed to the published version of the manuscript.

Funding: This research was funded by the São Paulo Research Foundation, FAPESP, grant numbers 19/27840-5, 16/50483-6, and 15/24207-9.

Institutional Review Board Statement: Not applicable.

Informed Consent Statement: Not applicable.

Data Availability Statement: Not applicable.

Conflicts of Interest: The authors declare no competing interest.

References

1. Howe, P.D.; Malcolm, H.M.; Dobson, S. *Manganese and Its Compounds: Environmental Aspects*; World Health Organization & International Programme on Chemical Safety: Geneva, Switzerland, 2004.
2. Luo, X.G.; Li, S.F.; Lu, L.; Liu, B.; Kuang, X.; Shao, G.Z.; Yu, S.X. Gene Expression of Manganese-Containing Superoxide Dismutase as a Biomarker of Manganese Bioavailability for Manganese Sources in Broilers. *Poult. Sci.* **2007**, *86*, 888–894. [CrossRef]
3. Kwakye, G.; Paoliello, M.; Mukhopadhyay, S.; Bowman, A.; Aschner, M. Manganese-Induced Parkinsonism and Parkinson's Disease: Shared and Distinguishable Features. *Int. J. Environ. Res. Public Health* **2015**, *12*, 7519–7540. [CrossRef] [PubMed]
4. Bowman, A.B.; Kwakye, G.F.; Herrero, E.; Aschner, M. Journal of Trace Elements in Medicine and Biology Role of Manganese in Neurodegenerative Diseases. *J. Trace Elem. Med. Biol.* **2011**, *25*, 191–203. [CrossRef]
5. Šarić, M.; Lucchini, R. *Handbook on the Toxicology of Metals*, 3rd ed.; Nordberg, G.F., Fowler, B.A., Nordberg, M., Friberg, L.T., Eds.; Academic Press Inc.: Burlington, MA, USA, 2007; pp. 645–674.
6. Hafeman, D.; Factor-Litvak, P.; Cheng, Z.; van Geen, A.; Ahsan, H. Association between Manganese Exposure through Drinking Water and Infant Mortality in Bangladesh. *Environ. Health Perspect.* **2007**, *115*, 1107–1112. [CrossRef]
7. Ljung, K.; Vahter, M. Time to Re-Evaluate the Guideline Value for Manganese in Drinking Water? *Environ. Health Perspect.* **2007**, *115*, 1533–1538. [CrossRef] [PubMed]
8. Jordão, C.P.; Pereira, J.L.; Jham, G.N.; Bellato, C.R. Distribution of Heavy Metals in Environmental Samples Near Smelters and Mining Areas in Brazil. *Environ. Technol.* **1999**, *20*, 489–498. [CrossRef]
9. Avnimelech, Y.; Kochba, M. Evaluation of Nitrogen Uptake and Excretion by Tilapia in Bio Floc Tanks, Using 15N Tracing. *Aquaculture* **2009**, *287*, 163–168. [CrossRef]
10. Bonne Hernández, R.; Oliveira, E.; Espósito, B.P. Distribution and Behavior of Manganese in the Alto Do Paranapanema Basin. *J. Environ. Monit.* **2009**, *11*, 1236. [CrossRef] [PubMed]
11. Templeton, D.M.; Ariese, F.; Cornelis, R.; Danielsson, L.-G.; Muntau, H.; van Leeuwen, H.P.; Lobinski, R. Guidelines for Terms Related to Chemical Speciation and Fractionation of Elements. Definitions, Structural Aspects, and Methodological Approaches (IUPAC Recommendations 2000). *Pure Appl. Chem.* **2000**, *72*, 1453–1470. [CrossRef]
12. Stueben, B.L.; Cantrelle, B.; Sneddon, J.; Beck, J.N. Manganese K-Edge XANES Studies of Mn Speciation in Lac Des Allemands as a Function of Depth. *Microchem. J.* **2004**, *76*, 113–120. [CrossRef]
13. Kenneth Klewicki, J.; Morgan, J.J. Kinetic Behavior of Mn(III) Complexes of Pyrophosphate, EDTA, and Citrate. *Environ. Sci. Technol.* **1998**, *32*, 2916–2922. [CrossRef]
14. Hernández, R.B.; Farina, M.; Espósito, B.P.; Souza-Pinto, N.C.; Barbosa, F.; Suñol, C. Mechanisms of Manganese-Induced Neurotoxicity in Primary Neuronal Cultures: The Role of Manganese Speciation and Cell Type. *Toxicol. Sci.* **2011**, *124*, 414–423. [CrossRef]
15. Queiroz, H.M.; Ying, S.C.; Abernathy, M.; Barcellos, D.; Gabriel, F.A.; Otero, X.L.; Nóbrega, G.N.; Bernardino, A.F.; Ferreira, T.O. Manganese: The Overlooked Contaminant in the World Largest Mine Tailings Dam Collapse. *Environ. Int.* **2021**, *146*, 106284. [CrossRef] [PubMed]
16. Hernández, R.B.; Nishita, M.I.; Espósito, B.P.; Scholz, S.; Michalke, B. The Role of Chemical Speciation, Chemical Fractionation and Calcium Disruption in Manganese-Induced Developmental Toxicity in Zebrafish (Danio Rerio) Embryos. *J. Trace Elem. Med. Biol.* **2015**, *32*, 209–217. [CrossRef] [PubMed]
17. Mehrifar, Y.; Bahrami, M.; Sidabadi, E.; Pirami, H. The Effects of Occupational Exposure to Manganese Fume on Neurobehavioral and Neurocognitive Functions: An Analytical Cross-Sectional Study among Welders. *EXCLI J.* **2020**, *19*, 372–386. [CrossRef] [PubMed]
18. Neves-Silva, P.; Heller, L. Rompimento Da Barragem Em Brumadinho e o Acesso à Água Das Comunidades Atingidas: Um Caso de Direitos Humanos. *Ciência Cult.* **2020**, *72*, 47–50. [CrossRef]
19. De Azevedo, D.C.B.; de Araujo Toledo, G.; Cohen, S.C.; Kligerman, D.C.; de Oliveira Cardoso, T.A. Desastre de Brumadinho: Contribuições Para Políticas Públicas e Gestão Do Saneamento Em Períodos Emergenciais. *Saúde Debate* **2020**, *44*, 221–233. [CrossRef]
20. De Freitas, C.M.; Barcellos, C.; Asmus, C.I.R.F.; da Silva, M.A.; Xavier, D.R. Da Samarco Em Mariana à Vale Em Brumadinho: Desastres Em Barragens de Mineração e Saúde Coletiva. *Cad. Saúde Pública* **2019**, *35*, 1–7. [CrossRef]
21. Mitchell, E.J.; Frisbie, S.H.; Roudeau, S.; Carmona, A.; Ortega, R. How Much Manganese Is Safe for Infants? A Review of the Scientific Basis of Intake Guidelines and Regulations Relevant to the Manganese Content of Infant Formulas. *J. Trace Elem. Med. Biol.* **2021**, *65*, 126710. [CrossRef]
22. Williams, B.B.; Kwakye, G.F.; Wegrzynowicz, M.; Li, D.; Aschner, M.; Erikson, K.M.; Bowman, A.B. Altered Manganese Homeostasis and Manganese Toxicity in a Huntington's Disease Striatal Cell Model Are Not Explained by Defects in the Iron Transport System. *Toxicol. Sci.* **2010**, *117*, 169–179. [CrossRef] [PubMed]
23. Pfalzer, A.C.; Wilcox, J.M.; Codreanu, S.G.; Totten, M.; Bichell, T.J.V.; Halbesma, T.; Umashanker, P.; Yang, K.L.; Parmalee, N.L.; Sherrod, S.D.; et al. Huntington's Disease Genotype Suppresses Global Manganese-Responsive Processes in Pre-Manifest and Manifest YAC128 Mice. *Metallomics* **2020**, *12*, 1118–1130. [CrossRef] [PubMed]

24. Martins, A.C.; Morcillo, P.; Ijomone, O.M.; Venkataramani, V.; Harrison, F.E.; Lee, E.; Bowman, A.B.; Aschner, M. New Insights on the Role of Manganese in Alzheimer's Disease and Parkinson's Disease. *Int. J. Environ. Res. Public Health* **2019**, *16*, 3546. [CrossRef]
25. Hernández, R.B.; Carrascal, M.; Abian, J.; Michalke, B.; Farina, M.; Gonzalez, Y.; Iyirhiaro, G.; Moteshareie, H.; Burnside, D.; Golshani, A.; et al. Manganese-Induced Neurotoxicity in Cerebellar Granule Neurons Due to Perturbation of Cell Network Pathways with Potential Implications for Neurodegenerative Disorders. *Metallomics* **2020**, *12*, 1656–1678. [CrossRef] [PubMed]
26. Gandhi, D.; Sivanesan, S.; Kannan, K. Manganese-Induced Neurotoxicity and Alterations in Gene Expression in Human Neuroblastoma SH-SY5Y Cells. *Biol. Trace Elem. Res.* **2018**, *183*, 245–253. [CrossRef] [PubMed]
27. Fernandes, J.; Chandler, J.D.; Liu, K.H.; Uppal, K.; Hao, L.; Hu, X.; Go, Y.-M.; Jones, D.P. Metabolomic Responses to Manganese Dose in SH-SY5Y Human Neuroblastoma Cells. *Toxicol. Sci.* **2019**, *169*, 84–94. [CrossRef] [PubMed]
28. Fernandes, J.; Chandler, J.D.; Lili, L.N.; Uppal, K.; Hu, X.; Hao, L.; Go, Y.-M.; Jones, D.P. Transcriptome Analysis Reveals Distinct Responses to Physiologic versus Toxic Manganese Exposure in Human Neuroblastoma Cells. *Front. Genet.* **2019**, *10*, 676. [CrossRef] [PubMed]
29. Kovalevich, J.; Langford, D. Considerations for the Use of SH-SY5Y Neuroblastoma Cells in Neurobiology. *Methods Mol. Biol.* **2013**, *1078*, 9–21. [CrossRef]
30. Altamirano, M.; Coates, C.W.; Grundfest, H. Mechanisms of direct and neural excitability in electroplaques of electric eel. *J. Gen. Physiol.* **1955**, *38*, 319–360. [CrossRef] [PubMed]
31. Korecka, J.A.; van Kesteren, R.E.; Blaas, E.; Spitzer, S.O.; Kamstra, J.H.; Smit, A.B.; Swaab, D.F.; Verhaagen, J.; Bossers, K. Phenotypic Characterization of Retinoic Acid Differentiated SH-SY5Y Cells by Transcriptional Profiling. *PLoS ONE* **2013**, *8*, e63862. [CrossRef]
32. Schlachetzki, J.C.M.; Saliba, S.W.; de Oliveira, A.C.P. Studying Neurodegenerative Diseases in Culture Models. *Rev. Bras. Psiquiatr.* **2013**, *35* (Suppl. 2), S92–S100. [CrossRef]
33. Xicoy, H.; Wieringa, B.; Martens, G.J.M. The SH-SY5Y Cell Line in Parkinson's Disease Research: A Systematic Review. *Mol. Neurodegener.* **2017**, *12*, 10. [CrossRef]
34. Hernández, R.B.; Moteshareie, H.; Burnside, D.; McKay, B.; Golshani, A. Manganese-Induced Cellular Disturbance in the Baker's Yeast, Saccharomyces Cerevisiae with Putative Implications in Neuronal Dysfunction. *Sci. Rep.* **2019**, *9*, 6563. [CrossRef] [PubMed]
35. Bonne Hernández, R.; Moteshareie, H.; Golshani, A. Manganese-Induced Disruption of Cross-Talking Pathways in Danio Rerio (Zebrafish) Is Potentially Linked to Toxicity and Neurodegeneration. *EC Pharmacol. Toxicol.* **2019**, *7*, 175–187.
36. Warner, J.R. The Economics of Ribosome Biosynthesis in Yeast. *Trends Biochem. Sci.* **1999**, *24*, 437–440. [CrossRef]
37. Thomson, E.; Ferreira-Cerca, S.; Hurt, E. Eukaryotic Ribosome Biogenesis at a Glance. *J. Cell Sci.* **2013**, *126*, 4815–4821. [CrossRef]
38. Liu, C.; Yan, D.; Wang, C.; Ma, Z.; Deng, Y.; Liu, W.; Xu, B. Manganese Activates Autophagy to Alleviate Endoplasmic Reticulum Stress–Induced Apoptosis via PERK Pathway. *J. Cell. Mol. Med.* **2020**, *24*, 328–341. [CrossRef]
39. Higashi, Y.; Asanuma, M.; Miyazaki, I.; Hattori, N.; Mizuno, Y.; Ogawa, N. Parkin Attenuates Manganese-Induced Dopaminergic Cell Death. *J. Neurochem.* **2004**, *89*, 1490–1497. [CrossRef] [PubMed]
40. Huang, Y.; Wen, Q.; Huang, J.; Luo, M.; Xiao, Y.; Mo, R.; Wang, J. Manganese (II) Chloride Leads to Dopaminergic Neurotoxicity by Promoting Mitophagy through BNIP3-Mediated Oxidative Stress in SH-SY5Y Cells. *Cell. Mol. Biol. Lett.* **2021**, *26*, 23. [CrossRef]
41. Rabin, O.; Hegedus, L.; Bourre, J.-M.; Smith, Q.R. Rapid Brain Uptake of Manganese(II) Across the Blood-Brain Barrier. *J. Neurochem.* **2006**, *61*, 509–517. [CrossRef]
42. Yokel, R.A. Manganese Flux Across the Blood–Brain Barrier. *NeuroMol. Med.* **2009**, *11*, 297–310. [CrossRef] [PubMed]
43. Stephenson, A.P.; Schneider, J.A.; Nelson, B.C.; Atha, D.H.; Jain, A.; Soliman, K.F.A.; Aschner, M.; Mazzio, E.; Reams, R.R. Manganese-Induced Oxidative DNA Damage in Neuronal SH-SY5Y Cells: Attenuation of Thymine Base Lesions by Glutathione and N-Acetylcysteine. *Toxicol. Lett.* **2013**, *218*, 299–307. [CrossRef] [PubMed]
44. Traverso, M.; Malnati, M.; Minetti, C.; Regis, S.; Tedeschi, S.; Pedemonte, M.; Bruno, C.; Biassoni, R.; Zara, F. Multiplex Real-Time PCR for Detection of Deletions and Duplications in Dystrophin Gene. *Biochem. Biophys. Res. Commun.* **2006**, *339*, 145–150. [CrossRef]
45. Jouannic, J.; Stieltjes, N.; Costa, J.; Girodon, E. Quantitative Real-Time PCR Assay for Rapid Identification of Deletion Carriers in Hemophilia. *Clin. Chem.* **2004**, *50*, 1269–1270. [CrossRef]
46. Samanfar, B.; Shostak, K.; Moteshareie, H.; Hajikarimlou, M.; Shaikho, S.; Omidi, K.; Hooshyar, M.; Burnside, D.; Márquez, I.G.; Kazmirchuk, T.; et al. The Sensitivity of the Yeast, Saccharomyces Cerevisiae, to Acetic Acid Is Influenced by DOM34 and RPL36A. *PeerJ* **2017**, *2017*, e4037. [CrossRef]
47. Alonso, R.; Salavert, F.; Garcia-Garcia, F.; Carbonell-Caballero, J.; Bleda, M.; Garcia-Alonso, L.; Sanchis-Juan, A.; Perez-Gil, D.; Marin-Garcia, P.; Sanchez, R.; et al. Babelomics 5.0: Functional Interpretation for New Generations of Genomic Data. *Nucleic Acids Res.* **2015**, *43*, W117–W121. [CrossRef] [PubMed]
48. Szklarczyk, D.; Morris, J.H.; Cook, H.; Kuhn, M.; Wyder, S.; Simonovic, M.; Santos, A.; Doncheva, N.T.; Roth, A.; Bork, P.; et al. The STRING Database in 2017: Quality-Controlled Protein-Protein Association Networks, Made Broadly Accessible. *Nucleic Acids Res.* **2017**, *45*, D362–D368. [CrossRef] [PubMed]
49. Davis, A.P.; Grondin, C.J.; Johnson, R.J.; Sciaky, D.; King, B.L.; McMorran, R.; Wiegers, J.; Wiegers, T.C.; Mattingly, C.J. The Comparative Toxicogenomics Database: Update 2017. *Nucleic Acids Res.* **2017**, *45*, D972–D978. [CrossRef]

50. Reimand, J.; Isserlin, R.; Voisin, V.; Kucera, M.; Tannus-Lopes, C.; Rostamianfar, A.; Wadi, L.; Meyer, M.; Wong, J.; Xu, C.; et al. Pathway Enrichment Analysis and Visualization of Omics Data Using g:Profiler, GSEA, Cytoscape and EnrichmentMap. *Nat. Protoc.* **2019**, *14*, 482–517. [CrossRef] [PubMed]
51. Tarale, P.; Daiwile, A.P.; Sivanesan, S.; Stöger, R.; Bafana, A.; Naoghare, P.K.; Parmar, D.; Chakrabarti, T.; Krishnamurthi, K. Manganese Exposure: Linking down-Regulation of MiRNA-7 and MiRNA-433 with α-Synuclein Overexpression and Risk of Idiopathic Parkinson's Disease. *Toxicol. In Vitro* **2018**, *46*, 94–101. [CrossRef]
52. Tinkov, A.A.; Paoliello, M.M.B.; Mazilina, A.N.; Skalny, A.V.; Martins, A.C.; Voskresenskaya, O.N.; Aaseth, J.; Santamaria, A.; Notova, S.V.; Tsatsakis, A.; et al. Molecular Targets of Manganese-Induced Neurotoxicity: A Five-Year Update. *Int. J. Mol. Sci.* **2021**, *22*, 4646. [CrossRef] [PubMed]
53. Dos Santos Vergilio, C.; Lacerda, D.; de Oliveira, B.C.V.; Sartori, E.; Campos, G.M.; de Souza Pereira, A.L.; de Aguiar, D.B.; da Silva Souza, T.; de Almeida, M.G.; Thompson, F.; et al. Metal Concentrations and Biological Effects from One of the Largest Mining Disasters in the World (Brumadinho, Minas Gerais, Brazil). *Sci. Rep.* **2020**, *10*, 5936. [CrossRef]
54. Thompson, F.; de Oliveira, B.C.; Cordeiro, M.C.; Masi, B.P.; Rangel, T.P.; Paz, P.; Freitas, T.; Lopes, G.; Silva, B.S.; Cabral, A.S.; et al. Severe Impacts of the Brumadinho Dam Failure (Minas Gerais, Brazil) on the Water Quality of the Paraopeba River. *Sci. Total Environ.* **2020**, *705*, 135914. [CrossRef]
55. Maddirala, Y.; Tobwala, S.; Ercal, N. N-Acetylcysteineamide Protects against Manganese-Induced Toxicity in SHSY5Y Cell Line. *Brain Res.* **2015**, *1608*, 157–166. [CrossRef]
56. Kemsheh, M.; Oblitey, R. Manganese Toxicity on Cultured SH-SY5Y Cells. In *All Zyzzogeton Presentations*; Augsburg University: Minneapolis, MN, USA, 2020; Volume 15.
57. Lehmkuhl, E.M.; Zarnescu, D.C. Lost in Translation: Evidence for Protein Synthesis Deficits in ALS/FTD and Related Neurodegenerative Diseases. *Adv Neurobiol.* **2018**, 283–301. [CrossRef]
58. Li, Y.; Sun, L.; Cai, T.; Zhang, Y.; Lv, S.; Wang, Y.; Ye, L. α-Synuclein Overexpression during Manganese-Induced Apoptosis in SH-SY5Y Neuroblastoma Cells. *Brain Res. Bull.* **2010**, *81*, 428–433. [CrossRef] [PubMed]
59. Younts, T.J.; Monday, H.R.; Dudok, B.; Klein, M.E.; Jordan, B.A.; Katona, I.; Castillo, P.E. Presynaptic Protein Synthesis Is Required for Long-Term Plasticity of GABA Release. *Neuron* **2016**, *92*, 479–492. [CrossRef] [PubMed]
60. Kano, M. Control of Synaptic Function by Endocannabinoid-Mediated Retrograde Signaling. *Proc. Jpn. Acad. Ser. B* **2014**, *90*, 235–250. [CrossRef]
61. Stoeckli, E.T. Understanding Axon Guidance: Are We Nearly There Yet? *Development* **2018**, *145*, dev151415. [CrossRef]
62. Liu, X.-A.; Rizzo, V.; Puthanveettil, S. Pathologies of Axonal Transport in Neurodegenerative Diseases. *Transl. Neurosci.* **2012**, *3*, 355–372. [CrossRef] [PubMed]
63. Cho, Y.; Shin, J.E.; Ewan, E.E.; Oh, Y.M.; Pita-Thomas, W.; Cavalli, V. Activating Injury-Responsive Genes with Hypoxia Enhances Axon Regeneration through Neuronal HIF-1α. *Neuron* **2015**, *88*, 720–734. [CrossRef]
64. Hou, C.; Wang, Y.; Liu, J.; Wang, C.; Long, J. Neurodegenerative Disease Related Proteins Have Negative Effects on SNARE-Mediated Membrane Fusion in Pathological Confirmation. *Front. Mol. Neurosci.* **2017**, *10*, 66. [CrossRef] [PubMed]
65. Corgnali, M.; Piconi, L.; Ihnat, M.; Ceriello, A. Evaluation of Gliclazide Ability to Attenuate the Hyperglycaemic 'Memory' Induced by High Glucose in Isolated Human Endothelial Cells. *Diabetes Metab. Res. Rev.* **2008**, *24*, 301–309. [CrossRef] [PubMed]
66. Fei, Q.; Ethell, D.W. Maneb Potentiates Paraquat Neurotoxicity by Inducing Key Bcl-2 Family Members. *J. Neurochem.* **2008**, *105*, 2091–2097. [CrossRef]
67. Srivastava, A.K.; Mishra, S.; Ali, W.; Shukla, Y. Protective Effects of Lupeol against Mancozeb-Induced Genotoxicity in Cultured Human Lymphocytes. *Phytomedicine* **2016**, *23*, 714–724. [CrossRef] [PubMed]
68. Kim, E.; Jung, H. Local Protein Synthesis in Neuronal Axons: Why and How We Study. *BMB Rep.* **2015**, *48*, 139–146. [CrossRef]
69. Hu, X.P.; Yang, Y.; Ma, B.G. Amino Acid Flux from Metabolic Network Benefits Protein Translation: The Role of Resource Availability. *Sci. Rep.* **2015**, *5*, 11113. [CrossRef]
70. Seo, Y.A.; Li, Y.; Wessling-Resnick, M. Iron Depletion Increases Manganese Uptake and Potentiates Apoptosis through ER Stress. *NeuroToxicol.* **2013**, *38*, 67–73. [CrossRef] [PubMed]
71. Putrament, A.; Baranowska, H.; Ejchart, A.; Prazmo, W. Manganese Mutagenesis in Yeast. A Practical Application of Manganese for the Induction of Mitochondrial Antibiotic-Resistant Mutations. *J. Gen. Microbiol.* **1975**, *90*, 265–270. [CrossRef] [PubMed]
72. Putrament, A.; Baranowska, H.; Ejchart, A.; Jachymczyk, W. Manganese Mutagenesis in Yeast-VI. Mn2+ Uptake, MitDNA Replication and ER Induction. Comparison with Other Divalent Cations. *MGG Mol. Gen. Genet.* **1977**, *151*, 69–76. [CrossRef] [PubMed]
73. Donaldson, S.G.; Fox, O.F.; Kishore, G.S.; Carubelli, R. Effect of Manganese Ions on the Interaction between Ribosomes and Endoplasmic Reticulum Membranes Isolated from Rat Liver. *Biosci. Rep.* **1981**, *1*, 727–731. [CrossRef] [PubMed]
74. Dambach, M.; Sandoval, M.; Updegrove, T.B.; Anantharaman, V.; Aravind, L.; Waters, L.S.; Storz, G. The Ubiquitous YybP-YkoY Riboswitch Is a Manganese-Responsive Regulatory Element. *Mol. Cell* **2015**, *57*, 1099–1109. [CrossRef]
75. Bray, M.S.; Lenz, T.K.; Haynes, J.W.; Bowman, J.C.; Petrov, A.S.; Reddi, A.R.; Hud, N.V.; Williams, L.D.; Glass, J.B. Multiple Prebiotic Metals Mediate Translation. *Proc. Natl. Acad. Sci. USA* **2018**, *115*, 12164–12169. [CrossRef]
76. Lee, C.-D.; Tu, B.P. Metabolic Influences on RNA Biology and Translation. *Crit. Rev. Biochem. Mol. Biol.* **2017**, *52*, 176–184. [CrossRef]

77. Cioni, J.-M.; Lin, J.Q.; Holtermann, A.V.; Koppers, M.; Jakobs, M.A.H.; Azizi, A.; Turner-Bridger, B.; Shigeoka, T.; Franze, K.; Harris, W.A.; et al. Late Endosomes Act as MRNA Translation Platforms and Sustain Mitochondria in Axons. *Cell* **2019**, *176*, 56–72. [CrossRef]
78. Giasson, B.I.; Lee, V.M.-Y. Are Ubiquitination Pathways Central to Parkinson's Disease? *Cell* **2003**, *114*, 1–8. [CrossRef]
79. Agbas, A. Trends of Protein Aggregation in Neurodegenerative Diseases. In *Neurochemistry*; IntechOpen: London, UK, 2018. [CrossRef]
80. Sidoryk-Węgrzynowicz, M.; Lee, E.-S.; Ni, M.; Aschner, M. Manganese-Induced Downregulation of Astroglial Glutamine Transporter SNAT3 Involves Ubiquitin-Mediated Proteolytic System. *Glia* **2010**, *58*, 1905–1912. [CrossRef]

MDPI
St. Alban-Anlage 66
4052 Basel
Switzerland
Tel. +41 61 683 77 34
Fax +41 61 302 89 18
www.mdpi.com

Toxics Editorial Office
E-mail: toxics@mdpi.com
www.mdpi.com/journal/toxics

www.ingramcontent.com/pod-product-compliance
Lightning Source LLC
Chambersburg PA
CBHW061554210926
51850CB00009B/2104
* 9 7 8 3 0 3 6 5 5 1 8 0 7 *